MICROBIOLOGICAL RISK ASSESSMENT SERIES

5

Risk assessment of *Listeria monocytogenes* in ready-to-eat foods

TECHNICAL REPORT

WORLD HEALTH ORGANIZATION
FOOD AND AGRICULTURE ORGANIZATION OF THE UNITED NATIONS

2004

WHO Library Cataloguing-in-Publication Data

Risk assessments of *Listeria monocytogenes* in ready-to-eat foods: technical report.

(Microbiological risk assessment series No. 5)
1. *Listeria monocytogenes* – pathogenicity 2. *Listeria monocytogenes* – growth and development 3. Food contamination 4. Food handling 5. Risk assessment – methods 6. Listeria infections I. World Health Organization II. Food and Agriculture Organization of the United Nations III. Series

ISBN 92 4 156262 5 (WHO) (LC/NLM classification: QW 142.5.A8)
ISBN 92 5 105127 5 (FAO)
ISSN 1726-5274

CONTENTS

PART 3. EXPOSURE ASSESSMENT

PART 4. EXAMPLE RISK ASSESSMENTS

PART 5. RISK CHARACTERIZATION: RESPONSE TO CODEX QUESTIONS

PART 6. KEY FINDINGS AND CONCLUSIONS

PART 7. REFERENCES CITED

APPENDICES

Acknowledgements

The Food and Agriculture Organization of the United Nations and the World Health Organization would like to express their appreciation to all those who contributed to the preparation of this report. Special appreciation is extended to the risk assessment drafting group for the time and effort that they freely dedicated to the elaboration of this risk assessment. Many people provided their time and expertise by reviewing the report and providing their comments, additional data and other relevant information. Both the risk assessment drafting group and the reviewers are listed in the following pages.

Appreciation is also extended to all those who responded to the calls for data that were issued by FAO and WHO and brought to our attention data which is not readily available in the mainstream literature and official documentation.

The risk assessment work was co-ordinated by the Joint FAO/WHO Secretariat on Risk Assessment of Microbiological Hazards in Foods. This included Sarah Cahill, Maria de Lourdes Costarrica and Jean Louis Jouve in FAO and Peter Karim Ben Embarek, Allan Hogue, Jocelyne Rocourt, Hajime Toyofuku and Jørgen Schlundt in WHO. During the development of the risk assessment additional support and feedback were provided by Lahsen Ababouch and Hector Lupin, Fishery Industries Division in FAO and Jeronimas Maskeliunas, Codex Secretariat. Publication of the report was coordinated by Sarah Cahill. Thorgeir Lawrence was responsible for editing the report for language and preparation for printing.

The preparatory work and the publication was supported and funded by the FAO Food Quality and Standards Service, the FAO Fishery Industries Division and the WHO Food Safety Department.

RISK ASSESSMENT DRAFTING GROUP

(in alphabetical order)

Robert Buchanan
Center for Food Safety and Nutrition, Food and Drug Administration,
United States of America

Roland Lindqvist
National Food Administration,
Sweden

Thomas Ross
University of Tasmania,
Australia

Mark Smith
Bureau of Biostatistics and Computer Applications, Health Products and Food Branch,
Health Canada, Canada

Ewen Todd
The National Food Safety & Toxicology Center, Michigan State University,
United States of America

Richard Whiting
Center for Food Safety and Nutrition, Food and Drug Administration,
United States of America

Assistance was provided to the drafting group by the following:

Elizabeth A. Junkins, **Michel Vigneault** and **William Ross**, Bureau of Biostatistics and Computer Applications, Health Products and Food Branch, Health Canada.

Jeff M. Farber, Bureau of Microbial Hazards, Health Products and Food Branch, Health Canada.

Sven Rasmussen, University of Tasmania.

Suzanne van Gerwen, Microbiology and Preservation Unit, Unilever Research, Vlaardingen, The Netherlands.

FDA/FSIS *Listeria Monocytogenes* Risk Assessment Team, Washington, DC, USA
(M. Bender, C. Carrington, S. Dennis, E. Ebel, A. Hitchens, N. Hynes, W. Long,
W. Schlosser, R. Raybourne, M. Ross, T. Rouse and R. Whiting)

Meat and Livestock Australia, for access to the report of Ross and Shadbolt (2001), and **Australian and New Zealand Food Authority,** for access to product composition data.

REVIEWERS

The risk assessment was reviewed on several occasions both during and after its elaboration, including through expert consultations and selected peer-reviews, and by members of the public in response to a call for public comments.

PARTICIPANTS IN EXPERT CONSULTATIONS

Amma Anandavally	Export Quality Control Laboratory, India.
Wayne Anderson	Food Safety Authority of Ireland, Ireland.
Olivier Cerf	Ecole Nationale Vétérinaire d'Alfort (ENVA), France.
Jean-Yves D'Aoust	Health Protection Branch, Health Canada, Canada.
Paw Dalgaard	Danish Institute for Fisheries Research, Ministry of Food Agriculture and Fisheries, Denmark.
Michael Doyle	Center for Food Safety, University of Georgia, United States of America.
Eric Ebel	United States Department of Agriculture, Food Safety and Inspection Service, United States of America.
Emilio Esteban	Centres for Disease Control and Prevention, United States of America.
Aamir Fazil	Population and Public Health Branch, Health Canada, Canada.
Lone Gram	Danish Institute of Fisheries Research, Technical University of Denmark, Denmark.
Steve Hathaway	MAF Regulatory Authority (Meat and Seafood), New Zealand
Matthias Hartung,	National Reference Laboratory on the Epidemiology of Zoonoses, Germany.
Inocencio Higuera Ciapara	Research Centre for Food and Development (CIAD), Mexico.
Andrew Hudson	The Institute of Environmental Science and Research Ltd, New Zealand.
David Jordan	New South Wales Agriculture, Wollongbar Agricultural Institute, Australia.

Fumiko Kasuga	National Institute of Infectious Diseases, Department of Biomedical Food Research, Japan.
Louise Kelly	Department of Risk Research, Veterinary Laboratories Agency, United Kingdom.
Julia A. Kiehlbauch	Microbiology Consultant, United States of America.
Günter Klein,	Division of Food Hygiene, Federal Institute for Health Protection of Consumers and Veterinary Medicine, Germany.
Susumu Kumagai	Graduate School of Agriculture and Life Sciences, The University of Tokyo, Japan.
Anna Lammerding	Population and Public Health Branch, Health Canada, Canada.
Xiumei Liu	Institute of Nutrition and Food Hygiene, Ministry of Health, China.
Carol Maczka	United States Department of Agriculture, Food Safety and Inspection Service, United States of America.
Patience Mensah	Noguchi Memorial Institute for Medical Research, University of Ghana, Ghana.
George Nasinyama	Department of Epidemiology and Food Safety, Makerere University, Uganda.
Gregory Paoli	Decisionalysis Risk Consultants Inc. , Canada.
Irma N.G. Rivera	Departamento de Microbiologia, Universidade de São Paulo, Brazil.
Son Radu	Department of Biotechnology, University Putra Malaysia, Malaysia.
Dulce Maria Tocchetto Schuch	Laboratorio Regional de Apoio Animal – Lara / RS Agriculture Ministry, Brazil.
Eystein Skjerve	Department of Pharmacology, The Norwegian School of Veterinary Science, Norway.
Bruce Tompkin	ConAgra Inc., United States of America.
Suzanne Van Gerwen	Microbiology and Preservation Unit, Unilever Research, The Netherlands.
Michiel Van Schothorst	Wageningen University, The Netherlands.
Kaye Wachsmuth	United States Department of Agriculture, Food Safety and Inspection Service, United States of America.

Helene Wahlström	National Veterinary Institute, Sweden.
Charles Yoe	Department of Economics, College of Notre Dame of Maryland United States of America.

REVIEWERS IN RESPONSE TO CALL FOR PUBLIC COMMENT

K.E. Aidoo	School of Biomedical and Biological Sciences,Glasgow Caledonian University, Scotland.
Scott Cameron	Department of Public Health, University of Adelaide, Australia.
Andrew Hudson	The Institute of Environmental Science and Research Ltd, New Zealand.
Industry Council for Development	United Kingdom
O.O. Komolafe	Department of Microbiology, College of Medicine, Malawi.
Douglas Marshall	Department of Food Science & Technology, Mississippi State University, United States of America.
Bruce Tompkin	ConAgra Inc., United States of America.
United States Food and Drug Administration	United States of America.

PEER-REVIEWERS

Paolo Aureli	Food Department, Istituto Superiore della Sanità, Italy.
Sava Buncic	School of Veterinary Science, University of Bristol, United Kingdom.
Carmen Buchrieser	Laboratoire de Genetique des Microorganismes Pathogenes, Institut Pasteur, France.
Marie Teresa Destro	Depto Alimentos e Nutrição Experimental, FCF – USP, Brazil.
Veronique Goulet	Institut de Veille Sanitaire, Hopital de Saint-Maurice, France.
Lone Gram	Danish Institute of Fisheries Research, Technical University of Denmark, Denmark.

Arie Havelaar Microbiological Laboratory for Health Protection, National Institute for Public Health and the Environment, The Netherlands.

Sophia Kathariou Department of Food Science, North Carolina State University, United States of America.

Heejeong Latimar United States Department of Agriculture, Food Safety and Inspection Service, United States of America.

Paul Martin Laboratoire des Listeria, Institut Pasteur, France.

Robert L. McMasters Department of Mechanical Engineering, Michigan State University, United States of America.

Thomas McMeekin Centre for Food Safety and Quality, University of Tasmania, Australia.

Birgit Norrung Danish Veterinary and Food Administration, Denmark.

Servè Notermans TNO Nutrition and Food Research Institute, The Netherlands.

Jocelyne Rocourt Food Safety Programme, World Health Organization, Switzerland.

Elliot T. Ryser Department of Food Science and Human Nutrition, Michigan State University, United States of America.

Don Schaffner Food Risk Analysis Initiative, Rutgers University, United States of America.

William Henry Sperber Cargill, United States of America.

David Vose David Vose Consulting, France.

Marion Wooldridge Department of Risk Research, Veterinary Laboratory Agency (Weybridge), United Kingdom.

FOREWORD

The Members of the Food and Agriculture Organization of the United Nations (FAO) and of the World Health Organization (WHO) have expressed concern regarding the level of safety of food both at national and international levels. Increasing foodborne disease incidence over the last decades seems, in many countries, to be related to an increase in disease caused by microorganisms in food. This concern has been voiced in meetings of the Governing Bodies of both Organizations and in the Codex Alimentarius Commission. It is not easy to decide whether the suggested increase is real or an artefact of changes in other areas, such as improved disease surveillance or better detection methods for microorganisms in foods. However, the important issue is whether new tools or revised and improved actions can contribute to our ability to lower the disease burden and provide safer food. Fortunately new tools, which can facilitate actions, seem to be on their way.

Over the past decade, Risk Analysis – a process consisting of risk assessment, risk management and risk communication – has emerged as a structured model for improving our food control systems with the objectives of producing safer food, reducing the numbers of foodborne illnesses and facilitating domestic and international trade in food. Furthermore, we are moving towards a more holistic approach to food safety, where the entire food chain needs to be considered in efforts to produce safer food.

As with any model, tools are needed for the implementation of the risk analysis paradigm. Risk assessment is the science-based component of risk analysis. Science today provides us with in-depth information on life in the world we live in. It has allowed us to accumulate a wealth of knowledge on microscopic organisms, their growth, survival and death, even their genetic make-up. It has given us an understanding of food production, processing and preservation, and of the link between the microscopic and the macroscopic world and how we can benefit from as well as suffer from these microorganisms. Risk assessment provides us with a framework for organizing all this data and information and to better understand the interaction between microorganisms, foods and human illness. It provides us with the ability to estimate the risk to human health from specific microorganisms in foods and gives us a tool with which we can compare and evaluate different scenarios, as well as to identify the types of data is necessary for estimating and optimizing mitigating interventions.

Microbiological risk assessment can be considered as a tool that can be used in the management of the risks posed by foodborne pathogens and in the elaboration of standards for food in international trade. However, undertaking a microbiological risk assessment (MRA), particularly quantitative MRA, is recognized as a resource-intensive task requiring a multidisciplinary approach. Yet foodborne illness is among the most widespread public health problems, creating social and economic burdens as well as human suffering, making it a concern that all countries need to address. As risk assessment can also be used to justify the introduction of more stringent standards for imported foods, a knowledge of MRA is important for trade purposes, and there is a need to provide countries with the tools for understanding and, if possible, undertaking MRA. This need, combined with that of the Codex Alimentarius for risk-based scientific advice, led FAO and WHO to undertake a programme of activities on MRA at the international level.

The Food Quality and Standards Service, FAO, and the Food Safety Department, WHO, are the lead units responsible for this initiative. The two groups have worked together to develop the area of MRA at the international level for application at both the national and international levels. This work has been greatly facilitated by the contribution of people from around the world with expertise in microbiology, mathematical modelling, epidemiology and food technology to name but a few.

This Microbiological Risk Assessment series provides a range of data and information to those who need to understand or undertake MRA. It comprises risk assessments of particular pathogen-commodity combinations, interpretative summaries of the risk assessments, guidelines for undertaking and using risk assessment, and reports addressing other pertinent aspects of MRA.

We hope that this series will provide a greater insight into MRA, how it is undertaken and how it can be used. We strongly believe that this is an area that should be developed in the international sphere, and have already from the present work clear indications that an international approach and early agreement in this area will strengthen the future potential for use of this tool in all parts of the world, as well as in international standard setting. We would welcome comments and feedback on any of the documents within this series so that we can endeavour to provide Member countries, Codex Alimentarius and other users of this material with the information they need to use risk-based tools, with the ultimate objective of ensuring that safe food is available for all consumers.

Ezzeddine Boutrif
Food Quality and Standards Service
FAO

Jørgen Schlundt
Food Safety Department
WHO

ABBREVIATIONS USED IN THE TEXT

AIDS	Acquired Immunodeficiency Syndrome
a_w	Water activity
BHI	Brain-heart infusion
CCFH	Codex Committee on Food Hygiene
CDC	Centers for Disease Control and Prevention (USA)
CFU	Colony-forming units
CNS	Central nervous system
CSFII	Continuing Survey of Food Intakes by Individuals (USA)
EGR	Exponential growth rate
FAO	Food and Agriculture Organization of the United Nations
FDA	Food and Drug Administration (USA)
FSIS	Food Safety and Inspection Service [USDA]
ID_{50}	Dose of an infectious organism required to produce infection in 50 percent of the experimental subjects or exposed population.
HIV	Human Immunodeficiency Virus
IV	Intravenous
LD_{50}	The amount of an infectious organism or toxic agent that is sufficient to kill 50 percent of the exposed population within a certain time.
LLO	Listeriolysin O
LMRA	*Listeria monocytogenes* Risk Assessment [FDA/FSIS]
MPD	Maximum population density
MPN	Most probable number
MRA	Microbiological risk assessment
MSE	Mean square error
NaCl	Sodium chloride
NHANES	National Health and Nutrition Examination Survey (USA)
RLT	Relative lag time
RTE	Ready-to-eat
USDA	United States Department of Agriculture
WHO	World Health Organization
WPS	Water phase salt

Executive Summary

This risk assessment on *Listeria monocytogenes* in ready-to-eat (RTE) foods was undertaken to (i) respond to the request of the Codex Committee on Food Hygiene (CCFH) for sound scientific advice as a basis for the development of guidelines for the control of *L. monocytogenes* in foods; and (ii) address the needs expressed by Member countries for adaptable risk assessments that they can use to support risk management decisions and to conduct their own assessments.

The risk assessment was tailored to address three specific questions posed by the 33rd session of the CCFH (CAC, 2000) namely:

1. Estimate the risk of serious illness from *L. monocytogenes* in food when the number of organisms ranges from absence in 25 grams to 1000 colony forming units (CFU) per gram or millilitre, or does not exceed specified levels at the point of consumption.
2. Estimate the risk of serious illness for consumers in different susceptible population groups (elderly, infants, pregnant women and immunocompromised patients) relative to the general population.
3. Estimate the risk of serious illness from *L. monocytogenes* in foods that support its growth and foods that do not support its growth at specific storage and shelf-life conditions.

By answering these questions, this risk assessment aims to assist risk managers in conceptualizing how some of the factors governing foodborne listeriosis interact, thereby assisting the development of strategies to reduce the rates of illness.

The risk assessment comprises the four steps of hazard identification, hazard characterization, exposure assessment and risk characterization. A quantitative approach was taken and mathematical modelling employed to estimate the risks per serving and risk to a population in a year from the selected foods. The risk assessment focused on four RTE foods in order to provide examples of how microbiological risk assessment techniques can be used to answer food safety questions at an international level. The study was limited to foods at retail and their subsequent public health impact at the time of consumption. The impact of post-retail factors that could influence the risk to a consumer, such as temperature and duration of refrigerated storage, was also examined. This was considered sufficient to address the questions posed by the CCFH within the time frame and resources available to the risk assessors, and also reflects the situation that most of the currently available exposure data for *L. monocytogenes* relate to the frequency and extent of contamination at the retail level.

HAZARD IDENTIFICATION

Foodborne listeriosis is a relatively rare but serious disease with high fatality rates (20–30%) compared with other foodborne microbial pathogens, such as *Salmonella*. The disease largely affects specific segments of the population who have increased susceptibilities. Basically,

L. monocytogenes is an opportunistic pathogen that most often affects those with a severe underlying disease or condition (e.g. immunosuppression, HIV/AIDS, chronic conditions such as cirrhosis that impair the immune system); pregnant women; unborn or newly delivered infants; and the elderly. *L. monocytogenes* is widely dispersed in the environment and foods. However, it was not until several large, common-source outbreaks of listeriosis occurred in North America and Europe during the 1980s that the significance of foods as the primary route of transmission for human exposure to *L. monocytogenes* was recognized (Broome, Gellin and Schwartz, 1990; Bille, 1990). An important factor in foodborne listeriosis is that the pathogen can grow to significant numbers at refrigeration temperatures when given sufficient time. Despite the fact that a wide variety of foods may be contaminated with *L. monocytogenes*, outbreaks and sporadic cases of listeriosis are predominately associated with RTE foods – a large, heterogeneous category of foodstuffs that can be subdivided in many different ways and vary from country to country according to local eating habits; availability and integrity of the chill chain; and regulations specifying, for example, the maximum temperature at retail level. Although listeriosis is a relatively rare disease, the severity of the disease and the very frequent involvement of industrially processed foods, especially during outbreaks, mean that the social and economic impact of listeriosis is among the highest of the foodborne diseases (Roberts, 1989; Roberts and Pinner, 1990). Listeriosis is mainly observed in industrialized countries and it is not known whether the differences in incidence rates between developed and developing countries reflect true geographical differences, differences in food habits and food storage, or differences in diagnosis and reporting practices.

HAZARD CHARACTERIZATION

The hazard characterization provides a description of the pathogen and host characteristics that contribute to an infection by *Listeria,* the public health outcomes of infection with this pathogen, the foods most commonly associated with listeriosis, and a description of the dose-response relationship. Various clinical manifestations are associated with listeriosis and these can be grouped in two categories: invasive listeriosis and non-invasive listeriosis. Invasive listeriosis are cases when initial infections of the intestinal tissue by *L. monocytogenes* leads to invasion of otherwise sterile body sites, such as the pregnant uterus, the central nervous system, or the blood, or combinations. Invasive listeriosis is characterized by a high case-fatality rate, ranging from 20 to 30% (Mead et al., 1999) and sequelae may follow listeriosis infections (McLauchlin, 1997), though their incidence is rarely estimated (Rocourt, 1996). Non-invasive listeriosis (referred to as febrile listerial gastroenteritis) has been observed during a number of outbreaks where the majority of cases developed symptoms of gastroenteritis, such as diarrhoea, fever, headache and myalgia, after a short period of incubation (Dalton et al., 1997; Salamina et al., 1996; Riedo et al., 1994; Aureli et al., 2000). These outbreaks have generally involved the ingestion of high doses of *L. monocytogenes* by otherwise healthy individuals. The incidence rate and factors that govern the onset of this non-invasive form are not known. As a result, this risk assessment only considered invasive listeriosis as the outcome of exposure.

Dose-response data from human volunteer studies with *L. monocytogenes* or from volunteer studies with a surrogate pathogen do not exist. Therefore dose-response relations have been developed and evaluated based on expert elicitations, epidemiological or animal data, or combinations of these. These dose-response relations, which were reviewed and summarized in this work, cover the spectrum of biological end-points, i.e. infection, morbidity and mortality, and have, to varying degrees of sophistication, been evaluated using human epidemiological data. All models assume that each microbial cell acts independently, and that a single bacterial cell has the potential to cause disease. However, none of the available models were fully able to meet the needs of the current risk assessment in relation to the parameters examined and simplicity of calculation. For these reasons, alternative approaches were developed and evaluated for this risk assessment.

The approach used took advantage of the epidemiological data and detailed exposure assessment available in the *Listeria* risk assessment developed in the United States of America (FDA/FSIS, 2001). The model contains one parameter, r, which is the probability that a single cell will cause invasive listeriosis. This parameter was estimated from the pairing of population consumption patterns (exposure) with epidemiological data on the number of invasive listeriosis cases in the population. The estimated r-value, which will vary with the data sets used and the assumptions made, was then used in the exponential model to estimate specific risks given the number of *L. monocytogenes* consumed.

EXPOSURE ASSESSMENT

A full farm-to-fork risk assessment was not required to address the questions posed by the CCFH. Thus, the focus of the exposure assessment models was to account for changes in the frequency and extent of contamination in the food between retail marketing and the point of consumption. This simplified the modelling and reduced the model uncertainties, thereby decreasing the ranges around the final risk estimates. The models developed describe the growth or decline of *L. monocytogenes* between the time of purchase and consumption, using information and models for the growth rate and the lag time of *L. monocytogenes* as affected by storage temperature and food composition, the maximum growth of *L. monocytogenes* supported by the food, and the distribution of retail and home storage times and temperatures. Calculating the numbers of *L. monocytogenes* actually consumed also required consideration of how much of and how often the food is eaten (i.e. the size and the number of servings).

RTE foods are a broad and diverse food category, prepared and stored in different ways and under different conditions, some of which support growth of *L. monocytogenes* and others that do not support growth at specific storage and shelf-life conditions. As it was therefore not possible to consider all RTE foods, four foods – pasteurized milk, ice cream, fermented meat and cold smoked fish – were selected to illustrate how the different factors mentioned above interact to affect the risk of acquiring listeriosis. Pasteurized milk is a food that is widely consumed, has very low frequencies and levels of contamination with *L. monocytogenes* but allows growth of the organism during storage. Ice cream is similar to milk but does not permit growth of *L. monocytogenes* during storage. Fermented meat products are often contaminated with *Listeria* and are produced without any lethal processing step, but their final composition prevents growth of the microbe during storage. Cold-smoked fish is frequently contaminated with *L. monocytogenes*, has no lethal processing step and permits growth during an extended storage period.

Several "what-if" scenarios were also considered in the case of milk and smoked salmon. These hypothetical scenarios have specific changes made to one or more of the exposure factors to demonstrate how the factors interact to affect the risk. In conducting the exposure assessments for these four foods, different databases were available and the modellers used slightly different techniques. These techniques are explained in the main risk assessment document and illustrate that there are numerous approaches that may be taken depending on the available data and the judgment of the risk assessors.

The outputs from the exposure assessment included a distribution of *L. monocytogenes* in the food at the point of consumption (frequency of contamination) and also the amount consumed (number of servings per year and size of servings).

RISK CHARACTERIZATION

The outputs from the exposure assessment were fed into the dose-response model to develop the risk characterization portion of the risk assessment to calculate the probability of contracting listeriosis. The outputs are described in terms of estimates of risk per million servings for the healthy and susceptible populations. The risk per serving and number of servings were used to estimate the number of illnesses in a specified population per year.

The mean risk estimates of the number of illnesses per 10 million people per year and the risk per serving for pasteurized milk, ice cream, fermented meats and smoked fish are shown in Table 1. For milk, for example, the risk per serving was low (5.0×10^{-9} cases per serving), but the very high frequency of consumption resulted in milk making substantial contributions to the total number of predicted cases of illness. In contrast, for smoked fish the risk per serving was estimated to be high (2.1×10^{-8} cases per serving). However, consumption of this product is modest (1 to 18 servings per year), and consequently the total number of cases of listeriosis was moderate.

Table 1

The mean risk estimates of the number of illnesses per 10 million people per year and the risk per serving for four ready-to-eat foods.

Food	Cases of listeriosis per 10 million people per year	Cases of listeriosis per 1 million servings
Milk	9.1	0.005
Ice cream	0.012	0.000014
Smoked fish	0.46	0.021
Fermented meats	0.00066	0.0000025

RESPONSE TO QUESTIONS POSED BY THE CCFH

These risk assessments were used to address the specific questions posed by the 33rd session of the CCFH. The replies to these questions are summarized below.

Question 1: *Estimate the risk of serious illness from* L. monocytogenes *in food when the number of organisms range from absence in 25 g to 1000 colony forming units (CFU) per gram or millilitre, or does not exceed specified levels at the point of consumption.*

Two approaches were taken: (i) the predicted risk per serving and predicted number of cases of listeriosis annually were estimated for a "worst-case" scenario by assuming that all servings had the maximum level being considered (0.04, 0.1, 1, 10, 100 and 1000 CFU/g); (ii) a more realistic, but also more complex, approach was to use a distribution of the levels of *L. monocytogenes* in foods when consumed rather than an absolute value to estimate the risk per serving and the predicted number of cases of listeriosis annually.

Comparisons between these two approaches indicated that there were vast differences in the estimated number of cases when one considers the worst-case scenario as opposed to a scenario that attempts to also consider the frequency and extent of contamination actually encountered in RTE foods. These two scenarios demonstrated that as either the frequency of contamination or the level of contamination increases, the risk and the predicted number of cases also increase. These scenarios assume that ingestion of a single cell has the possibility to cause illness. Thus, if all RTE foods went from having 1 CFU/serving to 1000 CFU/serving, the risk of listeriosis would increase 1000-fold (assuming a fixed serving size). Conversely, the effect of introducing into the food supply 10 000 servings contaminated with *L. monocytogenes* at a level of 1000 CFU/g would, in theory, be compensated by removing from the food supply a single serving contaminated at a level of 10^7 CFU/g.

In interpreting these results and the actual effect of a change in the regulatory limits for *L. monocytogenes* in RTE foods, one also has to take into account the extent to which non-compliance with established limits occurs. Based on data available for the United States of America, where the current limit for *L. monocytogenes* in RTE foods is 0.04 CFU/g, the estimated number of cases for listeriosis for that population was 2130 (baseline level used in the United States *Listeria* risk assessment). If a level of 0.04 CFU/g was consistently achieved, one could expect less than 1 case of listeriosis per year. This, in combination with available exposure data, suggests that a portion of RTE food contains a substantially greater number of the pathogen than the current limit and that the public health impact of *L. monocytogenes* is almost exclusively a function of the foods that greatly exceed the current limit. Therefore it could be asked if a less stringent microbiological limit for RTE foods could be beneficial in terms of public health if it simultaneously fostered the adoption of control measures that resulted in a substantial decrease in the number of servings that greatly exceeded the established limit.

To examine this concept further, a simple "what-if" scenario was developed describing the impact on public health of the level of compliance to a microbiological limit. Two often discussed limits, 0.04 CFU/g and 100 CFU/g, were examined in conjunction with different "defect rates" (a defect rate is the percentage of servings that exceed the specified limit). To simplify the model, a single level of *L. monocytogenes* contamination, 10^6 CFU/g, was assumed for all "defective" servings. This assumption focuses the scenario on the group of defective servings that is responsible for the majority of listeriosis cases. Data demonstrate that at 100% compliance, the number of predicted cases is low for both limits, with an approximate 10-fold difference between them, that is 0.5 cases versus 5.7 cases. As expected the number of cases increases with an increasing frequency of defective servings. However, it is possible that public health could be improved if an increase in the regulatory limit in RTE foods resulted in a substantial decrease in the number of servings that greatly exceeded the established limit, i.e. if the rate of compliance increased.

To summarize, the risk assessment demonstrates that the vast majority of cases of listeriosis result from the consumption of high numbers of *Listeria*, and foods where the level of the pathogen does not meet the current criteria, whatever they may be (0.04 or

100 CFU/g). The model also predicts that the consumption of low numbers of *L. monocytogenes* has a low probability of causing illness. Eliminating higher levels of *L. monocytogenes* at the time of consumption has a large impact on the number of predicted cases of illness.

Question 2: Estimate the risk of serious illness for consumers in different susceptible population groups (elderly, infants, pregnant women and immunocompromised patients) relative to the general population.

These results showed that the probability of becoming ill from ingesting *L. monocytogenes* was higher for susceptible populations (immunocompromised; elderly; and perinatal) than the general population. The probability of becoming ill was also shown to vary between the sub-groups of the susceptible population. Based on susceptibility information available from the United States of America, it was determined that the elderly (60 years and older) were 2.6 times more susceptible relative to the general healthy population, while perinatals were 14 times more susceptible. Conditions that compromise the immune system also affect susceptibility to varying extents (Table 2). These results are consistent with the physiological observation that, as an individual's immune system is increasingly compromised, the risk of listeriosis at any given dose increases.

Table 2 Relative susceptibilities for different sub-populations based on French epidemiological data.

Condition	Relative susceptibility
Transplant	2584
Cancer-Blood	1364
AIDS	865
Dialysis	476
Cancer-Pulmonary	229
Cancer-Gastrointestinal and liver	211
Non-cancer liver disease	143
Cancer-Bladder and prostate	112
Cancer-Gynaecological	66
Diabetes, insulin dependent	30
Diabetes, non-insulin dependent	25
Alcoholism	18
Over 65 years old	7.5
Less than 65 years, no other condition	1

Question 3: Estimate the risk of serious illness from L. monocytogenes *in foods that support its growth and foods that do not support its growth at specific storage and shelf-life conditions.*

The risk assessment provides three approaches for answering the question: (i) the general consideration of the impact of the ingested dose on the risk of listeriosis; (ii) a comparison of four foods that were selected (according to diversity of prevalence and level of contamination, food composition and consumption patterns), in part, to evaluate the effect of *L. monocytogenes* growth or non-growth on risk; and (iii) the ability to conduct "what-if scenarios" for the evaluated foods that support growth of *L. monocytogenes*.

The results of the risk assessment show that the potential for growth of *L. monocytogenes* strongly influences risk, though the extent to which growth occurs is dependant on the characteristics of the food and the conditions and duration of refrigerated storage. Using the selected RTE foods, their ability to support the growth of *L. monocytogenes* appears to increase the risk of listeriosis 100- to 1000-fold on a per-serving basis. While it is not possible to present a single value for the increased risk for all RTE foods, because of the divergent properties of the foods, the ranges of values estimated in the risk assessment provide some insight into the magnitude of the increase in risk that may be associated with the ability of food to support the growth of *L. monocytogenes.* Control measures that focus on reduction of both frequency and levels of contamination have an impact on reducing rates of listeriosis. Controlling growth post-processing is one of these measures.

KEY FINDINGS

The most important key findings of the risk assessment as a whole are:

- The probability of illness from consuming a specified number of *L. monocytogenes* is appropriately conceptualized by the disease triangle, where the food matrix, virulence of the strain and susceptibility of the consumer are all important factors.
- The models developed predict that nearly all cases of listeriosis result from the consumption of high numbers of the pathogen.
- Based on the available data, there is no apparent evidence that the risk from consuming a specific number of *L. monocytogenes* varies for the equivalent population from one country to another. Differences in manufacturing and handling practices in various countries may affect the contamination pattern and therefore the risk per serving for a food. The public health impact of a food can be evaluated by both the risk per serving and the number of cases per population per year.
- Control measures that reduce the frequencies of contamination will have a proportional reduction in the rates of illness, provided the proportions of high contaminations are reduced similarly. Control measures that prevent the occurrences of high levels of contamination at consumption would be expected to have the greatest impact on reducing rates of listeriosis.
- Although high levels of contamination at retail are relatively rare, improved public health could be achieved by reducing these occurrences at manufacture and retail in foods that do not permit growth. In foods that permit growth, control measures such as better temperature control or limiting the length of storage periods will mitigate increased risk due to increases in *L. monocytogenes*.
- The vast majority of cases of listeriosis are associated with the consumption of foods that do not meet current standards for *L. monocytogenes* in foods, whether that standard is zero tolerance or 100 CFU/g.

LIMITATIONS AND CAVEATS

- The risk assessment focuses on four RTE foods and only examines them from retail to consumption.
- The risk characterization results are subject to uncertainty associated with a modelled representation of reality involving simplification of the relationships among prevalence, cell number, growth, consumption characteristics and the adverse response to consumption of some number of *L. monocytogenes* cells. However, the modelling is appropriate to quantitatively describe uncertainty and variability related to all kinds of factors and attempts to provide estimates of the uncertainty and variability associated with each of the predicted levels of risk.
- The amount of quantitative data available on *L. monocytogenes* contamination was limited and restricted primarily to European foods.
- Data on the prevalence and number of *L. monocytogenes* in foods came from many different sources, which adds to uncertainty and variability. Also, assumptions had to be made with regard to distribution of the pathogen in foods.
- The data used for prevalence and cell numbers may not reflect changes in certain commodities that have occurred in the food supply chain during the past ten years.
- The consumption characteristics used in the risk assessment were primarily those for Canada or the United States of America.
- The r-values and their distributions were developed using epidemiological data on the current frequency of *L. monocytogenes* strain diversity observed, with their associated virulence. If that distribution of virulence were to change (as reflected by new epidemiological data), the r-values would have to be re-calculated.
- There is uncertainty associated with the form of the dose-response function used, and with the parameterization. Also, the dose-response section of the hazard characterization is entirely a product of the shape of the distribution of predicted consumed doses in the exposure assessment component of the *Listeria* risk assessment undertaken in the United States of America (FDA/FSIS, 2001). Therefore its validity is dependant on the validity of the FDA/FSIS exposure assessment, and changes to that exposure assessment should lead directly to changes in the parameter, r.
- Predictive modelling was used to model the growth of *L. monocytogenes* in RTE foods, between the point of retail and the point of consumption, and the exposure assessment was based on information derived from those models. It is known that models may overestimate growth in food, and so reliance on such a model can result in an overestimation of the risk.

CONCLUSION

This risk assessment reflects the state of knowledge on listeriosis and on contamination of foods with *L. monocytogenes* when the work was undertaken, in 2002. New data is constantly becoming available, but in order to complete this work it was not possible to incorporate the very latest data in the risk assessment. A future iteration of the work would incorporate such new data.

The risk assessment provides an insight into some of the issues to be addressed in order to control the problems posed by *L. monocytogenes,* and approaches for modelling a system to evaluate potential risk management options. It addresses the specific questions posed by the CCFH and provides a valuable resource for risk managers in terms of the issues to be considered when managing the problems associated with *L. monocytogenes,* and alternative or additional factors or means to consider when addressing a problem. For example, if a limit is being established, then the technical feasibility of achievable levels of compliance must also be considered. While the available data were considered adequate for the current purposes, the risk assessment could be improved with additional data of better quality for every factor in the assessment. For example, quantification provides new perspectives on the risk posed by exposure to different doses of *L. monocytogenes*. The gaps in the database have been identified and could be used as a basis for establishing priorities for research programmes. The risk assessment improves our overall understanding of this issue and can therefore pave the way for risk management action to address this problem at the international level.

REFERENCES CITED IN THE EXECUTIVE SUMMARY

Aureli, P., Fiorucci, G.C., Caroli, D., Marchiaro, B., Novara, O., Leone, L. & Salmoso, S. 2000. An outbreak of febrile gastroenteritis associated with corn contaminated by *Listeria monocytogenes*. *New England Journal of Medicine*, **342**: 1236–1241.

Bille, J. 1990. Epidemiology of listeriosis in Europe, with special reference to the Swiss outbreak. pp. 25–29, *in:* A.J. Miller, J.L. Smith and G.A. Somkuti (eds). *Topics in Industrial Microbiology: Foodborne Listeriosis*. New York NY: Elsevier Science Pub.

Broome, C.V., Gellin, B. & Schwartz, B. 1990. Epidemiology of listeriosis in the United States. pp. 61–65, *in:* A.J. Miller, J.L. Smith and G.A. Somkuti (eds). *Topics in Industrial Microbiology: Foodborne Listeriosis*. New York NY: Elsevier Science Pub.

CAC [Codex Alimentarius Commission]. 2000. Report of the 33rd Session of the Codex Committee on Food Hygiene (CCFH), Washington, DC, 23–28 October 2000.

Dalton, C.B., Austin, C.C., Sobel, J., Hayes, P.S., Bibb, W.F., Graves, L.M. & Swaminathan, B. 1997. An outbreak of gastroenteritis and fever due to *Listeria monocytogenes* in milk. *New England Journal of Medicine,* **336**: 100–105.

FDA/FSIS [U.S. Food and Drug Administration/Food Safety and Inspection Agency (USDA)]. 2001. Draft Assessment of the relative risk to public health from foodborne *Listeria monocytogenes* among selected categories of ready-to-eat foods. Center for Food Safety and Applied Nutrition (FDA) and Food Safety Inspection Service (USDA) (Available at: www.foodsafety.gov/~dms/lmrisk.html). [Report published September 2003 as: Quantitative assessment of the relative risk to public health from foodborne *Listeria monocytogenes* among selected categories of ready-to-eat foods. Available at: www.foodsafety.gov/~dms/lmr2-toc.html].

McLauchlin, J. 1997. The pathogenicity of *Listeria monocytogenes*: A public health perspective. *Reviews in Medical Microbiology*. **8**: 1–14.

Mead, P.S., Slutsker, L., Dietz, V., McCraig, L.F., Bresee, J.S., Shapiro, C., Griffin, P.M. & Tauxe, R.V. 1999. Food-related illness and death in the United States. *Emerging Infectious Diseases*, **5**: 607–625.

Riedo, F.X., Pinner, R.W., De Lourdes Tosca, M., Cartter, M.L., Graves, L.M., Reeves, M.W., Weaver, R.E. et al. 1994. A point-source foodborne outbreak: Documented incubation period and possible mild illness. *Journal of Infectious Diseases,* 170: 693–696.

Roberts, D. 1989. *Listeria monocytogenes* in foods – results of two PHLS [Public Health Laboratory Service] surveys. (In Annual General Meeting and Summer Conference, [see FSTA (1990) 22 6A42].) *Journal of Applied Bacteriology*, **67**(6): xix.

Roberts, T. & Pinner, R. 1990. Economic impact of disease caused by *L. monocytogenes*. pp. 137–149, *in:* A.J. Miller, J.L. Smith and G.A. Somkuti (eds). *Topics in Industrial Microbiology: Foodborne Listeriosis*. New York NY: Elsevier Science Pub.

Rocourt, J. 1996. Risk factors for listeriosis. *Food Control*, **7**: 192–202.

Salamina, G., Dalle Donne, E., Niccolini, A., Poda, G., Cesaroni, D., Bucci, M., Fini, R., Maldin, M., Schuchat, A., Swaminathan, B., Bibb, W., Rocourt, J., Binkin, N. & Salmasol, S. 1996. A foodborne outbreak of gastroenteritis involving *Listeria monocytogenes*. *Epidemiology and Infection*, **117**: 429–436.

Part 1.

Hazard Identification

1.1 HISTORICAL

Early reports suggest that *Listeria monocytogenes* may have been isolated from tissue sections of patients in Germany in 1891, from rabbit liver from Sweden in 1911, and from spinal fluid of meningitis patients in 1917 and again in 1920 (Reed, 1958; McCarthy, 1990). However, it was not until 1926 that the microorganism was fully described, when Murray, Webb and Swann (1926) isolated a small, Gram-positive rod bacterium that had caused an epizootic outbreak in 1924 among rabbits and guinea pigs. They named the organism *Bacterium monocytogenes*. This was a year after listeriosis in sheep was recognized in Germany as a disease syndrome, though the causative agent had not been isolated. At approximately the same time, Pirie (1927) isolated and described the same organism from gerbils in South Africa. He named the bacterium *Listerella hepatolytica*, and subsequently recommended in 1940 that the name be changed to *Listeria monocytogenes* (Reed, 1958; McCarthy, 1990). The first report of human listeriosis was in 1929, and the first perinatal case was reported in 1936 (Gray and Killinger, 1966). The microorganism has been reported to cause disease in a wide range of wild and domestic animals, and has been isolated from numerous species of mammals, birds, amphibians, fish, crustaceans, insects and reptiles (Hird and Genigeorgis, 1990; McCarthy, 1990; Ryser and Marth, 1991).

It is now widely recognized that human listeriosis is largely attributable to foodborne transmission of the microorganism. However, the first case of foodborne listeriosis was not reported until 1953, when the stillbirths of twins was linked to consumption by the mother of raw milk from a cow with listerial mastitis (Potel, 1953). It was not until several large, common-source outbreaks of listeriosis occurred in North America and Europe during the 1980s that the significance of foods as the primary route of transmission for human exposure to *L. monocytogenes* was recognized (Broome, Gellin and Schwartz, 1990; Bille, 1990). While the modes of transmission for *L. monocytogenes* can include vertical (mother to child), zoonotic (contact with animal to man), and nosocomial (hospital acquired), it is generally considered that most cases of human listeriosis involve foodborne transmission.

1.2 CHARACTERISTICS OF *LISTERIA MONOCYTOGENES*

L. monocytogenes is a Gram-positive, facultatively anaerobic, non-sporeforming rod, which expresses a typical tumbling motility at 20–25°C, but not at 35°C. The organism is psychrotrophic and grows over a temperature range of 0° to 45°C, with an optimum around 37°C. *L. monocytogenes* can grow at pH levels between 4.4 and 9.4, and at water activities ≥0.92 with sodium chloride (NaCl) as the solute (Miller, 1992). The effects of temperature, pH, water activity, oxygen availability and antimicrobials on the growth of *L. monocytogenes* have been studied extensively in both model systems and foods, and there are a number of mathematical models available for describing the interaction of these factors with the growth rate (Buchanan and Phillips, 2000).

L. monocytogenes is widely distributed in the environment and has been isolated from a variety of sources, including soil, vegetation, silage, faecal material, sewage and water. The bacterium is resistant to various environmental conditions, such as high salinity or acidity, which allows it to survive longer under adverse conditions than most other non-sporeforming bacteria of importance in foodborne disease (McCarthy, 1990; Ryser and Marth, 1991). *L. monocytogenes* occurs widely in food processing environments (Ryser and Marth, 1991, 1999), and can survive for long periods in foods, in processing plants, in households, or in the environment, particularly at refrigeration or frozen storage temperatures. The ability of *L. monocytogenes* to survive in foods and model systems has been studied extensively, and mathematical models are available that describe the effect of various environmental parameters on the microorganism's survival (Buchanan and Golden, 1994, 1995, 1998; Buchanan, Golden and Whiting, 1993; Buchanan et al., 1994; Buchanan, Golden and Phillips, 1997).

Although frequently present in raw foods of both plant and animal origin, it is also present in cooked foods due to post-processing contamination if the cooked food is handled post-cooking. *L. monocytogenes* has been often isolated from food processing environments, particularly those that are cool and wet. *L. monocytogenes* has been isolated in foods such as raw and pasteurized fluid milk, cheeses (particularly soft-ripened varieties), ice cream, raw vegetables, fermented raw meat and cooked sausages, raw and cooked poultry, raw meats, and raw and smoked seafood (Buchanan et al., 1989; Farber and Peterkin, 1991; FDA/FSIS, 2001; Ryser and Marth, 1991, 1999). Even when *L. monocytogenes* is initially present at a low level in a contaminated food, its ability to grow during refrigerated storage means that its levels are likely to increase during storage of those foods that can support the growth of the microorganism. A survey of a wide variety of foods from the refrigerators of listeriosis patients in the United States of America found *L. monocytogenes* in at least one food specimen in 64% of the patient's refrigerators. Food in 33% of the refrigerators had the same strain as the patient strain (Pinner et al., 1992). However, because the frequency at which people are exposed to *L. monocytogenes* is much higher than the incidence of listeriosis, there has been a public health debate about the significance of ingesting low levels of the pathogen, particularly for the portion of the population who are not immunologically compromised (Farber, Ross and Harwig, 1996; ICMSF, 1994).

1.3 OVERVIEW OF LISTERIOSIS

Listeriosis is a relatively rare disease. The reported yearly incidence of human listeriosis ranges from 0.1 to 11.3 cases per million persons (references cited in Notermans et al., 1998), with for example 0.3 to 7.5 cases per million people in Europe (EC, 2003), and 3 cases per million people in Australia. The data from the U.S. Centers for Disease Control and Prevention (CDC) active food surveillance programme, FoodNet, for the years from 1996 to 1998 indicate that there were about 5 reported cases of listeriosis per 1 000 000 population annually. Using the CDC 1996–97 surveillance data (CDC, 1998) and extrapolating to the 1997 total United States of America population, Mead et al. (1999) estimated that there were 2493 cases, including 499 deaths, due to foodborne listeriosis Although listeriosis is a relatively rare foodborne illness (Table 1.1), its severe nature makes it likely that individuals will seek medical care. In the United States of America, where listeriosis is a "reportable" disease, CDC estimates that it recognizes and identifies approximately half of all listeriosis cases, as compared to the 3% identification rate for most other foodborne pathogens (Mead et al., 1999).

One of the difficulties in characterizing the hazard associated with foodborne listeriosis is that there are no clear definitions for infection or cases in humans. In general, most cases that are reported to medical authorities are severe infections requiring medial intervention. Thus, for the purposes of the current hazard characterization, an infection in humans will be based on the colonization of the host, i.e. attachment and growth, which can include individuals that are asymptomatic, displaying febrile gastroenteritis, or suffering from severe symptoms or death. The terms "severe infection" or "invasive listeriosis" will be used to describe those infected individuals with life-threatening, systemic infections, such as perinatal listeriosis, meningitis or septicaemia, and where *L. monocytogenes* is present in normally sterile body tissues.

Table 1.1. Estimated incidence of foodborne disease from epidemiological surveillance.

Pathogen	Cases per 1 000 000 population
Vibrio	3
Listeria	5
Yersinia	10
E. coli O157:H7	28
Shigella	85
Salmonella	124
Campylobacter	217
All bacterial pathogens	472

SOURCE: FoodNet data for 1997 (CDC, 1998).

Invasive *L. monocytogenes* infections can be life threatening, with fatality rates of 20 to 30% being common among hospitalized patients. In 2000, the CDC (2000) reported that, of all the foodborne pathogens tracked by CDC, *L. monocytogenes* had the second-highest case fatality rate (21%) and the highest hospitalization rate (90.5%).

L. monocytogenes causes invasive listeriosis by penetrating the lining of the gastrointestinal tract and establishing infections in normally sterile sites within the body. Once *L. monocytogenes* penetrates the intestinal tissue it is taken up by cells of the immune system, the phagocytes. However, inside the phagocyte it is capable of escaping from the phagosome and subsequently growing. Phagocytes appear to be the means by which the bacterium can be transported to various parts of the body (Shelef, 1989; Farber and Peterkin, 1991).

The likelihood that *L. monocytogenes* will invade the intestinal tissue depends upon a number of factors, including the number of organisms consumed, host susceptibility, and virulence of the specific isolate (Gellin and Broome, 1989). Incubation periods can be long, e.g. typically 2-3 weeks, and up to three months (Gellin and Broome, 1989). *L. monocytogenes* can produce a wide range of symptoms. In non-pregnant adults, disease syndromes most commonly linked to *L. monocytogenes* include bacteraemia, meningitis and encephalitis (Rocourt and Cossart, 1997). In pregnant women, *L. monocytogenes* often causes an influenza-like bacteraemic illness, which leads to amnionitis and infection of the fetus and results in abortion, stillbirth or premature birth. Listeriosis occurs most often either very early in life or after 60 years of age. Figure 1.1 shows listeriosis incidence by age, using 1997 FoodNet data. The incidence of listeriosis in males and females is approximately equal.

foodborne listeriosis. Industry initiated HACCP programmes and increased sanitation efforts to eliminate contamination. Food control agencies expanded programmes to prevent contaminated foods from entering commerce. There were also consumer education campaigns that focused on food safety. In the United States of America, a reduction in listeriosis from 7.9 per million in 1989 to 4.4 per million in 1993 was observed (Tappero et al., 1995). Rates of listeriosis simultaneously declined in the United Kingdom after the British government issued health warnings regarding *L. monocytogenes* (Fyfe et al., 1991; McLauchlin et al., 1991). Similar declines have been reported as a result of public health initiatives in other parts of Europe and in Australia (Jacquet et al., 1999). For example, it is reported that preventative measures implemented by the French food industry played a substantial role in the 68% reduction observed in France between 1987 and 1997 (Goulet et al., 2001a). However, since that time, the incidence of listeriosis has remained relatively constant (CDC, 2000). The reported yearly incidence of human listeriosis in Europe ranges from 0.1 to 11.3 cases per 10^6 persons (references cited in Notermans et al., 1998). A more recent study within the European Union indicates a slight decrease, with the reported yearly incidence of listeriosis for 2000-2001 ranging from 0.3 to 7.8 cases per million persons (de Valk et al., 2003). However, the accuracy of these values is dependent on the vigour with which individual countries conduct national surveillance programmes for listeriosis.

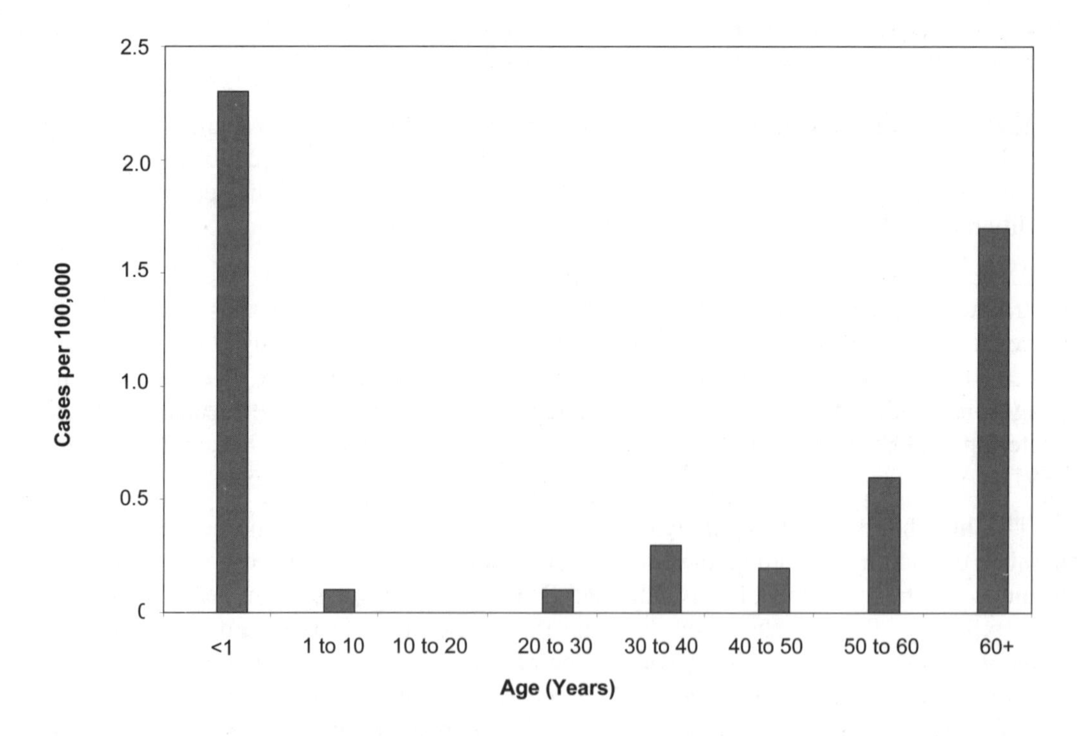

Figure 1.1. Estimated rate of listeriosis by age.
SOURCE: FoodNet 1997 data (CDC, 1998).

1.4 STATEMENT OF PROBLEM AND SCOPE OF RISK ASSESSMENT

Foodborne listeriosis represents a relatively rare but clinically serious disease that largely affects specific higher-risk segments of the population. The microorganism is widely dispersed in the environment and foods and it appears to be ingested in low numbers by consumers on a routine basis. Despite the fact that a wide variety of foods may be contaminated with *L. monocytogenes*, outbreaks and sporadic cases of listeriosis appear to be predominately associated with ready-to-eat (RTE) products. A number of risk assessments and related evaluations of foodborne listeriosis have been conducted by investigators and national governments. The current risk assessment was undertaken to determine, in part, how previously developed risk assessments done at a national level could be adapted or expanded to address concerns related to *L. monocytogenes* in RTE foods at an international level. This included an international team conducting a risk assessment to answer questions posed by an international organization. Data from different countries were used. This does not imply that the risk assessment reflects the global food supply or that the specific results are universally applicable, as different countries will have differences in contamination levels, processing and consumption patterns that are not addressed in this risk assessment. In addition, after initiation of the risk assessment, the risk assessment developers were asked by the Codex Committee on Food Hygiene (CCFH), through FAO/WHO, to consider three specific points related to RTE foods in general, namely:

- Estimate the risk of serious illness from *L. monocytogenes* in food when the number of organisms ranges from absence in 25 grams to 1000 colony forming units (CFU) per gram or millilitre, or does not exceed specified levels at the point of consumption.
- Estimate the risk of serious illness for consumers in different susceptible population groups (elderly, infants, pregnant women and immunocompromised patients) relative to the general population.
- Estimate the risk of serious illness from *L. monocytogenes* in foods that support its growth and foods that do not support its growth at specific storage and shelf-life conditions.

Considering the resources available and the time constraints placed on the risk assessment developers, it was impossible to consider all RTE foods that could be contaminated with *L. monocytogenes*. Accordingly, it was decided to limit the risk assessment to a small number of RTE foods selected to represent various classes of product characteristics. These foods were selected to provide realistic examples of how microbiological risk assessment techniques could be used to evaluate food safety questions at an international level. This educational component is a stated goal of the FAO/WHO microbiological risk assessment programme under the auspices of which the current risk assessment was developed. It was also decided to limit the scope of the risk assessment to foods at retail and their subsequent public health impact at time of consumption. This was done for two reasons. The first is that such a scope was sufficient to address the charge provided by the requestors of the risk assessment within the time frames and resources made available to the developers. Second, most of the exposure data for *L. monocytogenes* that are currently available are frequencies and extents of contamination at the retail level. More detailed examination of factors contributing to the levels found at retail as a result of manufacturing parameters would have either restricted evaluation to a much smaller range of foods, or required that substantially greater resources and data be made available. Accordingly, the assessment does not evaluate the risks associated with different means of manufacturing the products selected. However, the risk assessment does consider several post-retail factors that could influence the consumers' risk of acquiring foodborne listeriosis, such as the temperature and duration of refrigerated storage.

Part 2.

Hazard Characterization

2.1 LISTERIOSIS

Most cases of human listeriosis appear to be sporadic, although a portion of these cases may represent previously unrecognized common-source clusters (Broome, Gellin and Schwartz, 1990; Farber and Peterkin, 1991). The source and route of infection is usually unknown, but contaminated food is considered to be the principal route of transmission, and estimated to be the source in as high as 99% of the cases (WHO, 1988; Mead et al., 1999).

L. monocytogenes appears to be a frequent transitory resident of the intestinal tract in humans. The proportion of individuals whose faecal samples have been positive for *L. monocytogenes* range from a low 0.5% to a high 29% (Farber and Peterkin, 1991). On average, 2 to 10% of the general population are carriers of the organism without any apparent adverse consequences (Farber and Peterkin, 1991; Rocourt and Cossart, 1997; Skidmore, 1981; Slutsker and Schuchat, 1999; Mascola et al., 1992; Schuchat, Swaminathan and Broome, 1991). Because of the high rate of clinically healthy carriers, Farber and Peterkin (1991) suggested that the presence of *L. monocytogenes* in the faeces is not necessarily an indication of infection. The role of healthy carriers is not clear, but investigations during an outbreak in California in 1985 suggested that community-acquired outbreaks might be amplified through secondary transmission by stool carriers (Rocourt, 1996). Pregnancy, while predisposing to listeriosis, does not seem to predispose women to carriage of the organism (Lamont and Postlethwaite, 1986). Healthy pregnant women may be carriers of *L. monocytogenes* and still give birth to healthy infants.

2.1.1 Manifestations of listeriosis

The pregnant uterus, the central nervous system (CNS) or the blood are the locations where bacteria are most often found when initial infections of the intestinal tissue by *L. monocytogenes* leads to invasion of otherwise sterile body sites. A summary of 782 cases of listeriosis reported from 20 countries in 1989 showed that 43% were perinatal (prenatal + neonatal) infections, 29% were septicaemic infections, 24% were CNS infections and 4% were atypical forms (Rocourt, 1991). However, changes in the epidemiology of listeriosis over the past ten years have been noted. For example, more recent data from England and Wales, France, Denmark and the United States of America show that the proportion of pregnancy related cases ranges from 11 – 31%: 17% in England and Wales in 1995-99 (Smerdon et al., 2001), 11% in Denmark in the period 1999-2000, 24% in France in 1999 (Goulet et al., 2001b) and 31% in the United States of America in 1993 (Tappero et al., 1995). In the non–pregnancy related group, the proportion of bacteraemic forms has increased and represents at least two thirds of the cases. This form nearly always occurs in patients with an underlying disease, whereas CNS infection also occurs in previously healthy persons. Sequelae may follow listeriosis infections (McLauchlin, 1997), but their incidence is rarely estimated (Rocourt, 1996). Up to 11% of neonates and 30% of survivors of CNS infection suffer from residual symptoms, and psychiatric sequelae have also been reported (references cited in Rocourt, 1996). However, for survivors of CNS infection 30% is considered unusually high and the rate of occurrence of sequelae is normally lower. For example, in a study of 225 patients in France in 1992, neurological sequelae were observed in 12% of patients and 15% of survivors (Goulet and Marchetti, 1996).

A classification scheme has been proposed for differentiating the manifestations of syndromes associated with *L. monocytogenes* that takes into consideration host status, route of transmission, severity and incubation period (Table 2.1). It has been estimated that as much as 20% of the population may belong to groups with a greater risk for developing listeriosis (Buchanan et al., 1997; Lindqvist and Westöö, 2000). These higher-risk people can be divided into non-perinatal and perinatal groups. When severe infection occurs in adults and children, listeriosis is usually superimposed on another illness (Lorber, 1990; Broome, Gellin and Schwartz, 1990; Schuchat, Swaminathan and Broome, 1991; Shelef, 1989; Gray and Killinger, 1966; Linnan et al., 1988; WHO, 1988.). The high-risk cases primarily consist of persons with chronic debilitating illnesses that impair their immune system, such as cancer, diabetes or alcoholism; HIV/AIDS; persons taking immunosuppressive medication (e.g. immune suppressors taken by transplant patients); and persons over the age of 60–65, particularly individuals with pre-existing, debilitating medical conditions. Healthy children and immunocompetent adults have a low risk of severe infection from *L. monocytogenes*.

There have also been a number of outbreaks where the majority of cases developed mild symptoms (Dalton et al., 1997; Salamina et al., 1996; Riedo et al., 1994; Aureli et al., 2000), such as diarrhoea, fever, headache and myalgia. These outbreaks have generally involved the ingestion of high doses of *L. monocytogenes* by otherwise healthy individuals and these gastroenteritis symptoms generally self-resolve within a few days.

A summary of epidemiological information from some foodborne listeriosis outbreaks is shown in Table 2.2.

2.1.1.1 Systemic listeriosis

Non-perinatal

In non-pregnant humans, systemic listeriosis usually presents as either CNS infections, with or without bacteraemia, or bacteraemia alone. Cases of bacteraemia alone are often confined to the immunocompromised or elderly (McLauchlin, 1996).

In addition to these clinical manifestations, less common manifestations include peritonitis (Polanco et al., 1992; Nguyen and Yu, 1994), hepatitis and liver abscess (Bourgeois et al., 1993; Braun et al., 1993), endocarditis (Gallagher and Watanakunakorn, 1988), arterial infections (Gauto et al., 1992), myocarditis (Stamm et al., 1990), lung and pleural fluid infection (Mazzulli and Salit, 1991), septic arthritis and osteomyelitis (Louthrenoo and Schumacher, 1990; Ellis et al., 1995), and chorioretinitis, endophtalmitis and corneal ulcer (Ballen, Loffredo and Painter, 1979; Huismans 1986; Holland et al., 1987).

Table 2.1 Classification of illness caused by *Listeria monocytogenes.*

Type of Listeriosis	Mode of transmission	Severity	Time to onset
Occupational infection	Primary cutaneous listeriosis after direct contact with infected animal tissues.	Usually mild and self-resolving.	1–2 days.
Neonatal infection	Infection of newborn babies from infected mother during birth or due to cross-infection from one neonate in the hospital to other babies.	Can be extremely severe, resulting in meningitis and death.	1–2 days (early onset), usually from congenital infection prior to birth. 5–12 days (late onset), following cross-infection from another infant.
Infection during pregnancy (prenatal)	Acquired following consumption of contaminated food.	Mild flu-like illness or asymptomatic in the mother, but serious complications for unborn infant, including spontaneous abortion, fetal death, stillbirth and meningitis. Infection is more commonly reported in third trimester.	
Infection of non-pregnant adults (non-perinatal)	Acquired following consumption of contaminated food.	Asymptomatic or mild illness, which may progress to CNS infections such as meningitis. Most common in immunocompromised or elderly.	Illness may occur within 1 day or up to 3 months, but commonly within 20–30 days.
Listeria food poisoning (febrile gastroenteritis)	Consumption of food with exceptionally high levels of *L. monocytogenes*, > 10^7/g.	Vomiting and diarrhoea, sometimes progressing to bacteraemia but usually self-resolving.	<24 h after consumption.

SOURCE: Modified from Bell and Kyriakides, 1998, as described in EC, 1999.

Despite the fact that infections can be treated successfully with antibiotics, between 20 and 40% of cases are fatal (Gellin and Broome, 1989; McLauchlin, 1996). In severely immunocompromised patients, the case-fatality rate may approach 75% (Nørrung, Andersen and Schlundt, 1999).

Perinatal (prenatal/neonatal) infections

The perinatal group consists of pregnant women and their fetuses or newborns. About two-thirds of *L. monocytogenes*-infected pregnant women will present with a prodromal influenza-like illness, which includes fever, chills and headache. About three to seven days after the onset of prodromal symptoms, a woman may abort the fetus or have premature labour (Gellin and Broome, 1989). Sepsis or fever is reported in about 30% of pregnant women with listeriosis (Gellin and Broome, 1989). Women may get listeriosis at any time during pregnancy, but most cases are reported in the third trimester (Slutsker and Schuchat, 1999). In the first trimester, listeriosis may result in spontaneous abortion. In later stages of pregnancy the result may be stillbirth or a critically ill newborn. Listeriosis is rarely severe or life threatening to the mother and is not known to cause increased risk in subsequent pregnancies (Skidmore, 1981; Farber and Peterkin, 1991). The epidemiological records for prenatal cases are incomplete in that the rate of occurrence during early pregnancy and recovery of fetuses from infection are unknown. Neonatal cases of listeriosis are better documented and the rate of prenatal listeriosis was estimated to be 1.5 times that of neonates (FDA/FSIS, 2001).

Table 2.2 Summary of epidemiological information from some published foodborne listeriosis outbreaks.

Country	Number of cases				Number of deaths			Percentage of Manifestations					Source
	Total (exposed)	Healthy	Materno -fetal	Immunocom-promised	Total (%)	Adults	Perinatal	Septi-caemia	Meningitis	Other CNS	Other	GI	
Australia	9	–	–	–	6 (67)	–	–	–	–	–	–	–	[10]
Australia	4	4	0	0	0	0	0	–	–	–	100	75	[11] [12]
Australia	5	0	0	5	1 (20)	1	0	100	–	–	–	–	[24]
Denmark	26	10	3	13	6 (23)	–	–	26(4)	65(4)	–	–	–	[9]
Canada	41	7	34	0	18 (44)	2	16	14(4)	86(4)	–	–	–	[3]
Finland	25	–	0	24	6 (24)	6	0	80	16	–	4	–	[27]
France	279	62	92	125	88 (32)	59	29	–	–	–	–	–	[15] [16] [17] [28]
France	36(7)	8	18	19	9 (25)	4	5	22	67(4)	11(4)	–	–	[23] [18]
France	38	2	31	5	11 (29)	1	10	28(4)	57(4)	14(4)	45–93(1)	3(1)	[18] [19]
France	10	1	3	6	3 (30)	2	1	4	57(4)	43(4)	–	–	[29]
France	32	12	9	11	9 (28)	5	4	7	30(4)	70(4)	–	–	[29]
Italy	1566 (2930)	–	0	–	0	0	0	0	0	0	6-82	19-72	[25]
Italy	18 (39)	18	0	0	0	0	0	–		–	22–100	78	[20]
New Zealand	22	0	22	0	6 (27)	0	6	27(2), 55(1)	28(2)	–	82(1)(3)	45(1)	[2]
New Zealand	4	0	4	0	2 (50)	0	2	–	25(2)	–	–	–	[13] [14]
Sweden	8	0	3	5	2 (25)	1	1	50	25	–	–	–	[22]
Switzerland	122	33/17(5)	65	24/40(5)	34 (28)	18	16	21(4)	40(4)	39(4)	56(6)	46(4)	[6] [7]
UK	>350	–	–	–	–	–	–	–	–	–	–	–	[8]
USA	20	10	0	10	5 (25)	5	0	90	50	30	–	65	[1]
USA	49	0	7	42	14 (29)	12	2	69(4), 29(2)	31(4), 42(2)	–	–	–	[4]
USA	142	1	93	48	48 (34)	18	30	52(1), 71(4)	0(1), 14(4)	–	–	–	[5]
USA	45 (60)	44	1	0	0	0	0	–	–	–	3-72	79	[21]
USA	101	–	–	–	21	–	–	–	–	–	–	–	[26]

KEY: CNS = central nervous system. GI = gastrointestinal. – information not available

NOTES: (1) refers to the pregnant women; (2) refers to the fetus or the baby; (3) flu like illness or urinary tract symptoms; (4) refers to adults (not including pregnant women); (5) Including age >65 year as predisposing factor; (6) Including meningismus and altered mental status; (7) information given on only 20 cases.

SOURCES: [1] Ho et al., 1986. [2] Lennon et al.,1984. [3] Schlech et al., 1983. [4] Fleming et al., 1985. [5] Linnan et al., 1988. [6] Bille, 1990. [7] Büla, Bille and Glausser, 1995. [8] McLauchlin et al., 1991. [9] Jensen, Frederiksen and Gerner-Smidt, 1994. [10] Kittson, 1992. [11] Mitchell, 1991. [12] Misrachi, Watson and Coleman, 1991. [13] Baker et al., 1993. [14] Brett, Short and McLauchlin, 1998. [15 Rocourt et al., 1993. [16] Salvat et al., 1995. [17] Jacquet et al., 1995a. [18] Jacquet et al., 1995b. [19] Goulet et al., 1998. [20] Salamina et al., 1996. [21] Dalton et al., 1997. [22] Ericsson et al., 1997. [23] Goulet et al., 1995. [24] Hall et al., 1996. [25] Aureli et al., 1998. [26] Mead, 1999. [27] Lyytikäinen et al., 2000. [28] Goulet, 1995. [29] de Valk et al., 2001.

Neonates may present with an early-onset or late-onset form of listeriosis. Early onset (infected in utero) is defined as a case of listeriosis (Granulomatosis Infantisepticum) in a neonate less than 7 days old. Early-onset listeriosis is characterized by premature birth, respiratory distress and circulatory failure. Most early-onset cases present with sepsis and about 20% have meningitis. Late-onset is defined as listeriosis in a neonate between 8 to 28 days of life. Usually, late-onset neonates are born healthy and at full term. Meningitis is more common in late-onset babies (Farber, 1991). The mothers of late-onset babies usually had an uneventful pregnancy without prodromal illness. *L. monocytogenes* is rarely isolated from the mother and the source of listeriosis is often not identified in late-onset cases (Farber and Peterkin, 1991; Slutsker and Schuchat, 1999). While a number of alternative sources of *L. monocytogenes* could be hypothesized for the purposes of the current hazard characterization, it will be assumed that neonatal infections are the result of *in utero* exposure. About 25% of neonates with listeriosis die (Gellin and Broome, 1989; McLaughlin, 1990a), with the mortality rate being 15–50% in early-onset listeriosis and 10–20% in late-onset listeriosis (Farber and Peterkin, 1991).

2.1.1.2 Febrile gastroenteritis

Typical signs and symptoms associated with febrile listerial gastroenteritis include chills, fever, diarrhoea, headache, abdominal pain and cramps, nausea, vomiting, fatigue, joint and muscle pain, and myalgia. *L. monocytogenes* infection manifestation may be limited to these symptoms in otherwise healthy individuals. Although mild symptoms associated with listeriosis have been reported in several countries, and a variety of foods have been implicated as the vehicle of infection, there is a high potential for underreporting of mild illness due to *L. monocytogenes* because of the general nature of the symptoms. Table 2.3 summarizes reported outbreaks where most of the cases reported only mild symptoms (Aureli et al., 2000; Miettinen et al., 1999; Heitmann, Gerner-Smidt and Heltberg, 1997; Dalton et al., 1997; Salamina et al., 1996; Riedo et al., 1994). The reports from Italy (1997), Denmark (1996) and the United States of America (1994) are of particular note because they show that listeriosis can be limited to mild symptoms even if blood cultures are positive for *L. monocytogenes*.

Table 2.3 Reports of mild illness associated with *Listeria monocytogenes.*

Location	Year	Cases	Vehicle	Ref.
Denmark	1996	3	Unknown	[1]
Finland	1999	5	Smoked rainbow trout	[2]
Italy	1997	1 566	Maize and tuna salad	[3]
Italy	1993	18	Rice salad	[4]
USA	1994	45	Chocolate milk	[5]
USA	1989	10	Shrimp	[6]

SOURCES: [1] Heitmann, Gerner-Smidt and Heltberg, 1997. [2] Miettinen et al., 1999. [3] Aureli et al., 2000. [4] Salamina et al., 1996. [5] Dalton et al., 1997. [6] Reido et al., 1994.

There are insufficient data available about the incidence of the milder symptoms to allow the impact of this biological end point on public health to be assessed in the current exercise.

2.1.2 Foods associated with foodborne listeriosis

Food is the principal route of transmission of listeriosis (WHO, 1988). Listeriosis cases are observed in conjunction with both common-source outbreaks and individual sporadic cases. Foods of most concern include RTE products that (i) support growth of *L. monocytogenes,* (ii) have a long refrigerated shelf-life, and (iii) are consumed without further listericidal treatments (Pinner et al., 1992; Rocourt, 1996; FDA/FSIS, 2001; Nørrung, Andersen and Schlundt, 1999). This includes products that receive a listericidal treatment but are subject to post-processing recontamination. This also includes cross-contamination in both the retail and home setting. For example, in the French outbreak in 1992, cross-contamination was suspected at the distribution level (Rocourt, 1996). Similar cross-contamination is likely to occur in the home (Schwartz, Pinner and Broome, 1990).

Common-source outbreaks have been associated or linked epidemiologically with the consumption of Hispanic-style soft cheeses (*queso fresco*); soft, semi-soft and mould-ripened cheeses; hot dogs; pork tongue in jelly; processed meats; paté; salami; pasteurized chocolate-flavoured milk; pasteurized milk; unpasteurized milk; butter; cooked shrimp; smoked salmon; maize and rice salad; maize and tuna salad; potato salad; raw vegetables; and cole slaw (see FDA/FSIS, 2001). In addition, sporadic cases have been linked to the consumption of raw milk; unpasteurized ice cream; ricotta cheese; goat, sheep and feta cheeses; soft, semi-soft and mould-ripened cheeses; Hispanic-style cheese; salami; hot dogs; salted mushrooms; smoked cod roe; smoked mussels; undercooked fish; pickled olives; raw vegetables; and cole slaw.

In general, the levels of *L. monocytogenes* in the implicated food have been greater than 10^3 CFU/g (EC, 1999), but there have been instances where the observed level of *L. monocytogenes* in the implicated food has been substantially lower. However, there is a great deal of uncertainty concerning these estimates because the actual level of the pathogen in a serving of food consumed by an individual could have varied considerably from that observed in other portions of the food during a subsequent investigation.

2.2 DOSE-RESPONSE RELATIONS

2.2.1 Characterization of severity and the selection of appropriate biological end points to be modelled

The severity of a hazard can be evaluated by qualitative, semi-quantitative and quantitative approaches. Roberts, Ahl and McDowell (1995) summarized approaches used to rank or prioritize different foodborne illnesses in terms of their severity or consequences. Different criteria used to evaluate severity included:

- The number of acute illness cases.
- The number of deaths.
- The number of chronic illness cases.
- The quality-adjusted life-years lost due to the illness.
- The damage to society in terms of medical costs and loss of productivity.
- The willingness of the society to pay for reducing the risk of illness (Roberts, Ahl and McDowell, 1995).

Other work to assess or describe the severity of microbial hazards has tried to relate the dose to the severity of the disease (Glynn and Bradley, 1992). Due to the difficulty of obtaining the relevant information during outbreaks, case fatality rate and hospitalization rate have been used for assessing severity, while attack rate, incubation period, amount of contaminated food and the vehicle involved have been used as proxy measures of infecting dose. At least for *Salmonella,* there appears to be an association between the dose and the incubation period. A correlation between dose and severity was found also for some of the food poisoning salmonellae, but not for S*almonella. typhi* (Glynn and Bradley, 1992; Glynn and Palmer, 1992). No similar evaluation for the relation between the dose and severity of illness, or the dose and incubation period for *L. monocytogenes* was found in the literature.

In this risk assessment, characterization of the severity of listeriosis is limited to a description of the manifestations of the disease and a summary of epidemiological information from outbreaks. The quantitative relationship between the dose and the severity is addressed by selection of the biological end-points to be modelled, i.e. infection, morbidity or mortality. However, this is complicated because infection has been differently defined and estimated in different studies, e.g. faecal-positive versus an infected spleen. Similarly, the morbidity endpoint may cover the whole range from mild to severe manifestations. If the probability of morbidity after an infection were low, the use of the infection endpoint would be excessively conservative for a risk assessment model. The shape of the dose-response relationship, and thus the appropriate dose-response model, for these two biological end-points may also be different (FDA/FSIS, 2001).

2.2.2 Factors that affect dose-response relations for *L. monocytogenes*

The response of a human population to exposure of a foodborne pathogen is highly variable. This reflects the fact that the incidence of disease is dependent on a variety of factors, such as the virulence characteristics of the pathogen, the numbers of cells ingested, the general health and immune status of the host, and any attributes of the food that alter microbial or host status. Thus, the likelihood that any individual will become ill due to an exposure to a foodborne pathogen depends on the integration of host, pathogen and food matrix effects. These interactions are often referred to as the infectious disease triangle. Each of these classes of factors and how they affect the dose-response relations for *L. monocytogenes* will be discussed briefly.

2.2.2.1 Virulence of L. monocytogenes *isolates*

The traditional taxonomic scheme for the genus *Listeria* differentiates the species *L. monocytogenes* and *Listeria innocua* based on their ability to produce listeriolysin. Otherwise, the two species have nearly identical cultural and biochemical characteristics. Listeriolysin is a haemolysin (i.e. an enzyme capable of lysing red blood cells) produced by *L. monocytogenes* that is associated with the microorganism's ability to cause disease. Thus taxonomically, all *L. monocytogenes* were presumed to be pathogenic. However, the relative virulence of individual *L. monocytogenes* isolates can vary substantially (at least in animal models), presumably from different forms of other virulence factors (Hof and Rocourt, 1992). This variability influences the number of microorganisms required to produce an infection, the potential for an infection to become symptomatic, the severity or manifestations of illness, and the population at greatest risk.

Invasive listeriosis is characterized by bacterial dissemination to the CNS and the feto-placental unit, due to the capacity of *L. monocytogenes* to cross the intestinal barrier, the blood-brain barrier and the feto-placental barrier. Recent advances in the study of virulence factors have improved our understanding of the steps in the infection process at the cellular level, although much remains unknown (Lecuit et al., 1999; Vazquez-Boland et al., 2001). *L. monocytogenes* are facultative intracellular parasites and one important feature of this bacterium is its ability to induce its own internalization into cells that are normally nonphagocytic. *L. monocytogenes* readily invades many types of cells *in vitro*, which suggests that there may be multiple routes by which the bacterium invades the host, but animal experiments suggest that the small intestine acts as the primary site of invasion (McLauchlin, 1997). Two invasion proteins of *L. monocytogenes* have been characterized – internalin A (InlA) and B (InlB) – that mediate entry into different cell types. The bacterial surface protein InlA is necessary for *L. monocytogenes* entry into human gut epithelial cells in the small intestine through binding to a human host receptor, a protein called E-cadherin (Lecuit et al., 2001). This appears to be a host-specific process and it was recently shown that InlA interacts with human and guinea pig E-cadherin, but not with mouse and rat E-cadherin (Lecuit et al., 1999, 2001). Thus, *L. monocytogenes* readily invades human and guinea pig gut epithelial cells but not mouse and rat epithelial cells. Instead, in mice it has been demonstrated that *L. monocytogenes* may colonize the Peyers's patches of the host through the M cells (Vazquez-Boland et al., 2001). These results indicate that the mouse, which has been the most widely used animal model for the study of listeriosis, is inappropriate to study specific features of human listeriosis, i.e. the crossing of the intestinal barrier following exposure via the oral route. This is in contrast to the guinea pig (Lecuit et al., 1999) or a newly developed transgenic mouse model expressing human E-cadherin (Lecuit et al., 2001). The impact of this limitation for other aspects of listeriosis studied in a mouse or rat model is unclear at present, but this example illustrates the complexity of the pathogenesis of many bacterial diseases and the necessity for careful evaluation of results from surrogate animal studies.

After invasion of gut epithelial cells, *L. monocytogenes* bacteria are carried to the lymph nodes and then other tissues, including the spleen and the liver, by dendritic cells, phagocytes or as free cells (Pron et al., 2001). Based on experimental infection of mice via the intravenous route, it appears that most of the *L. monocytogenes* bacteria accumulate in the liver and that most of the ingested bacteria are killed by resident macrophages in the spleen and liver (Vazquez-Boland et al., 2001). However, surviving *L. monocytogenes* cells start multiplying in the liver, the principal site being the hepatocytes (Vazquez-Boland et al., 2001). In the majority of individuals, *L. monocytogenes* invasion may be successfully cleared, but if the infection is not controlled by an adequate immune response, proliferation of *L. monocytogenes* may result in the release of bacteria into the circulation system and a successive invasion of other sites, such as the uterus, fetus or CNS (McLauchlin, 1997; Vazquez-Boland et al., 2001).

From a mechanistic perspective, the virulence of *L. monocytogenes* has been studied extensively. Most studies of *L. monocytogenes* virulence have used genetically inbred mouse varieties as the surrogate animal model, and have, of necessity, been conducted using well-characterized strains of *L. monocytogenes* selected – or in some cases genetically modified – for the presence or absence of the specific virulence genes. These studies have discovered a large number of virulence determinants involved in the entry and colonization of host tissue. Examples of steps in the infection process include internalization by eucaryotic cells, lysis of the resulting phagosome, replication as well as movement within the host cytoplasm, direct cell-to-cell spread, and lysis of a double-membrane vacuole when entering neighbouring cells (Brehm et al., 1996). As discussed above, internalin is required for *L. monocytogenes* entry into epithelial cells (Lebrun et al., 1996). The production of superoxide dismutase by *L. monocytogenes* may aid in the survival in the macrophages (Farber and Peterkin, 1991). Bacteria that survive or are in a non-activated phagocyte then dissolve the phagosome by means of listeriolysin O (LLO) or possibly by phospholipase C (McLauchlin, 1997). *Listeria* strains lacking LLO are avirulent, failing to colonize liver or spleen in gastric infection studies in mice (Gaillard, Berche and Sansonetti, 1986; Roll and Czuprynski, 1990; Tabouret et al., 1991; Erdenlia, Ainsworth and Austin, 2000). The production of the enzyme phospholipase C by virulent *L. monocytogenes* is important for its ability to survive the early host neutrophil-mediated defence mechanism (Conlan and North, 1992). The listerial surface protein ActA is required for actin polymerization and confers intracellular mobility and enables the bacterium to invade an adjacent host cell (Kocks et al., 1992). The surface-bounded protein actin A mediates the contact to the actin filament system of the host cell. The cell-to-cell spread is also mediated by phospholipase and lecithinase (Schwarzkopf, 1996). Most virulence genes are activated by the transcriptional regulator *prfA* (Mengaud et al., 1991; Renzoni, Cossart and Dramsi, 1999). The expression of *pfrA* and *prfA*-dependent proteins is under the control of several environmental parameters, such as temperature, pH, stress conditions and composition of the medium (Brehm et al., 1996).

While the use of tightly defined systems (i.e. clonal bacteria and genetically identical hosts) is needed to study the pathogen's virulence mechanisms, the frequency of naturally occurring strains that are deficient in one or more virulence markers appears to be relatively rare among populations of foodborne isolates of *L. monocytogenes*. Various cell cultures have been proposed as a means for differentiating virulent and non-virulent isolates of *L. monocytogenes*. While the methods have had varying degrees of success, most have indicated that the majority of isolates from foods have a complete array of virulence-associated genes and are virulent (del Corral et al., 1990; Pine et al., 1991; Wang et al., 1998). Accordingly, it is generally assumed that, except for atypical isolates such as listeriolysin-deficient mutants, all *L. monocytogenes* isolates are potentially pathogenic (Rocourt, 1996).

Testing with surrogate animal models (mice) has demonstrated substantial variation among isolates in relation to the differential levels of the microorganisms needed to induce morbidity or death after oral or intraperitoneal administration. For example, del Corral et al.

Table 2.4 The lethality of *Listeria monocytogenes* food and clinical isolates for immunocompromised mice.

Strain	LD_{50} (CFU)	Source
MF2-L-P	6	Food
V3-VT	13	Food
GV2-VS	29	Food
F3-VJ-G	31	Food
HO-V6-G	31	Food
LG4-VS	42	Food
VS2-VJ	74	Food
Scott A	93	Clinical
H4-V-G	100	Food
GVG-VS	110	Food
GLB1-LS	200	Food
CCR8-V-G	1 000	Food
S9-VJ-G	1 400	Food
F-4259	2 000	Clinical
GVN4-VG	3 100	Food

SOURCE: Adapted from del Corral et al., 1990.

(1990) found a 3-log range for LD_{50} values when immunocompromised mice were administered *L. monocytogenes* by an intraperitoneal route (Table 2.4). However, virtually all listeriolysin-positive clinical and food isolates were pathogenic for immunocompromised mice (del Corral et al., 1990; Pine, Malcolm and Plikaytis, 1990; Notermans et al., 1998). The level of the pathogen required to produce infections and morbidity declined by several orders of magnitude when the mice were immunocompromised (Golnazarian et al., 1989). Inhibition of gastric acid production also decreases the levels of *L. monocytogenes* needed to produce infections and morbidity after oral administration (Golnazarian et al., 1989). Ribotyping in combination with allelic analysis of virulence genes and DNA sequencing has identified disease-associated sub-types of *L. monocytogenes* (Wiedman et al., 1997). Resistance to arsenite has been reported to occur at a higher rate in clinical isolates of *L. monocytogenes* (Buchanan et al., 1991; McLauchlin, 1997). Thus, a substantial heterogeneity in virulence has been observed in several *in vivo* (mice) and *in vitro* (cell culture) studies. However, no consistent pattern of increased virulence associated with any specific serotype or subtype in animal or *in vitro* studies has emerged (Pine et al., 1991; Tabouret et al., 1991; Hof and Rocourt, 1992; Wiedman et al., 1997) and none of the present methods have consistently identified strains that are non-pathogenic or less virulent (McLauchlin, 1997).

Nevertheless, there is evidence for variation in virulence among foodborne isolates of *L. monocytogenes*. Although human listeriosis may be caused by all 13 serotypes of *L. monocytogenes*, most listeriosis cases are associated with a restricted number of serotypes: 1/2a (15–25%); 1/2b (10–35%); 1/2c (0–4%); 3 (1–2%); 4b (37–64%); and 4 not b (0–6%) (McLauchlin, 1990b; Farber and Peterkin, 1991). The frequency of serotype 4b was significantly greater in pregnancy cases, whereas serovar 1/2b was most commonly associated with non-pregnancy cases (McLauchlin, 1990b). However, the frequency with which these serotypes can be isolated from foods does not closely parallel the disease distribution (Pinner et al., 1992). Contamination of hot dogs by two serotypes of *L. monocytogenes* (1/2a and 4b) resulted in disease associated only with the 4b serotype, which was present at apparently much lower concentrations (FDA/FSIS, 2001). This suggests that the 4b isolate was either more virulent, better able to survive transport through the stomach or grew at a greater rate in the food.

The difference in the distribution of strains isolated from foods and human clinical cases, does not necessarily reflect a difference in virulence, but may also be a reflection of the adaptations by this bacterium to different ecological niches (Boerlin and Piffaretti, 1991). It may also be a reflection of the methodology used. MacGowan et al. (1991) investigated faeces from different categories of patients and detected more than one *Listeria* species or serovar in 40% of the positive samples. Similarly, a direct plating method recovered two serovars from a gravad rainbow trout (Loncarevic, Tham and Danielsson-Tham, 1996).

The observed variability in virulence of different *L. monocytogenes* isolates reflects the number of microorganisms required to produce an infection, the potential for an infection to become symptomatic, the severity or manifestations of illness, and which individuals in the population are at greatest risk. In addition to the factors that directly influence the ability of *L. monocytogenes* to produce a systemic infection, the microorganism's virulence can also be influenced by characteristics that increase its likelihood of reaching the intestinal tract. For example, *L. monocytogenes* does have adaptive acid-resistance mechanisms that, when induced, increases the likelihood that it will survive passage through the stomach (Kroll and Pratchett, 1992; Buchanan et al., 1994). This will be discussed more fully below.

2.2.2.2 Host susceptibility

Human populations are highly diverse in their response to infectious agents, reflecting the population's diversity in genetic background, general health and nutrition status, age, immune status, stress level and prior exposure to infectious agents. For certain foodborne diseases, it appears that prior exposure to the agent renders the individual resistant to subsequent exposures to the pathogen (e.g. for *Cyclospora cayetanensis*). However, for many infectious and toxico-infectious foodborne pathogens, immunity is of limited importance, due to either the presence of the pathogen being restricted to the intestinal tract (e.g. enterohaemorrhagic *Escherichia coli*), great diversity of serotypes (e.g. *Salmonella*), or mechanisms for avoiding or overcoming the host's defences (e.g. *L. monocytogenes*).

Severe listeriosis most often affects those with severe underlying illness, the elderly, pregnant women and both unborn or newly delivered infants (McLauchlin, 1996). Infection in healthy adults is typically asymptomatic. The rate of *L. monocytogenes* carriage among these asymptomatic individuals is not known (Slutsker and Schuchat, 1999). The majority of human cases of severe listeriosis occur in individuals who have an underlying condition that suppresses their T-cell mediated immunity (Farber and Peterkin, 1991; Rocourt, 1996). A summary of listeriosis cases in 1989 from 16 countries showed that 31% of the cases occurred in patients older than 60 years, and 22% occurred in patients younger than 1 month (Rocourt, 1991). In addition to age (elderly and the neonates) and pregnancy, risk factors include cancer and immunosuppressive therapy, AIDS, and chronic conditions such as cardiovascular disease, congestive heart failure, diabetes, cirrhosis and alcoholism (Nieman and Lorber, 1980; McLauchlin, 1990a; Paul et al., 1994; Goulet and Marchetti, 1996; Rocourt, 1996). A review of 98 cases of non-pregnancy associated sporadic listeriosis in the United States of America revealed that 98% of individuals involved had at least one underlying condition (Schuchat et al., 1992). Most, but not all, of these were associated with probable immunosuppression. Antacid therapy (Ho et al., 1986) and iron overload (Lorber, 1990) were also reported as risk factors. However, in surveillance data from France for 1999 no identified immunosuppressive condition was noted in up to 15% of cases (Goulet et al., 2001b). Thus, individuals without any of the risk factors mentioned above have occasionally become severely infected.

As discussed earlier, *L. monocytogenes,* at high numbers, can cause febrile gastroenteritis in healthy persons (Salamina et al., 1996; Miettinen et al., 1999; Aureli et al., 2000). The course of the disease appears to be similar to more classical foodborne pathogens, such as *Salmonella*, where infections are generally limited to gastroenteritis, but, for a small percentage of the population, particularly those with an underlying condition, systemic, life threatening infections may occur.

For the purposes of this hazard characterization, the elderly will be considered to be individuals aged 60 years or older, and the very young are ≤ 28 days of age.

While susceptibility in these groups is thought to be related primarily to an impaired or undeveloped immune function, another physiological parameter thought to be relevant to susceptibility is a reduced level of gastric acidity. As previously mentioned, antacid use has been identified as a risk factor for severe listeriosis. Reduced gastric acidity may be associated with aging or with drug treatment for gastric hyperacidity. An increasing portion of the population suffers from achlorhydria as age increases above 50 years. Two dose-response studies dealing with this issue involved treatment of mice or rats with the acid suppressor Cimetidine concurrent with oral infection with *L. monocytogenes*. The mouse study showed no significant effect with the drug treatment (Golnazarian et al., 1989), while the rat study showed increased infectivity of *L. monocytogenes* at the lowest dose (Schlech, Chase and Badley, 1993). Another factor that can reduce gastric acidity is infection with *Helicobacter pylori*. Basal gastric acidity was found to be increased in individuals following successful eradication of *H. pylori* compared with subjects whose infection persisted after antibiotic therapy (Feldman et al., 1999). The subjects in this study were asymptomatic for *H. pylori* infection, as are the majority of infected individuals. While this population may be more at risk for infection with *L. monocytogenes* (and other pathogenic bacteria) by reduction of the stomach acid barrier, no studies were found that focused on this relationship.

With respect to immune function, specific human dose-response information must be gleaned from surveillance data. However, much of our understanding of the effect of immune status on the pathogenicity of *L. monocytogenes* comes from research with surrogate animals. Thus, an underlying assumption is that human and animal resistance mechanisms are similar. The mouse is the most thoroughly characterized with respect to the role that specific immune defects have on susceptibility to *L. monocytogenes*. Host resistance mechanisms against *L. monocytogenes* have been studied primarily using a variety of immunocompromised mouse models. These models include gene knockout models, depletion of cytokines or immune cells with monoclonal antibodies, and mouse strains with genetic defects related to macrophage-mediated killing of *L. monocytogenes* (Czuprynski, Theisen and Brown, 1996; Stevenson, Rees and Meltzer, 1980; Cheers and McKenzie, 1978).

Within some susceptible human populations, immune system defects that correlate with resistance in mouse models have been identified. In pregnancy, there is a characteristic inhibition of natural killer (NK) cell activity in the placenta (Schwartz, 1999). During the early phase of resistance in the mouse, NK cells, stimulated by interleukin 12, are the primary source of gamma-interferon, a key component of resistance (Unanue, 1997; Tripp et al., 1994). Pregnancy is also associated with development of a Th-2 cytokine environment that favours the production of interleukins 4 and 10 (Schwartz, 1999). Using gnotobiotic pregnant mice, Lammerding et al. (1992) observed cellular immune response in the mother's liver and spleen, but a similar response was not observed in the placenta or fetus. Immune defects in the mouse that reflect these changes have a negative effect on resistance (Nakane et al., 1996; Genovese et al., 1999) while cytokines characteristic of a Th-1 response (e.g. gamma-interferon) are critical for resistance (Unanue, 1997; Tripp et al., 1994; Huang et al., 1993). Listeriosis symptoms in pregnancy are often mild (Slutsker and Schuchat, 1999), suggesting that pregnancy may not predispose mothers to more severe illness. However, it is possible that immunosuppression as a consequence of pregnancy results in increased likelihood that even small numbers of *Listeria* in the circulation can colonize placental tissues, increasing the chances of fetal exposure. The consequences of fetal exposure are severe, often resulting in stillbirth or neonatal infection.

At the extremes of age – neonates and the elderly – changes in both innate and acquired immunity have been observed. Numerous biomarkers of immune responsiveness have been measured in the elderly, including decreased gamma-interferon production and NK cell activity, and increased IL-4 and IL-10 production (Rink, Cakman and Kirchner, 1998;

Mbawuike et al., 1997; Di Lorenzo et al., 1999). The effects on IL-4 and IL-10 are suggestive of a predominant Th-2 versus Th-1 response. A similar imbalance, characterized by decreased gamma-interferon production and down regulation of IL-10, may occur in neonates (Lewis, Larsen and Wilson, 1986; Genovese et al., 1999). Thus, in pregnancy, as well as in elderly and neonatal immune systems, there are changes in the immune system and biomarkers can be documented in mouse models that correlate with decreased resistance. Relatively few mouse studies investigated dose-response in an oral infection model in immunocompromised mice (Czuprynski, Theisen and Brown, 1996; Golnazarian et al., 1989).

Because the experimental studies summarized above all involve highly controlled manipulation of the immune system, it is very difficult to interpret the results with respect to a highly variable human population. Furthermore, the results of studies involving knockout mice or treatment with monoclonal antibodies reflect a nearly complete abrogation of the immune parameter in question – a condition that is probably seldom the case in humans. In addition, most studies were not conducted using oral administrations of *L. monocytogenes* that might have an impact on targeted immune mechanisms locally in the gut.

2.2.2.3 Food matrix effects

Traditionally, food had been viewed as a neutral vehicle for the pathogen, and as such had little impact on dose-response relations. However, recently there has been increasing awareness of the impact that the food matrix can have on the likelihood of disease. Much of the focus has been on the impact that microbial adaptation has on the acid resistance of enteric pathogens. The stomach acts as the body's first defence against foodborne pathogens via their inactivation by the pH of gastric fluids (Gianella, Broitman and Zamcheck, 1973; Peterson et al., 1989). The key factors influencing the extent of inactivation of ingested pathogens by this barrier are the pH of the stomach, the residence time of the bacteria in the stomach, and the pathogen's inherent acid resistance. Since the inactivation of *L. monocytogenes* due to adverse pH values follows first order kinetics (Buchanan, Golden and Phillips, 1997), the extent of inactivation will also be dependent on the initial numbers of bacterial cells (i.e. dose) ingested. Exposure times of between 15 and 30 minutes were required to achieve more than a 5-log inactivation of three strains of *L. monocytogenes* in simulated gastric juice (Roering et al., 1999). Anything that reduces the contact between bacteria and gastric acid could potentially have the effect of reducing the number of bacterial cells needed to produce an infection. For example, outbreaks of salmonellosis involving water and other liquids have often been associated with low levels of the pathogen. Mossel and Oei (1975) demonstrated that a liquid bolus of less than 50 ml could pass rapidly through the stomach because the pyloric sphincter fails to constrict when challenged with such a small bolus.

Fatty food vehicles can protect bacteria from the gastric acid during passage through the stomach (Blaser and Newman, 1982). This was illustrated by an outbreak of *Salmonella* Typhimurium present in chocolate at very low levels (Kapperud et al., 1990). However, a reduced intestinal colonization and diarrhoea in rats fed milk with a high fat content as opposed to milk with a low fat content was recently reported for *L. monocytogenes* (Sprong, Hulsterin and Van der Meer, 1999), whereas *Salmonella* Enteritidis infection was apparently unaffected. *L. monocytogenes* were killed mainly in the stomach by free fatty acids and monoglycerides resulting from digestion of fat, whereas the Gram-negative cell wall was suggested as protecting the *Salmonella* (Sprong, Hulsterin and Van der Meer, 1999). In agreement with these results, Schlech (1993) reported a lower proportion of infection in rats arising from *L. monocytogenes* grown in Brain Heart Infusion (BHI) broth and administered in milk than when administered in BHI alone, which suggested that milk might have an inhibitory effect on the number of organisms available for colonization. In mice, however, Notermans et al. (1998) reported that the ID_{50} was the same when *L. monocytogenes* was orally administered in water or in milk with 1% or 3% fat. However, given that two different animal models were used (rats and mice), the proper interpretation for humans of these findings is difficult.

Food vehicles with high buffering capacity may also protect bacteria from gastric acid, although the gastric response to exogenous buffers may be complex (Blaser and Newman, 1982). Volunteers either quickly secreted more acid to overcome the effect of the buffer or experienced a prolonged buffering effect. Volunteers with a prolonged buffering had a higher attack rate for *Vibrio cholerae* than those who overcame the effect (references cited in Blaser and Newman, 1982).

While there is a clear indication that food matrix effects could influence dose-response relations associated with *L. monocytogenes*, there are insufficient data to allow this to be considered currently as a variable within the hazard characterization.

2.2.2.4 Interaction of pathogen, host and matrix variables

Based on the observation that serovars 1/2a, 1/2b and 4b dominate among the strains isolated from human cases, whereas a wider range of serovars have been isolated from foods, it has been suggested that this is a reflection of their different potential for causing disease. Schlech (1991) suggested that the sporadic nature of outbreaks is more consistent with changes in the virulence of strains than in host susceptibility, since the population at risk may not vary greatly. In fact, an indirect vaccination due to the presence of strains with reduced or no virulence has been suggested as an explanation for the low incidence of listeriosis, despite the frequent exposure due to contaminated food (Chakraborty et al., 1994; Schwarzkopf, 1996). McLauchlin (1996), in contrast, commented that the explanation for the wide variation observed in incubation periods after oral ingestion is unknown but it may be dose dependent, strain dependent, or perhaps reflect unknown differences in host susceptibility.

One of the adaptive mechanisms in *L. monocytogenes* is its ability to develop acid resistance (Buchanan et al., 1994; Patchett et al., 1996; Phan-Thank and Montagne, 1998). An acid-tolerant mutant demonstrated an increased lethality in mice following intraperitoneal inoculation (O'Driscoll, Gahan and Hill, 1996). Conversely, a mutant that was acid-tolerant deficient had decreased lethality to mice (Marron et al., 1997). Acid adaptation was reported to lead to an enhanced resistance to a number of other environmental stresses, including heat treatment (Farber and Pagotto, 1992), lactoperoxidase (Ravishankar, Harrison and Wicker, 2000), bacteriocins (van Schaik, Gahan and Hill, 1999) and other preservatives (Lou and Yousef, 1997), and to increased survival in acidic foods (Gahan, O'Driscoll and Hill, 1996).

The effect of growth temperature on the subsequent pathogenicity of *L. monocytogenes* has been examined by several investigators. Growth at 4°C was reported to increase the virulence in mice infected by the intravenous route but not by the intragastric route (Czuprynski, Brown and Roll, 1989; Stephens et al., 1991). The effect was suggested to be dose dependent because it was observed only at levels greater than 10^4 viable

L. monocytogenes cells (Stephens et al., 1991). This suggested that there might be variants within the population with increased virulence as a result of the non-optimal growth conditions. Clinical strains were demonstrated to be more resistant to cold storage in terms of lag phase duration and the degree of pathogenicity to chick embryos compared with strains isolated from meat (Avery and Buncic, 1997). Cells grown at 10°C were less acid resistant than cells grown at 30°C (Patchett et al., 1996) and as such would be less likely to survive passage through the stomach. Differences in pathogenicity of cells grown at 5° and 10°C were not observed when the pathogen was grown in crabmeat or microbiological media (Brackett and Beuchat, 1990).

An apparent variation in the virulence of *L. monocytogenes* can also reflect a change in the health status of the host. Clinical and epidemiological investigations of an outbreak of listeriosis in 1987 in Philadelphia in the United States of America led to the suggestion that individuals colonized by *L. monocytogenes* but previously asymptomatic for listerial infection became symptomatic because of a co-infecting disease (Schwartz et al., 1989; Rocourt, 1996).

2.2.3 Approaches to modelling dose-response relations

2.2.3.1 General approaches and limitations to modelling dose-response relations for foodborne pathogens

When modelling dose-response relations, the number of microorganisms entering the digestive tract per exposure may be expressed as a mean number of functional particles of the pathogenic organism, CFU, spores, oocysts, etc. (Teunis et al., 1996; Vose, 1998). This is the dose, a quantitative measure of the intensity of the exposure. At a certain dose, certain effects in the host occur. The frequency within the exposed population of hosts at which this occurs constitutes the response. The response may be more or less well defined, but generally there will not be a one-to-one relationship between the size of the dose and the specific kind and frequency of the biological effect it produces. Furthermore, pathogenic microorganisms generally produce an array of effects or conditions within an affected host. Thus, instead of a single dose-effect relation there will be a series of dose-response relations that describe the relationship between the various biological effects and the magnitude of the dose (Teunis et al., 1996). The effects commonly considered, which are also referred to as biological end points, include infection (for *L. monocytogenes* this is often measured by the presence of bacteria in the spleen or the liver of an animal model), various forms of morbidity, or death (Vose, 1998).

The response of a human population to an exposure to a foodborne pathogen is highly variable, in terms of both the duration and the severity of the symptoms observed. The variability is a reflection of the dependency of the frequency and extent of disease on a variety of factors, such as the virulence characteristics of the pathogen, the number of bacterial cells ingested, the general health and immune status of the host, and attributes of the food that may alter microbial or host status. Thus, the relationship between the dose and the response is a function of the *L. monocytogenes* strain in terms of its virulence properties and its survival characteristics, the food in which it resides, and the susceptibility of the host. A mathematical relationship between the dose and the response would ideally be able to describe the interactions between all these factors. It is important to note that such mathematical relations describe the dose-response relationship on a population basis and cannot describe the likelihood of illness for any specific individual.

Sources of data and general considerations

An appreciation of the factors described above is critical to a scientifically rigorous consideration of dose-response relations. Equally as important is an appreciation of the uncertainty and variability associated with the different sources of dose-response data.

Human volunteer feeding studies

The primary source of microbiological dose-response data for other pathogens has been human volunteer feeding studies. Such trials provide the most direct measure of human response to pathogens and have been the data of choice for quantitative microbial risk assessments. However, these data do have limitations that must be considered when these dose-response relations are used to estimate the susceptibility of the entire population. Volunteers for these studies have been almost exclusively limited to healthy adult males. Information on the susceptibility of higher-risk populations or potential gender effects is generally not available. Of necessity, volunteer studies are limited to foodborne diseases that are not considered life threatening for the test subjects. Thus, volunteer feeding studies are unlikely to be conducted for pathogens or diseases that are either life threatening (e.g. enterohaemorrhagic *E. coli*) or that almost exclusively affect higher-risk populations (e.g. *L. monocytogenes*). Volunteer studies have often been conducted in conjunction with vaccine trials, which tend to focus on higher dose levels. Typically, there are relatively few test subjects per dose, and because of the small size of the test population, dose levels are used that produce relatively high rates of infection or morbidity. It is usually not possible to evaluate doses that are directly pertinent to the pathogen levels most often associated with human exposures via food. Thus, most dose-response determinations rely on extrapolations of the dose-response relations based on high doses. This leads to a high degree of uncertainty at the low-dose levels. Human volunteer feeding studies are not available for *L. monocytogenes*.

Surrogate animals

Because *L. monocytogenes* primarily affects specific, high susceptibility populations, human feeding studies are ethically not feasible. Animal models have been used as the primary alternative means of studying its dose-response relations. The successful use of animal models is dependent on a number of factors, not the least of which is the need for a "conversion factor" that allows the quantitative relations observed in the animal to be correlated with human response to the pathogen. Success is highly dependent on the selection of an appropriate animal model. This can be a significant challenge with many foodborne pathogens. It assumes that the pathogen causes disease by the same mechanism of pathogenicity in both the human and surrogate animal, that the animal's physiological and immune responses are similar to that of humans, and that quantitative relationships between infectivity, morbidity and mortality are similar for the two species. Further, animal feeding studies have many of the same limitations as human volunteer studies. For example, most studies are conducted using only healthy animals that are similar in age and weight. In fact, most laboratory animals are so highly inbred that genetic diversity among the animals is negligible. This reduces the variability associated with the testing but brings into question the data's applicability to the general population.

While the disease characteristics of *L. monocytogenes* have been examined in a wide range of animals, the primary animal model for dose-response studies with this microorganism has been the mouse, with death being the primary biological end point measured. Care must be taken in reviewing these studies since the dose-response relations vary substantially depending on both route of entry and the variety of mouse employed as the surrogate for humans. This caution is reinforced by the recent finding, discussed in the section on virulence of *L. monocytogenes*, that the mouse may not be an appropriate model for study of at least some aspects of human listeriosis (Lecuit et al., 1999).

Epidemiological approaches

Potentially, epidemiological investigations could be a source for human dose-response information, particularly for outbreaks involving RTE foods that do support the growth of the pathogenic bacterium. However, to be useful for risk assessments, the investigations would have to be expanded beyond their usual scopes. In addition to detailed information about who became ill, the investigations also have to acquire information about a variety of other factors such as who consumed the food and did not become ill, the amounts of food consumed by both groups, and the frequency and extent of contamination. Regrettably, few epidemiological investigations have been conducted in a manner that provided such data.

An alternative approach for pathogens that are not appropriate for human volunteer feeding studies have been suggested by Buchanan et al. (1997). These authors proposed using data on the annual national incidence for a disease, and food survey data on the frequency and extent of contamination of an RTE food, to produce an estimate of the microorganism's dose-response relationship. Assuming that all cases of listeriosis were due to a single food, this approach was used to generate a conservative estimate of the dose-response relations for *L. monocytogenes* in higher-risk populations.

Mathematical models

The relation between ingestion of a certain number, N, of a pathogenic microorganism and the possible outcomes has been quantitatively described in a number of ways (Table 2.5).

Models may be classified or distinguished in different ways. Depending on the assumptions and parameter values chosen, some models may be special cases of other models (Haas, 1983; Holcomb et al., 1999). One important distinction is between models describing infection as a deterministic or a as stochastic process (Haas, 1983). The deterministic view assumes that for each microorganism there is an inherent minimum dose, i.e. there is a threshold level, below which no response is seen. Thus, in a deterministic threshold model, the risk below the threshold is zero. However, the threshold level (i.e. the minimum dose) may potentially vary across individuals in a population. If this variation is incorporated into the model it becomes a stochastic model. The stochastic view on infection holds that the actions of individual cells of pathogenic microorganisms are independent from other cells and that a single microorganism has the potential to infect and provoke a response in the individual, i.e. a single-hit, non-threshold model (Haas, 1983).

Table 2.5 Summary of some dose-response models used for foodborne pathogens. The table is adapted and modified from Holcomb et al. (1999) and other sources.

Model name	Function Probability (P) =	Parameter definitions	Comments and sources
Log-Normal	$\phi[b_0 + b_1 * \log_{10}(N)]$	ϕ = cumulative normal distribution function b_0 = intercept b_1 = log10(dose) slope parameter	[1]
Log-Logistic	$\beta/1+[(1-p)/p] * e^{-\varepsilon\{\log_{10}(N) - \chi\}}$	β = Asymptotic value of probability of infection as dose approaches ∞. β =1 in Holcomb et al. (1999). χ = Predicted dose at specified value of p where p = P ε = Curve rate value affecting spread of curve along dose axis	[2]
Simple Exponential	$1- e^{-r*\log_{10}(N)}$	r = Reflects host/microorganism interaction probability	[3] Note (1)
Flexible Exponential	$\beta * [1-p*e^{-\varepsilon\{\log_{10}(N) - \chi\}}]$	β = Asymptotic value of probability of infection as dose approaches ∞. β=1 in Holcomb et al. (1999). χ = Predicted dose at specified value of p where p = 1 – P ε = Curve rate value affecting spread of curve along dose axis	[2]
Beta-Poisson	$1 - (1 + N/\beta)^{-\alpha}$	α, β = Parameters affecting the shape of the curve	[4] Note (2)
Beta-Binomial	$1-(1-P_I(1))^N$	$P_I(1)$ = probability of illness from exposure to one organism. $P_I(1)$ assumed to be Beta(α, β) distributed	[5]
Weibull-Gamma	$1 – [1 + (N)^b/\beta]^{-\alpha}$	α, β, b = parameters affecting shape of curve	[6]
Gompertz	1 – exp[-exp(a + bf(x))]	a = model (intercept) parameter; b = model (slope) parameter; f(N) = function of dose.	[7]

KEY: N = Ingested dose of microorganisms; P = Probability of infection.

NOTES: (1) Rose, Haas and Regli (1991) used the form $1- e^{-r*dose}$.
(2) See Vose (1998) for a discussion on the interpretation of α, β.

SOURCES: [1] Dupont et al., 1972. [2] Levine et al., 1973. [3] Rose, Haas and Regli, 1991 [4] Haas, 1983. [5] Cassin et al., 1998. [6] Todd and Harwig, 1996. [7] Coleman and Marks, 1998.

Models can also be differentiated on the basis of whether they are mechanistic or empirical. Buchanan, Smith and Long (2000) suggested that most dose-response models used currently are empirical and are limited because they attempt to extrapolate beyond the limits for which there are data. Potentially, mechanistic models would be more flexible since they focus on specific physiological or chemical attributes; however, there have been few attempts to develop such models. Buchanan, Smith and Long (2000) encouraged the development of mechanistic dose-response models, and outlined a simple three-compartment dose-response model for foodborne salmonellosis. The model compartments were survival in the stomach; attachment and colonization in the intestine; and invasion of body tissues or production of toxins. No mechanistic dose-response models are currently available for *L. monocytogenes*.

Threshold models assume a minimum threshold dose before the response occurs. In a given population, the variation in the minimal dose can be described by a distribution. For instance, the Log-Normal model (Probit model) assumes that the minimal dose is lognormally distributed (Haas, 1983), and the Log-Logistic model assumes that $\log_{10}$ of the infectious dose follows a logistic distribution (Holcomb et al., 1999).

Threshold models

Marks et al. (1998) compared a Beta-Poisson model with a combined Beta-Poisson model that also employed a threshold level (3 bacteria) in a risk assessment for *E. coli* O157 in hamburgers. The introduction of a threshold means that, at low doses, the location of the dose-response curve is shifted along the x-axis by the threshold amount. The differences between these models were significant only in the low dose range. The resulting estimates of risk were 100- to 1000-fold larger, depending on cooking temperature, using the non-threshold model. The authors concluded that the two-parameter Beta-Poisson model appeared insufficient for describing the complexity of dose-response interactions and that it was inadequate as a default model for microbial risk assessment, especially in cooked foods (Marks et al., 1998). They also concluded that the consideration of threshold models as alternative dose-response models is of great importance and that additional research was needed in this area.

Stochastic – single-hit models

Other researchers have favoured the use of single hit models, which in many instances have described data quite well (Haas, 1983; Teunis et al., 1996). For instance, dose-response data for protozoan parasites can be well described by the exponential model (Teunis, 1997), and bacterial infection data are generally well described by Beta-Poisson models (Teunis, 1997; Teunis, Nagelkerke and Haas, 1999), or by the Weibull-Gamma model (Holcomb et al., 1999). The same model may not be equally effective for all biological end points caused by the pathogen. For example, the FDA/FSIS *L. monocytogenes* risk assessment reported that the exponential model did not fit mouse infection data (i.e. isolation of *L. monocytogenes* from the spleen and liver), but was among the best models for describing the relationship between dose and the frequency of death (FDA/FSIS, 2001).

Exponential model

In the derivation of this model it is assumed that all of the ingested organisms have the same probability, r, of being individually capable of causing an infection to a specific consumer. Further, the probability of a single-hit, r, is independent of the size of the inoculum. Then the probability of infection after ingesting N organisms is the probability of one or more hits:

$$P_{inf} = 1 - (1\text{-}r)^{N}$$

Assuming that the distribution of organisms follows a Poisson process, with a mean number of organisms N per portion, the exponential dose-response relation follows (Haas, 1983; Vose, 1998):

$$P = 1\text{-} \exp^{-r*N}$$

In a few cases, notably for the pathogenic protozoa *Cryptosporidium parvum* and *Giardia lamblia*, this model provides an acceptable fit, but the slope of the model is generally steeper than what is observed from data (Teunis et al., 1996). Holcomb et al. (1999) modified the form of the exponential model and termed it the Simple Exponential model by using the $\log_{10}$ of the dose instead of the dose directly (Table 2.5), the reason being that the Simple Exponential model fitted more of the investigated data sets. Holcomb et al. (1999) also used a model similar to the Simple Exponential model, which they termed the Flexible Exponential model (Table 2.5). According to the authors, the benefit of this model was that it could be applied to experimental data where doses much greater than 1 still resulted in zero percent infection.

Beta-Poisson model

In this model, heterogeneity in the microorganism–host interaction is introduced. The r-value, the probability of an organism initiating infection given a successful introduction into the host, is assumed to follow a Beta-distribution. Haas (1983) suggested that this variation reflected the variation in virulence of the individual pathogens or in the sensitivity of the host, or both. In contrast, Vose's (1998) interpretation was that the Beta-distribution characterized by its α and β values describes the expected probability of each of the consumed microorganisms causing infection, averaged over all volunteers.

A complex function results from the derivation of this model. However, assuming that β is much larger than both α and 1, the following approximation can be used:

$$P = 1- [1 + N/\beta)]^{-\alpha}$$

In some cases the use of this model to fit dose-response data has not fulfilled this condition (Teunis et al., 1996). Initially the authors proposed using an approximated function in all cases because the influence of using the more rigorous function was considered relatively insignificant. However, Teunis and Havelaar (2000) recently showed that the discrepancies between the models were largest in the low-dose region, which is the region of interest for many risk applications, and that errors may become very large in the results of uncertainty analysis or when the data contain little low-dose information. Vose (1998) criticised using α and β just as fitted parameters without any consideration of their interpretation in the beta distribution. For instance, values between 0 and 1 of these parameters mean that the distribution for the probability of infection will peak at both 0 and 1. This could be interpreted as a partition among volunteers into susceptible and non-susceptible populations. Teunis et al. (1996) concluded that the Beta-Poisson model appears to fit most available dose-response data well and has the desired property of being conservative when extrapolated to low doses.

Beta-Binomial model

Cassin et al. (1998) developed a Beta-Binomial dose-response model to assess the risk of *E. coli* O157:H7 in hamburgers. The model reflected the same assumptions used in the original Beta-Poisson model. However, the Beta-Binomial model yields variability for probability of illness from a particular dose in contrast to the original model, which only specifies a mean population risk.

$$P = 1- (1-P_I(1))^N$$

$P_I(1)$ is the probability of illness from ingestion of one microorganism, and this probability was assumed to be Beta-distributed with parameters α and β. By fitting the model to data from human feeding studies with *Shigella,* it was possible to generate a dose-response curve showing the estimated uncertainty in the average probability of illness verses the ingested dose. The variability between feeding studies was used as a proxy for the uncertainty in the parameters α and β.

Weibull-Gamma model

This model was chosen by Farber, Ross and Harwig (1996) because of its flexibility, i.e. it is possible to accommodate the available qualitative dose-response information for *L. monocytogenes* and to adapt to both healthy and higher-risk groups. The starting point for the derivation is the Weibull model:

$$P = 1 - e^{-a*N^b}$$

where N is the dose ingested and *a* and *b* are parameters. The parameter *a* is related to the probability of illness given exposure to a single organism and *b* determines the shape of the individual dose-response curve. In this model, host–pathogen heterogeneity is considered by assuming that *a* follows a Gamma distribution characterized by the parameters α and β. The resulting equation, the Weibull-Gamma model becomes:

$$P = 1 - [1 + (N^b)/ \beta]^{-\alpha}$$

Depending on the parameter values, the Weibull-Gamma model can be reduced to both the Beta-Poisson and the Log-Logistic models (Farber, Ross and Harwig, 1996; Holcomb et al., 1999).

Gompertz model

Recognizing that a number of empirical models may fit observed data adequately, Coleman and Marks (1998), in addition to Logistic and Beta-Poisson models, used a Gompertz model to describe the results of human feeding studies:

$$P = 1 - \exp [-\exp(a + b f(N))]$$

where *a* is a model (intercept) parameter, *b* is a model (slope) parameter, and f(N) is a function of dose.

Choice of dose-response model

The issue of which functional form truly describes reality, i.e. the interactions between the pathogen, the food vehicle and the host, remains an open question needing additional research. For example, an equally good fit for *Shigella* dose-response data was provided by a Gompertz function as by the Beta-Poisson model. However, outside the data range, the predictions differed greatly (Marks et al., 1998). The choice of dose-response model may depend on its applicability, e.g. how well it fits the available data, the simplicity of the model in relation to parsimony in the number of parameters used, and the range of conditions over which the model gives good predictions (Holcomb et al., 1999). Holcomb et al. (1999) emphasized the flexibility of the dose-response model to fit data from different organisms, thereby allowing direct comparisons of infectious doses for use in risk assessment. In their comparison of how well the models in Table 2.5 fitted different experimental data, they reported differences of up to nine orders of magnitude in the predicted dose affecting the most sensitive 1-percentile of the population. This illustrates the difficulty of extrapolating from high to low doses. They also concluded that the three-parameter Weibull-Gamma model was the only model capable of describing all data sets. However, this flexibility is achieved through the use of three parameters, which generally increases the extent of uncertainty in the prediction.

Although results of dose-response experiments fit a single-hit model well, e.g. the Beta-Poisson model, some serious shortcomings have been noted (Teunis, 1997; Teunis, Nagelkerke and Haas, 1999). The models ignore the incubation period and there is no opportunity for generalization with regard to microorganism, host and vector. Furthermore, the experimental evidence suggests that the probability of illness changes with dose in a manner that differs from that of the probability of infection. For instance, a decrease in the probability of illness was noted with a higher dose of *Campylobacter jejuni* (Black et al., 1988). Teunis, Nagelkerke and Haas (1999) developed a hazard function for the probability of illness, given successful infection, occurring in the time between onset and clearing of the infection. The duration of the infection period was assumed to follow a Gamma distribution. The scale parameter, λ, representing the time scale for the primary events responsible for clearing the infection (a Poisson process) was the authors' primary choice for dose dependence. Three possible scenarios were modelled: increased likelihood of illness with dose; decreased likelihood of illness with dose; and the likelihood of illness being independent of dose. Examples of each of these possible scenarios were illustrated using volunteer data from the literature. The different alternatives were suggested to reflect the balance of the interactions between the pathogen and the host (Teunis, Nagelkerke and Haas, 1999).

In their risk assessment of *L. monocytogenes*, FDA/FSIS (2001) assumed that there is no *a priori* means of determining which is the "correct" model to fit to a data set. Accordingly, they employed an alternative approach, namely that of fitting several of the dose-response models described above to dose-response data. An integrated dose-response relationship was then derived by combining the individual dose-response curves after weighting for how well each model fitted the data and for the parsimony of each model. The differences in the response value at any single dose that were predicted by the individual models were used as a means of estimating the uncertainty related to model selection.

2.2.3.2 Listeria monocytogenes *dose-response models developed from epidemiological data and expert elicitations*

The models of Farber, Ross and Harwig (1996) and of Bemrah et al. (1998)

Citing the lack of volunteer feeding studies and the tenuous extrapolation of animal data to the human situation, Farber, Ross and Harwig (1996) evaluated different dose-response models to determine which had the flexibility to use qualitative data. They proposed the Weibull-Gamma model as having advantageous attributes. As a means of demonstrating how the model would be used, they estimated, based on the literature (Farber and Peterkin, 1991; McLauchlin, 1993), that the ID_{10} and ID_{90} for *L. monocytogenes* are 10^7 and 10^9 CFU for healthy adults and 10^5 and 10^7 CFU for high-risk individuals, respectively. Farber, Ross and Harwig (1996) used this dose-response relationship to estimate the probability of illnesses based on data for both the consumption of soft cheeses and their contamination by *L. monocytogenes*. However, in so doing they did not clearly define the case definition for infection and assumed that all individuals that become infected also become symptomatic. This is an assumption that is not supported by work with surrogate animals. The results did demonstrate how these techniques could be used to develop quantitative microbial risk assessments. However, the study also demonstrated that care must be taken in ensuring that the advice of experts provides estimates of dose-response relations that are realistic in terms of the incidence of disease in the population and are consistent with the biological end point of concern. This initial estimate of the dose-response relationship is generally considered to predict substantially higher rates of illness than actually occur in human populations. Farber, Ross and Harwig (1996) also assumed that only 1% to 10% of *L. monocytogenes* strains are pathogenic, an assumption that is not in keeping with the observations described earlier, namely that most isolates are pathogenic.

Based on the work of Farber, Ross and Harwig (1996), Bemrah et al. (1998) used the Weibull-Gamma model for the dose-response relations for the risk assessment of human listeriosis from the consumption of soft cheese made from raw milk. They used values of $\alpha = 0.25$ and $b = 2.14$ for both the general population and the more highly susceptible population, and β values of $10^{15.26}$ and $10^{10.98}$ for these two groups, respectively. Based on these parameter values, the doses that would lead to 50% of the general and of the more highly susceptible populations becoming ill would be 48 000 000 CFU and 480 000 CFU, respectively.

The models of Buchanan et al. (1997) and of Lindqvist and Westöö (2000)

Epidemiological investigations of listeriosis outbreaks have not generally been useful in elucidating dose-response relations. This is because the levels of *L. monocytogenes* in the suspect food and the percentage of the individuals that consumed that food but that did not become ill are seldom quantified adequately. As an alternative approach to using epidemiological data, Buchanan et al. (1997) explored whether a purposefully conservative dose-response relation for *L. monocytogenes* could be developed using the annual incidence of listeriosis in combination with food survey data. Taking advantage of the fact that the exponential model is a single-parameter model, as discussed above, they used data for the incidence of listeriosis in Germany and the levels of *L. monocytogenes* in smoked fish as a means of deriving the r-value for the exponential model. Based upon *L. monocytogenes* prevalence and food consumption data, smoked fish was the likely source for most of the illnesses in Germany. They further assumed that symptomatic cases of listeriosis were largely restricted to that portion of the population that was immunocompromised. Based on this, Buchanan et al. (1997) estimated that the dose that would be expected to produce severe illnesses in half of a population of immunocompromised individuals was 5.9×10^{9} CFU, based on a r-value of 1.179×10^{-10}. The validity of this approach relies on several assumptions, including the percentage of individuals susceptible to severe *L. monocytogenes* infections, the uniformity of consumption patterns, the suitability of the exponential model to describe the pathogen's dose-response relationship in humans, and the accuracy of the statistics on the annual rate of severe listeriosis cases. The model was purposefully precautionary in relation to each of its underlying assumptions. However, the approach has several advantages, including "anchoring" the dose-response relationship to values that are based upon observed incidences of disease and using data based on the entire population instead of the small sample in human volunteer studies.

Using data for the consumption of smoked fish in Sweden and the incidence of listeriosis in that country, Lindqvist and Westöö (2000) used a similar approach to derive an r-value for *L. monocytogenes*. They reported an r-value of 5.6×10^{-10} for the 20% of the population at greater risk of listeriosis. Lindqvist and Westöö (2000) subsequently compared the exponential model of Buchanan et al. (1997) and the Weibull-Gamma model of Farber, Ross and Harwig (1996), assuming that this product was the primary source of listeriosis in that population. If all *L. monocytogenes* were assumed to be equally pathogenic, the exponential model predicted 168 cases and the Weibull-Gamma predicted 95 000. The reported number of cases per year in Sweden was 37. If it was assumed that only 1% to 10% of the *L. monocytogenes* isolates were pathogenic, the predicted number of cases was 9 and 5200 for the exponential and Weibull-Gamma models, respectively.

Evaluation of outbreak data

While epidemiological investigations of listeriosis outbreaks have generally been insufficient to allow calculation of dose-response relations or even attack rates, the US FDA/FSIS *L. monocytogenes* risk assessment (FDA/FSIS LMRA) team (FDA/FSIS, 2001) did acquire sufficient information related to two outbreaks that permitted a more detailed evaluation. These were the outbreak associated with Hispanic-style cheese that occurred in the United States of America in 1985, and the outbreak associated with butter among patients at a hospital in Finland in 1998–99. These evaluations and similar considerations of outbreaks of febrile gastroenteritis were used in the current document to estimate dose-response curves using the exponential model.

FDA/FSIS LMRA estimates for attack rates and dose ranges: Pregnant females and United States of America Hispanic-style cheese outbreak

Archival data from the Hispanic-style cheese outbreak in Los Angeles County, California, in 1985 were re-examined to determine if the attack rate and dose range could be estimated. The original report did not contain information on the amount consumed by individuals or the attack rate. Fortunately, consumption data by individuals had been collected and records from the outbreak were saved such that an attack rate could be estimated (FDA/FSIS, 2001).

The strategy used to estimate the dose-response for pregnant females was to assume a very high percent of pregnant Hispanic females ate the implicated cheese. Using the outbreak odds ratio table, the number of controls that were exposed to the implicated food was derived. This number of exposed controls, divided by the total number of controls results in a quotient (Q = 11/31 = 35%) that is an estimate of the proportion of the population that consumed the implicated food. The population giving rise to the cases was identified from the outbreak data and the common feature for cases and controls in this outbreak was pregnancy in Hispanic females. The total number of pregnant Hispanic females within the cheese marketing area during the time interval of interest provides (P). The proportion (Q) of the population that consumed the implicated food was multiplied by the total number of people (P), and this product, $Q \times P$, is an estimate of the number in the population of interest that ate the implicated food.

Laboratory data provided the total number of food samples qualitatively tested (T = 85) and the number of samples that were positive (T+ = 22). An estimate of the proportion of food contaminated was obtained by dividing the number of positive tests by the total number of tests (T+/T = 22/85 = 0.26). Multiplication of $(T+/T) \times Q \times P$ ($0.26 \times 11\ 775 = 3061$) provided an estimate of the total number of exposed persons in the population. Based on 21 cases answering a questionnaire on the frequency and amount of Hispanic cheese consumed, an average of 219 servings during the critical time of 20 weeks for a pregnant woman was estimated (R.C. Whiting, pers. comm., 2001). There were several brands of Hispanic cheese on the market. Assuming that 50% of the servings were the outbreak brands, an average of 110 servings of the implicated brand of cheese were consumed per person. From the total number of exposed persons and the number of servings, the attack rate was estimated. A second estimate of T+/T was based on 56 qualitatively-positive samples of 665 samples tested (56/665 = 0.084).

The cases caused by the implicated food were defined as those cases infected with the outbreak phage type. The estimated attack rate then equalled the number of cases that were infected with the outbreak phage type divided by the total number of exposed persons in the population. The proportion of actual cases that were identified during the outbreak was then estimated. Using this strategy, the estimated attack rate (i.e. the percentage of exposed pregnant women (or their fetuses) who became cases during the pregnancy) during the Hispanic-style cheese outbreak was between a low of 2.1–2.7% and a high of 6.4–8.5% pregnant Hispanic females with the epidemic phage type. Sample calculations used by the FDA/FSIS LMRA are shown in Table 2.6.

The dose of microorganisms consumed was based on the most likely consumption of cheese for each case, multiplied by the estimated number of organisms (CFU/g) of food. From the outbreak records, the estimated one-day consumption of the implicated cheese by 39 of 63 pregnant Hispanic females infected by the epidemic phage of *L. monocytogenes* serotype 4b ranged from 0.5 ounces/day [15 g/day] to 21 ounces/day [650 g/day] (median about 5.5 ounces/day [170 g/day]). In addition to reporting consumption for one day, about 38% of the females reported their usual consumption of cheese for more than one day. Contamination of the cheese by *L. monocytogenes* has been reported to be 10^3 to 10^4 *L. monocytogenes* CFU/g (NACMCF, 1991) and 140 000 to 500 000 *L. monocytogenes* CFU/g (Ryser and Marth, 1999). Thus, about 2.1–8.5% of pregnant Hispanic females that consumed between 1.5×10^4 and 5.0×10^7 *L. monocytogenes* serotype 4b organisms in a single day became ill. The effect of cumulative doses on the attack rate and pathogenesis was not estimated.

Table 2.6 FDA/FSIS LMRA (2001) estimates of attack rates and dose ranges: Pregnant women and United States of America Hispanic-style cheese outbreak.

Hispanic births (January – June, 1985), LA County	33 628
Hispanic fetal and neonatal deaths (January – June, 1985)	+ 350
Proportion of multi-gestational births (1%)	- 336
Population giving rise to cases (Total Hispanic pregnant females, January – June, 1985)	33 642
Total Hispanic pregnant females that ate the implicated cheese (based on an estimate that 35% of controls ate the implicated cheese)	11 775
High estimate of Hispanic pregnant females that ate contaminated cheese (based on an estimate of 26% product contamination × 11 775)	3 061
Average number of servings consumed	110
Total listeriosis cases among Hispanic pregnant females	81
Cases with outbreak phage type	63
Attack rate if all cases were identified (63/(3 061 × 110))	1.9×10^{-4}
Attack rate if 75% of cases identified (63/(3 061 × 110))/0.75	2.5×10^{-4}
Low estimate of Hispanic pregnant females that ate contaminated cheese (based on an estimate of 8.4% product contamination × 11 775)	989
Total listeriosis cases among Hispanic pregnant females	81
Cases with outbreak phage type	63
Attack rate per serving if all cases were identified (63/989 × 110)	5.8×10^{-4}
Attack rate per serving if 75% of cases identified (63/989 × 110)/0.75	7.7×10^{-4}

FDA/FSIS LMRA estimates for attack rates and dose ranges: Immunocompromised individuals and Finnish butter outbreak

The strategy used to calculate the Finland butter outbreak attack rate and dose range was similar to the strategy described above to calculate the attack rate and dose range for the Hispanic-style cheese outbreak (FDA/FSIS, 2001). Between December 1998 and February 1999, an increase in cases of listeriosis due to *L. monocytogenes* serotype 3a in Finland was recognized (O. Lyytikäinen, pers. comm., 1999; Lyytikäinen et al., 2000; Maijala et al., 2001). Review of national laboratory surveillance data from 1 June 1998 to 31 March 1999 identified a total of 25 *L. monocytogenes* serotype 3a cases, including six deaths. Cases of listeriosis were identified by cultures of blood, cerebrospinal fluid and samples from other sterile sites. Most of the cases were haematological or organ transplant patients. The median age of cases was 53 years (range 12–85). There were ten males and no pregnant females or newborns. The median hospital stay within 70 days prior to positive *Listeria* culture was 31 days for cases, and 10 days for controls.

Butter was implicated, and the isolates of *L. monocytogenes* serotype 3a from the butter and from 15 of the cases were indistinguishable. In the tertiary-care hospital where most cases occurred the implicated butter brand was the only brand consumed during the outbreak period. The hospital is the only site for organ transplantation and is also where most bone marrow transplants are performed. The total number of butter samples obtained at the hospital kitchen was 13 (pooled samples of 7-g packages), with an additional 139 samples (packages of 25 kg, 500 g, 7 g and 10 g) obtained from the dairy and from retail outlets. In addition, there were three estimates of the proportion of product contamination. There were 13 positives in 13 samples from the hospital kitchen (100%), and 4 positives in 5 samples (80%) and 3 positives in 5 samples (60%) from the retail outlets. In all positive hospital kitchen samples, the number of *L. monocytogenes* was <100 CFU/g (range 7–79). One wholesale supplier butter sample from a 7-g package contained 11 000 CFU/g.

It was possible to estimate butter consumption for five patients. The estimated consumption was divided by 31 days (median hospital stay) to estimate daily butter consumption. To determine the most likely dose range, the minimum butter consumption (1.1 g/day) was multiplied by the minimum contamination level for the hospital kitchen samples (7 CFU/g), and the maximum butter consumption (55 g/day) was multiplied by the maximum contamination level for the hospital samples (79 CFU/g). Using quantitative levels from the hospital samples, the consumed dose would be 8×10^0 to 4.3×10^3 CFU/day. If the maximum contamination level found in the wholesale samples (11 000 CFU/g) was the contamination actually consumed by those who became ill, then the daily dose consumed would range between 1.21×10^4 and 6.05×10^5 CFU/day.

Table 2.7 shows the attack rate calculations for the 1999 Finland butter outbreak. Approximately 6.4–10.7% of the haematological and transplant patients at the hospital that consumed between 8×10^0 and 4.3×10^3 *L. monocytogenes* serotype 3a organisms in a single day developed listeriosis. It is assumed that hospitalized patients ate 2.5 servings per day of the implicated butter on each of 31 days (median hospital stay), implying a total of 77.5 servings while hospitalized. The majority of the illnesses were associated with severe symptoms. The effect of cumulative doses on the attack rate and pathogenesis was not estimated.

Table 2.7 FDA/FSIS LMRA calculation of attack rate for an outbreak of *Listeria monocytogenes* serotype 3a infections from butter in Finland for haematological and transplant patients.

Annual number of new diagnoses for acute leukaemias or lymphomas plus annual number of kidney or liver transplants performed at the hospital.	410
Total persons at risk (time interval × annual new diagnoses; time interval was June 1998 to February 1999 = 9/12 months)	308
Estimated number of haematological and transplant patients in the population that ate the butter (proportion of controls that ate implicated butter, 76%)	234
Average number of servings during hospital stay	77.5
Number of cases during the outbreak	25
Number of cases with the same phage type	15
High estimate of product contamination (100%)	
Total number of contaminated servings (1 × 234 × 77.5)	18 135
Attack rate per serving (15/18135)	8.3×10^{-4}
Mid-estimate of product contamination (80%)	
Total number of contaminated servings (0.8 × 234 × 77.5)	14 508
Attack rate per serving (15/14 508)	1.0×10^{-3}
Low estimate of product contamination (60%)	
Total number of contaminated servings (0.6 × 234 × 77.5)	10 881
Attack rate per serving (15/10 881)	1.4×10^{-3}

SOURCE: FDA/FSIS, 2001.

Dose-response relation for L. monocytogenes *based on attack rates from the Hispanic-style cheese and butter outbreaks*

While the attack rate calculations for the Hispanic-style cheese and butter outbreaks were not used to generate a dose-response model in the FDA/FSIS LMRA (FDA/FSIS, 2001), for this risk assessment an attempt was made to use that information to estimate a relationship. For the Hispanic-style cheese outbreak, it was assumed that (1) the attack rate was the average of the estimates (4.5×10^{-4}) (see Table 2.6); (2) the contamination level was the higher estimate of 5×10^{5} CFU/g; and (3) a median portion size was 34 g (ca 1 ounce). These dose and response values were then used in combination with the exponential model. The derived r-value was 2.6×10^{-11}. This r-value, in turn, leads to an estimate that if a dose of 1×10^{6} CFU was consumed by a population of pregnant women, 0.0026% of their perinates/neonates would acquire listeriosis ($P = 2.6 \times 10^{-5}$). The FDA/FSIS LMRA team (FDA/FSIS, 2001) used a significantly more sophisticated dose-response model (see section 2.3.3.4 below). Their dose-response curve estimated that a 1×10^{6} dose would lead to an attack rate of 1.6×10^{-6} for neonates (or 4.0×10^{-6} for all pregnancy-associated cases). These dose and attack rate values would yield an r-value of 4.0×10^{-12}. Considering the uncertainties regarding the numbers of *L. monocytogenes* that were consumed, the one-log difference between the calculated r-values from the Hispanic cheese outbreak and the FDA/FSIS model suggests the models were in reasonable agreement.

In the Finnish outbreak, the estimated attack rates varied between 8.3×10^{-4} and 1.3×10^{-3}, with a median of 1.0×10^{-3}. The estimated dose ranged from 8×10^{0} to 4.3×10^{3} CFU (hospital samples), and 1.21×10^{4} to 6.05×10^{5} CFU (wholesale samples, see earlier). Obviously, the large uncertainty in the estimation of the dose consumed leads to a large uncertainty in the estimated r-value for the exponential model. It was assumed that the attack rate was 1.03×10^{-3} (median attack rate), and the dose was 8.2×10^{3} CFU (the median of the dose range values). The derived r-value was 3.15×10^{-7}, which leads to an estimate that if a dose of 1×10^{6} CFU were consumed by haematological and organ transplant patients, 27% of them would acquire listeriosis.

Although the virulence of this strain has not been compared to other strains, the susceptibility of these patients to listeriosis is clearly greater than other groups. An estimate based on French epidemiological data of relative risk of transplant patients versus people less than 65 years old with no immunosupression was 2854 to 1. Estimates based on the FDA/FSIS data for the relative risk for immunocompromised persons within the intermediate age group versus the rest of the intermediate age group was 1584 to 1.

Dose-response relation for L. monocytogenes *based on attack rates from two outbreaks of febrile gastroenteritis among immunocompetent individuals*

- Chocolate milk - United States of America: Gastroenteritis in healthy adults

  An outbreak of gastroenteritis and fever occurred among persons who had attended a picnic in Elizabeth, Illinois, in the United States of America in 1994 (Dalton et al., 1997; Proctor et al., 1995). By both epidemiological and laboratory findings, the outbreak was linked to consumption of contaminated chocolate milk. None of those attending the picnic were reported to have an underlying chronic illness or immunodeficiency. Forty-five of the 60 people (75%) who consumed chocolate milk at the picnic reported illness that met the case definition, compared with none of the 22 people who did not drink chocolate milk. Nine other people who consumed chocolate milk had an illness in the week after the picnic that did not meet the case definition. This indicates that the attack rate was between 75% and 90%.

  Based on laboratory investigations of milk containers and the estimated consumption, the median dose was estimated to be 2.9×10^{11} CFU per person.

- Maize-tuna salad – Italy: Febrile gastroenteritis in immunocompetent students and staff

  Of those interviewed, 72% reported symptoms and 18.6% of those individuals had been hospitalized. The symptoms included headache, abdominal pain, diarrhoea, nausea, vomiting and joint and muscular pain, but no sepsis or deaths were reported. The investigation indicated that sweet corn and tuna salad were associated with the highest relative risk. The food-specific attack rate for sweet corn and tuna salad was reported to be 83.9%.

  The level of *L. monocytogenes* contamination in the salad was $>10^{6}$ CFU/g (Aureli et al., 2000). No estimate of consumption among the students was reported. Assuming that the average consumption was 100 g of sweet corn-tuna salad and the level of *L. monocytogenes* was 10^{6} CFU/g, the estimated dose becomes 10^{8} CFU.

The estimated ingested dose and attack rates of these outbreaks were then used in combination with the exponential model. The derived r-values were estimated to be 5.8×10^{12} (chocolate milk) and 1.8×10^{-8} (sweet corn-tuna salad), respectively.

Using the r-values described above, the exponential model dose-response curves derived from epidemiological data are shown in Figure 2.1 and compared with the Weibull-Gamma model developed using expert estimates for the low risk group (Farber, Ross and Harwig, 1996). It should be emphasized that while the end-points in these relationships are the same, namely morbidity, they are based on data reflecting a wide range of symptoms in terms of severity.

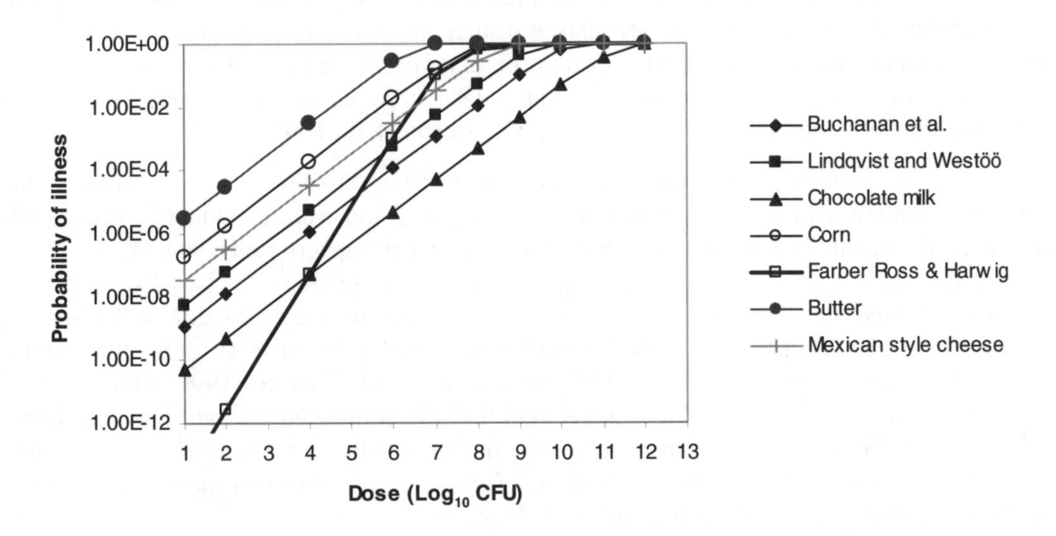

Figure 2.1 A comparison of the dose-response curves for morbidity derived from epidemiological data or expert elicitations. The models include outbreaks where the primary symptoms included serious illness (smoked fish: Buchanan et al., 1997; smoked fish: Lindqvist and Westöö, 2000, and Farber, Ross and Harwig, 1996; butter, current study), perinatal/neonatal infections (Hispanic-style cheese, current study), and febrile gastroenteritis (sweet corn-tuna salad and chocolate milk, current study).

2.2.3.3 Listeria monocytogenes *dose-response models developed from data derived from surrogate pathogens or surrogate animals*

Surrogate pathogens

L. monocytogenes is unusual among foodborne pathogens in terms of its pathogenicity, susceptible populations and clinical manifestations. As such, there is no microorganism that can serve as a surrogate for *L. monocytogenes* in relation to its ability to cause disease. The genus *Listeria* is differentiated into two very closely related species, *L. monocytogenes* and *L. innocua.* The differentiation is based on the ability to produce listeriolysin, a protein that causes lysis of red blood cells. This toxin is considered a key virulence determinant for *L. monocytogenes*. The original goal of the taxonomic scheme was to differentiate between pathogenic and non-pathogenic isolates. However, listeriolysin does not appear to be the only difference between these species, as shown in the recent comparison of the genomes of *L. monocytogenes* and *L. inoccua,* which indicated the presence of 270 and 149 strain-specific genes, respectively (Glaser et al., 2001). *L. innocua* is non-pathogenic and as such has been used extensively as a surrogate microorganism to study the growth and survival characteristics of *L. monocytogenes* in foods.

Surrogate animal data

Various feral, domestic and laboratory animals are susceptible to infections by *L. monocytogenes*, but rabbits, mice and rats have been used most extensively since these animals die within 1 to 7 days following intravenous (IV) or intraperitoneal inoculation (McLauchlin, 1997). Using these inoculation routes, the LD_{50} for mice ranged from 10^2 to 10^9 CFU (Table 2.8), depending on the strain of *L. monocytogenes*, the strain of mouse and the route of inoculation (Audurier et al., 1980; Mainou-Fowler, MacGowan and Postlethwaite, 1988; Golnazarian et al., 1989; Notermans et al., 1998).

Mice are susceptible to oral infection although the LD_{50} ratios for oral and intraperitoneal administration are not consistent among different strains (Pine, Malcolm and Plikaytis, 1990). However, the results of Lecuit et al. (1999) indicate that the specific mechanism for crossing of the intestinal barrier is not the same as in human listeriosis since mice lack E-cadherin, the receptor for internalin A. Conflicting results appear in the literature concerning the levels of *L. monocytogenes* necessary to induce infection or mortality by an oral route, and reported LD_{50} values range from 50 to $>10^9$ CFU (Pine, Malcolm and Plikaytis, 1990; Audurier et al., 1980; Notermans et al., 1998). Notermans et al. (1998) concluded that an intestinal barrier and a specific immune defence mechanism act independently in protecting mice inoculated orally with *L. monocytogenes* from infection. These results mean that mouse data must be used with caution when making inferences for humans.

The approximate LD_{50} for mice was not statistically different for five *L. monocytogenes* strains when grown in milk or suspended in milk, compared with suspension in phosphate-buffered saline (Pine, Malcolm and Plikaytis, 1990). Similarly, growth in milk did not enhance the virulence of *L. monocytogenes* for Sprague-Dewley rats but instead was suggested to have an inhibitory effect on the number of bacteria available for colonization (Schlech, 1993). The ID_{50} for these rats following gastric inoculation with *L. monocytogenes* was 10^6 CFU. This was not affected by pregnancy, although the invasive infection led to abnormal reproductive outcomes (Schlech, 1993).

The pathogenicity of a *L. monocytogenes* isolate injected intraperitoneally into mice did not differ when grown in crabmeat or tryptose phosphate broth (Brackett and Beuchat, 1990). The ID_{50} of pregnant mice inoculated orally with *L. monocytogenes* appeared to be lower than for normal control mice, although the difference was not statistically significant due to the small number of mice used in the trial (Golnazarian et al., 1989). The authors also compared the ID_{50} of normal mice and mice immunocompromised other than by pregnancy. The immunocompromised mice were beige mutants (deficient in lysosome production within their monocytes and granulocytes), cimetidine-treated mice (decreases gastric acidity) and hydrocortisone acetate-treated mice. With the exception of treatment with hydrocortisone acetate (an observed response with 2.5 mg/day but not with 0.25 mg/day), the responses of predisposed mice were not significantly different from the response of normal mice (Golnazarian et al., 1989). Similarly, decreasing gastric acidity with an antacid had no substantial effect on the infective dose in a non-human primate model (Farber et al., 1991). Immunosuppression by cyclosporin A did not alter the ID_{50} but led to more prolonged infections (Schlech, Chase and Badley, 1993). In contrast to the results of Golnazarian et al. (1989), treating rats with cimetidine lowered the infective dose significantly (Schlech, Chase and Badley, 1993). However, the interpretation of the effects of these treatments should be treated with caution since the animals did not have the underlying physiological condition that required this cimetidine treatment (Golnazarian et al., 1989).

In the absence of human clinical data, various animal and *in vitro* (e.g. tissue culture) surrogates have been used to acquire experimental dose-response data. For foodborne listeriosis, the model with the greatest apparent similarity to human infections comes from dose-response studies that use oral infection of mammals. The primary animal surrogate used has been the mouse.

Table 2.8 Summary of some *Listeria monocytogenes* dose-response studies using animal models.

Animal model	Route	ID_{50} (CFU)	LD_{50} (CFU)	Other	Source
Monkey	IG			10^5, shedding 2 days 10^7, shedding 3 wks 10^9, shedding 3 wks, symptoms	[1]
Outbred mice	AS		$10^{3.1}$ –$10^{5.5}$		[2]
	GI			10^{10}, 20–100% mortality	
Mice	IV	<$2.7x10^2$	$2.6x10^5$		[3]
	SC	<$2.1x10^2$	>$2.1x10^8$		
	O	$9.9x10^6$	$7.0x10^9$		
C57BL/6 mice	IV		0.8–$6.2x10^6$		[4]
BALB/c mice	IV		0.04–$0.6x10^6$		[4]
Mice	IP		$10^{2.57}$; $10^{2.69}$; $10^{4.96}$; $10^{5.08}$; $10^{5.75}$; $10^{5.91}$		[5]
	IG		$10^{5.47}$		[5]
Mice	IP		$10^{2.68}$; $10^{3.62}$; $10^{4.56}$; $10^{4.57}$; $10^{4.73}$; $10^{4.95}$; $10^{5.47}$; $10^{6.00}$; $10^{6.23}$; $10^{8.88}$; $10^{9.70}$		[6]
S-D Rats	O	10^6			[7]
Stelma	IP		$10^{6.04}$; $10^{6.80}$; $10^{7.28}$; $10^{7.30}$; $10^{7.54}$		[8]
Mice	IP	$10^{3.2}$	$10^{4.79}$; $10^{4.52}$		[9]
	O	$10^{4.57 (1)}$; $10^{4.00 (2)}$; $10^{3.30 (3)}$; $10^{2.48 (4)}$	$10^{4.77}$; $10^{4.24}$; $10^{0.94 (1)}$		[9]
Mice	IP		$10^{0.77}$ **; $10^{1.11}$ **; $10^{1.46}$ **; $10^{1.49}$ **; $10^{1.49}$ **; $10^{1.62}$ **; $10^{1.87}$ **; $10^{1.97}$ **; $10^{2.00}$ **; $10^{2.04}$ **; $10^{2.30}$ **; $10^{3.00}$ **; $10^{3.15}$ **; $10^{3.30}$ **; $10^{3.49}$ **		[10]
Mice	IV	$10^{1.8}$; $10^{5.6}$*; $10^{1.0}$**	$10^{3.2}$; $10^{5.8}$*; $10^{2.3}$**		[11]
	O	$10^{6.5}$; >$10^{9.0}$*; $10^{6.3}$**	>$10^{9.0}$; >$10^{9.0}$*; >$10^{8.0}$**		[11]

KEY: IG = Intragastric. IV = Intravenous. IP = intraperitoneal. AS = Aerosol. SC = Subcutaneous, GI = Gastric intubation. O = Oral. * = Previously exposed to *L. monocytogenes.* ** = Immunosuppressed by carrageenan. S-D = Sprague-Dewley.

NOTES: (1) = Hydrocortisone acetate treated. (2) = Lysosome deficient. (3) = Cimetidine treated. (4) = Pregnant.

SOURCES: [1] Farber et al., 1991. [2] Bracegirdle et al., 1994. [3] Audurier et al., 1980. [4] Mainou-Fowler, MacGowan and Postlethwaite, 1988. [5] Pine, Malcolm and Plikaytis, 1990. [6] Pine et al., 1991. [7] Schlech, 1993. [8] Stelma et al., 1987. [9] Golnazarian et al., 1989. [10] del Corral et al., 1990. [11] Notermans et al., 1998.

Endpoints in studies with animal surrogates have usually been infection or death. Because infection in mice is based on the recovery of *L. monocytogenes* from normally sterile internal organs (e.g. spleen, liver), it is difficult to relate this to data for humans where infection has been assessed largely on the basis of the colonization of the intestinal tract by the microorganism.

One study that determined both endpoints following oral dosing of inbred mice (Golnazarian et al., 1989) is useful for determining the relationship between these endpoints. No dose-response studies of *L. monocytogenes* in animal surrogates that used host physiological endpoints or biomarkers other than infection or lethality appeared to have been reported. Other animal surrogates, such as rats (Schlech, Chase and Badley, 1993) and primates (Farber et al., 1991), have also been used for oral dose-response studies, but are not as developed as the mouse system, lacking the extensive genetic and immunological tools that are available in the mouse model. A study with pregnant primates was underway in the United States of America, but results from this investigation were not yet available at the time of preparing this report. There is also a paucity of human data to directly correlate relevant biomarkers of exposure in mice to the frequency and severity of listeriosis in humans.

The work of Notermans et al. (1998)

Notermans et al. (1998) examined dose-response relations for a single strain of *L. monocytogenes* in normal and immunocompromised (i.e. carrageenan-treated) mice using both intravenous and oral dosing (Table 2.9). Both infection and lethality were used as biological end points. They also tested mice previously exposed to *L. monocytogenes* in order to elicit immune protection. Both infectivity and lethality were greater when the pathogen was administered by intravenous injection, and lethality was not observed with any of the orally dosed animals. Immunosuppression decreased the ID_{50} when *L. monocytogenes* was administered by intravenous injection, but did not affect the oral ID_{50}. Prior exposure of the mice decreased the infectivity and lethality of the *L. monocytogenes* isolate.

Table 2.9 Effect of immunosuppression, route of entry and prior exposure on the ID_{50} and LD_{50} values for mice exposed to *Listeria monocytogenes* strain EGD (serotype 1/2a).

Condition of mice	ID_{50}		LD_{50}	
	IV[1]	Oral	IV	Oral
Normal Immunocompetency, naive (Unexposed)	1.8[2] (1.10×10^{-2})[3]	6.5 (2.00×10^{-7})	3.2 (4.37×10^{-4})	>9.0
Normal Immunocompetency, protected (Prior exposure)	5.6 (1.70×10^{-6})	>9.0	5.8 (1.10×10^{-6})	>9.0
Immunosuppressed (carrageenan-treated), naive (Unexposed)	1.0 (6.93 × 10-2)	6.3 (3.00×10^{-7})	2.3	>8.0
Immunosuppressed (carrageenan-treated), protected (Prior exposure)	0.8 (1.10×10^{-1})	7.9 (8.73×10^{-9})	3.2 (4.37×10^{-4})	>8.0

NOTES: (1) IV = intravenous. (2) Log_{10} CFU. (3) r-value from fitted exponential model.
SOURCE: Adapted from Notermans et al., 1998.

Notermans et al. (1998) found the slope of the various dose-response curves to be steep, with both the infection and lethality data being described well using the exponential model. Using an oral LD_{50} value of $\log_{10} = 8.0$, the minimum value for immunosuppressed, protected mice, in conjunction with the yearly human exposure estimates of Notermans et al. (1998), the number of human cases predicted by the exponential dose-response model is <2054 deaths per 1 000 000 persons. This incidence is substantially higher than the 4 to 7 cases per 1 000 000 actually observed. Moreover, these models have limited usefulness since the data for oral administration did not actually establish a dose-response relation for lethality, and intravenous administration is questionable in relation to dose-response relations in humans.

The work of Haas et al. (1999)

The dose-response relation for *L. monocytogenes* infectivity was evaluated by Haas et al. (1999) using the data of Audurier et al. (1980) and Golnazarian et al. (1989). Both data sets represent mice that were orally administered *L. monocytogenes*. The data were fitted using the exponential model and the Beta-Poisson model. The exponential model did not adequately fit the data, whereas the Beta-Poisson model could describe the data sets. In comparing the dose-response curves for strains 10401 (serovar 4b) and F5817 (serovar 4b), used by Audurier et al. (1980) and Golnazarian et al. (1989), respectively, Haas and co-workers concluded that there were significant differences in the strains' infectivity, and that one strain could not be used to describe the dose-response relation of the other. The α and ID_{50} values were 0.17 and 2.1×10^6 CFU for strain 10401, and 0.25 and 2.76×10^2 for strain F5817. Haas and co-workers speculated that the difference in the infectivity of the strains might reflect the method and vehicle of administration. Golnazarian et al. (1989) dosed animals by gavage using milk, whereas Audurier et al. (1980) dosed through drinking water.

Using these models, Haas et al. (1999) compared predicted values with the burden of disease, both in relation to annual incidence and for several outbreaks of febrile gastroenteritis. They concluded that the model based on the data of Golnazarian et al. (1989) greatly over-predicted the infectivity of *L. monocytogenes* when compared with the attack rates reported for outbreaks associated with rice salad (Salamina et al., 1996) and chocolate milk (Dalton et al., 1997). Predictions based on the dose-response model from the Audurier et al. (1980) data were more realistic in comparison with the observed data. When the two dose-response models were used to evaluate the exposure estimates of Notermans et al. (1998), Haas et al. (1999) concluded that the predicted infection rate was unrealistically high.

These observations provide a good example of the care that must be exercised in developing and interpreting dose-response models based on surrogate animal data. First, care must be taken to ensure that the models are based on the same biological end point as the disease's manifestation in humans. It is not surprising that a dose-response model based on infectivity does not provide predictions that match the data on reported cases of illness. The public health data is based largely on cases of meningitis, septicaemia and other severe symptoms, rather than febrile gastroenteritis. Lethality would be a more consistent biological end point and there is a relatively constant ratio in human cases between severe cases and fatalities (i.e. approximately 20% to 30% of hospitalized patients die). The dose differential between the LD_{50} and ID_{50} for the Audurier et al. (1980) and Golnazarian et al. (1989) studies was approximately 1000-fold and 10-fold, respectively. Second, there is no assurance that the dose-response relations for mice and humans are the same. There is a need to correlate or "anchor" the response in a surrogate animal with that in humans. Traditionally, this has been done with human volunteer feeding studies. However, this is not possible with *L. monocytogenes*, so an alternative means is needed, such as annual disease statistics.

2.2.3.4 Listeria monocytogenes *dose-response models developed from data derived from a combination of surrogate animal and epidemiological data*

The FDA/FSIS LMRA team (FDA/FSIS, 2001) employed an approach that combined the use of surrogate animal data with epidemiological findings. The exposure assessment provided an estimate of the frequency and distribution of consuming *L. monocytogenes*. Surrogate animal data were used to establish the shape of the dose-response curve and the epidemiological data to anchor the results, i.e. the results were constrained so that the predicted incidence of disease approximated the incidence of severe infections noted in a population. A dose-response adjustment factor was created so the exposure data and dose-response model would calculate the estimated number of cases per year in the United States of America. The dietary consumption surveys did not indicate any major difference between population groups. Separate dose-response models were calculated for three populations: pregnant women and their unborn; the elderly (> 60 years); and intermediate-aged (everyone else). The numbers of deaths per year from the epidemiological surveillance were 50 neonatal, 250 elderly and 200 intermediate-aged. The estimated total number of deaths per year for the entire perinatal group (prenatal and neonatal) was 125.

The risk assessment examined two different biological end points, infection (defined as serious illness) and lethality. The incidences of infection, i.e. the incidence of cases requiring hospitalization, were derived from the lethality data using the established ratios between human infection and lethality. The models were developed to account for the variability and uncertainty both in the biological phenomena being modelled and the modelling approaches employed. Accordingly, the dose-response models were designed to run as Monte Carlo simulations. The models factored-in the differences in the virulence of *L. monocytogenes* isolates and the differences in the susceptibility among the three groups of humans (i.e. the general population, the elderly, and perinates/neonates).

Dose-response model based on studies with mice

The relationship between the number of *L. monocytogenes* consumed and the occurrence of either infection (serious illness) or death (mortality) was modelled using data obtained for immunocompetent mice orally administered *L. monocytogenes* F5817. Because of the effects of strain variation, host susceptibility and the differences between a surrogate animal in a controlled environment versus humans in an uncontrolled environment, the mouse model served primarily to establish the shape or steepness of the dose-response curve. In actuality, with the added uncertainties and the linear shape of the curve in the probability of illness range of interest (log dose-log probability plot), the shape of the mouse curve had relatively little influence on the final dose-response curve.

The data used to develop a dose-response curve for mortality was taken from Golnazarian et al. (1989). Data were fitted to six different models using an iterative, least-squares curve-fitting procedure. The best four models (Beta-Poisson, exponential, logistic and Gompertz-Log) were used to characterize the uncertainty in the shape of the dose-response curve. The Gompertz-Log and Weibull-Gamma models were discarded for lack of fit. The Exponential model provided the best fit and received the most weight (Figure 2.2). Notermans et al. (1998) also reported the exponential model to be effective for depicting the relationship between dose and lethality in mice. The resultant FDA/FSIS LMRA dose-response relations for infection and mortality in mice are summarized in Table 2.10.

Because mortality is a more consistent measure than infection or illness in mice and human data, the FDA/FSIS LMRA used lethality as their primary model to define the *L. monocytogenes* dose-response relations. The number of serious illnesses that did not lead to death was estimated to be four times the number of deaths based upon epidemiological data. An exception to this was the dose-response relation for perinatal infections, where the primary public health impact was considered to be death. The frequency of perinatal deaths was estimated to be 1.5 times the observed frequency of neonatal deaths. For the United States of America in recent years, this was approximately 500 deaths and 2000 additional cases of serious illness.

Modelling the variability in virulence among strains of L. monocytogenes

Since there appears to be substantial variability in the pathogenicity of *L. monocytogenes* strains, based on animal model data, the FDA/FSIS LMRA included a model for variability in virulence. The model is based on data acquired using mice. Specifically, the range of LD_{50} values observed in mice was also used to characterize the range of variation expected in humans. Adjustment of the dose-response relationship relative to the LD_{50} presumes that the shape of the population dose-response function is the same for different strains.

The data used was from three studies (Stelma et al., 1987; Pine, Malcolm and Plikaytis, 1990; Pine et al., 1991) wherein *L. monocytogenes* was administered to immunocompetent mice by intraperitoneal injection (Table 2.11). The LD_{50} values encompassed a range of over 7 orders of magnitude. Although some of the strains were obtained directly from food, most of the strains tested were clinical isolates, which may have biased the model towards strains that are more virulent. Conversely, while a range of strains were used, there are no definitive studies that have attempted to examine the relative virulence of foodborne *L. monocytogenes* strains on the basis of their relative occurrence.

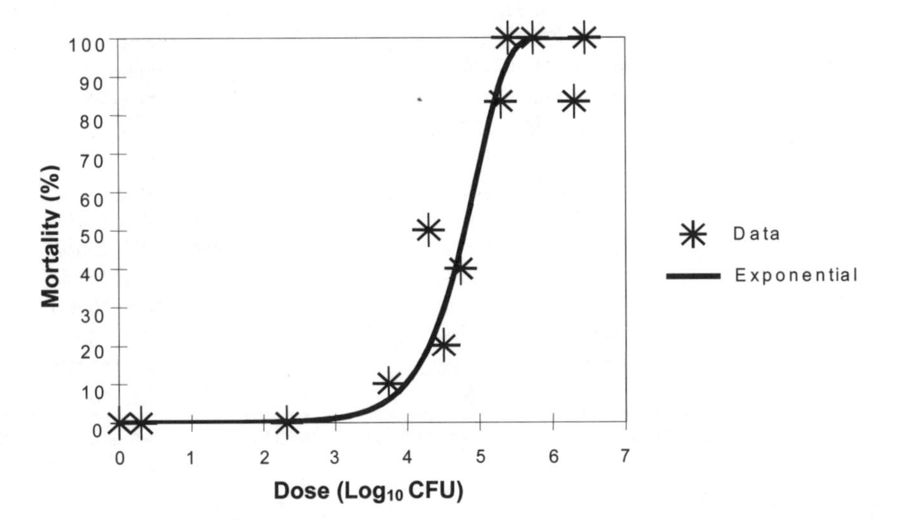

Figure 2.2 Dose versus frequency of mortality in mice administered *Listeria monocytogenes.*

SOURCE: Adapted from FDA/FSIS, 2001.

The strains examined in the laboratory investigations were selected to provide a wide virulence range, and not on the basis of their relative occurrence in foods. Because most food isolates do not appear to cause outbreaks, the array of strains in Table 2.11 could have a disproportionate number of strains with reduced virulence. No large or obvious trends in the LD_{50} values relative to either serotype or strain source were apparent.

Because the three studies used to estimate the variability in virulence among *L. monocytogenes* strains were based on studies using intraperitoneal administration and the dose-response model was based on orally dosed mice, there was concern that this could introduce a bias. To avoid this potential source of error, the FDA/FSIS LMRA developed a correction factor for adjusting the virulence model. The correction factor was based on a study of Pine, Malcolm and Plikaytis (1990) that compared the LD_{50} values for *L. monocytogenes* strains administered both by intragastric gavage and by intraperitoneal injection.

Table 2.10 Dose-response functions for infection and mortality in mice resulting from oral exposure to *Listeria monocytogenes* strain F5817.

Effective Dose (Log_{10} CFU)	Infection	Mortality
0	9.11%[1] (0.87%, 9.38%)[2]	0.001% (0.000%, 0.007%)
0.5	12.2% (2.6%, 12.5%)	0.003% (0.000%, 0.018%)
1	16.4% (7.3%, 16.5%)	0.011% (0.000%, 0.045%)
1.5	21.7% (17.1%, 21.8%)	0.035% (0.000%, 0.117%)
2	28.4% (28.4%, 31.3%)	0.110% (0.000%, 0.301%)
2.5	36.7% (36.5%, 46.0%)	0.347% (0.002%, 0.775%)
3	46.4% (46.2%, 58.7%)	1.09% (0.06%, 1.99%)
3.5	57.4% (57.1%, 68.7%)	3.42% (1.06%, 5.11%)
4	68.8% (68.6%, 76.3%)	10.4% (8.3%, 12.9%)
4.5	79.6% (79.5%, 82.1%)	30.8% (29.4%, 32.0%)
5	88.6% (86.5%, 88.6%)	66.7% (63.6%, 67.6%)
5.5	94.8% (89.8%, 95.0%)	95.2% (90.7%, 96.9%)
6	98.2% (92.3%, 98.4%)	100% (98%, 100%)
6.5	99.6% (94.2%, 99.7%)	100% (100%, 100%)
7	99.9% (95.6%, 100.0%)	100% (100%, 100%)
7.5	100% (97%, 100%)	100% (100%, 100%)
8	100% (97%, 100%)	100% (100%, 100%)
8.5	100% (98%, 100%)	100% (100%, 100%)
9	100% (99%, 100%)	100% (100%, 100%)
9.5	100% (99%, 100%)	100% (100%, 100%)
10	100% (99%, 100%)	100% (100%, 100%)
10.5	100% (99%, 100%)	100% (100%, 100%)
11	100% (100%, 100%)	100% (100%, 100%)
11.5	100% (100%, 100%)	100% (100%, 100%)
12	100% (100%, 100%)	100% (100%, 100%)

NOTES: (1) Median estimate. (2) Confidence intervals representing the 5th and 95th percentiles of the uncertainty distribution.

SOURCE: Based on data from Golnazarian et al. (1989).

Table 2.11 LD_{50} values for various *Listeria monocytogenes* strains administered by intraperitoneal injection to immunocompetent mice.

Strain	Serotype	Source	LD_{50} (Log_{10} CFU)	Study
G9599	4	Clinical	*2.57*	[1]
G1032	4	Clinical	*2.69*	[1]
G2618	1/2a	Food	*2.89*	[2]
F4244	4b	Clinical	3.62	[2]
F5738	1/2a	Clinical	3.67	[1]
F6646	1/2a	Clinical	4.49	[1]
15U	4b	Clinical	4.56	[2]
F4246S	1/2a	Clinical	4.57	[2]
F7208	3a	Clinical	4.61	[1]
G2228	1/2a	Clinical	*4.66*	[1]
F2381	4b	Food	4.73	[2]
G2261	1/2b	Food	*4.95*	[2]
F2380	4b	Food	*4.96*	[1]
F2392	1/2a	Clinical	5.08	[1]
1778+H1b	1/2a	Clinical	*5.47*	[2]
F7243	4b	Clinical	*5.75*	[1]
F7245	4b	Clinical	*5.91*	[1]
SLCC 5764	1/2a	Clinical	6.00	[2]
V37 CE	–	Food	6.04	[3]
F7191	1b	Clinical	6.23	[2]
V7	–	Food	6.80	[3]
Brie 1	–	Food	7.28	[3]
Murray B	–	Clinical	7.30	[3]
Scott A	4b	Clinical	7.54	[3]
G970	1/2a	Clinical	8.88	[2]
NCTC 5101	3a	Clinical	9.70	[2]

NOTE: (1) The italicized LD_{50} values are averages from multiple experiments. (2) The original studies were [1] Pine, Malcolm and Plikaytis, 1990. [2] Pine et al., 1991. [3] Stelma et al., 1987.

SOURCE: FDA/FSIS, 2001: Table IV-2.

Table 2.12 Effect of administration route (intraperitoneal vs intragastric gavage) on mouse LD_{50} values.

Strain	Serotype	Source	Log_{10} ratio (intragastric/intraperitoneal)
F2380	4b	Food	-1.81
F7243	4b	Clinical	-0.75
F7245	4b	Clinical	-0.47
G2228	1/2a	Clinical	0.00
G2261	2/1b	Food	0.00
NCTC 7973	1/2a	Food	0.04
F6646	1/2a	Clinical	0.21
F2380	4b	Food	0.71
G9599	4	Clinical	0.96
G1032	4	Clinical	1.60
F5738	1/2a	Clinical	1.81
G2618	1/2a	Food	2.00

Source: FDA/FSIS, 2001: Table IV-3., based on data from Pine, Malcolm and Plikaytis (1990).

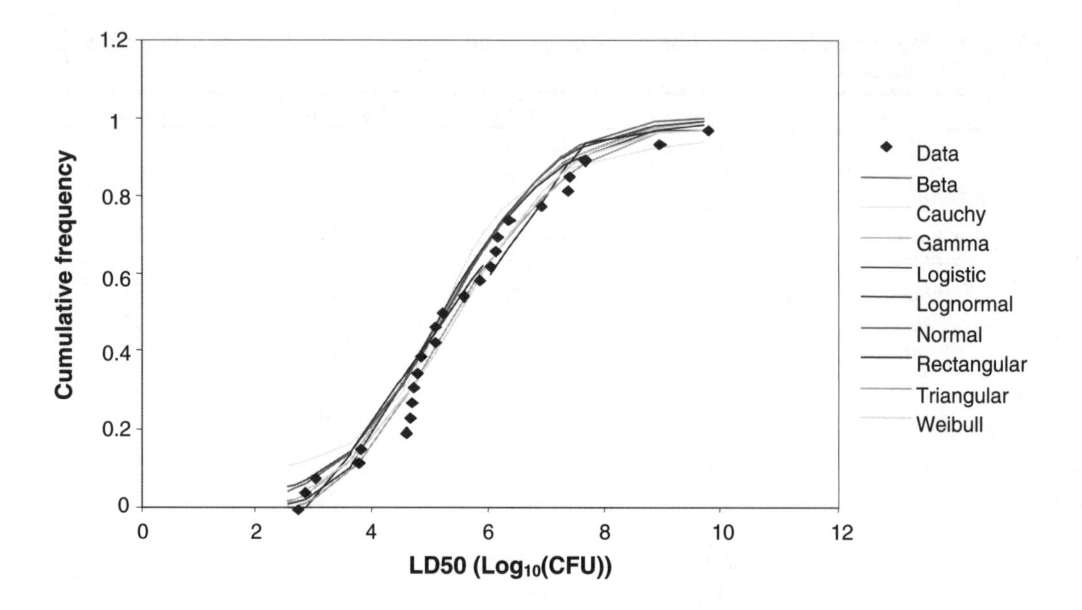

Figure 2.3 Variation in *Listeria monocytogenes* Strain Virulence: Nine Distributions.

SOURCE: FDA/FSIS, 2001. Figure IV-2.

Table 2.13 Model output for *Listeria monocytogenes* strain virulence

Percentile	LD_{50} (Log_{10} CFU)
1st	2.55 (0.97, 2.80)
5th	3.12 (2.47, 3.32)
10th	3.53 (3.18, 3.66)
25th	4.28 (4.20, 4.39)
Median	5.25 (5.15, 5.34)
75th	6.35 (6.23, 6.48)
90th	7.45 (7.25, 7.67)
95th	8.06 (7.84, 8.54)
99th	9.47 (8.52, 10.59)

NOTE: Values in parentheses are the 5th and 95th percentiles for the uncertainty about the distribution in virulence.

SOURCE: FDA/FSIS, 2001: Table IV-4.

Although there was up to a 100-fold difference in the LD_{50} values by the two routes, the intragastric or the intraperitoneal route was the most effective depending upon the strain (Table 2.12). The median value of the ratio between the LD_{50} determined using the intragastric and intraperitoneal routes, respectively, was greater than 1.0 (i.e. [LD_{50}oral/ LD_{50}ip]), indicating that the correction factor for virulence could overestimate the virulence of *L. monocytogenes* (by approximately half a log).

In Table 2.12, a $\log_{10}$ ratio of 0 indicates that the LD_{50} by the two routes were identical. A negative number indicates a lower LD_{50} by the intragastric route, while a positive number indicates a greater LD_{50} by the intragastric route. The data in Table 2.12 were modelled by fitting nine distributions. The best five were used to characterize model uncertainty associated with distribution (Figure 2.3). The resulting distribution in LD_{50} is given in

Table 2.13. This distribution was used to describe the extent of virulence variability in determining dose-response. Because the virulence was estimated from the distribution of intraperitoneal administered doses, the estimated LD_{50} was increased by zero to one log (uniform uncertainty range) to adjust the virulence value to more accurately predict the estimated oral LD_{50}.

Dose-response model for mortality in humans

Intermediate-Age Population

The FDA/FSIS LMRA considered three age-related populations: perinatal cases (mothers, and fetus and newborns from 16 weeks of gestation to 30 days of age); elderly cases (>60 years of age); and an intermediate age group (>30 days, <60 years). Considering that it appears that humans are commonly exposed to low levels of foodborne *L. monocytogenes*, direct application of the mouse dose-response model would greatly overestimate the incidence of lethal infections in humans from *L. monocytogenes*. The LD_{50} in the mouse study from which the curve was derived was about $\log_{10} = 4.3$, or about 20 000 CFU. The periodic exposure of humans to such numbers of *L. monocytogenes* is frequent (FDA/FSIS, 2001; Buchanan et al., 1997; Notermans et al., 1998). Based on the described consumption, contamination and growth data, it was estimated that if the mouse dose-response model were used directly, it would overestimate the number of illnesses and deaths due to listeriosis by a factor of over 10^9. If the estimates of the occurrence of *L. monocytogenes* in food (developed in the exposure assessment) are reasonable, then human beings are much less susceptible than laboratory mice to *L. monocytogenes*. Therefore, the mouse-derived models had to be adjusted to reflect human susceptibility. A dose-response adjustment factor was applied that allowed the models to predict serious illness and death occurrences roughly consistent with surveillance data reported to FoodNet. Thus, while the shape of the curve was initially derived from mice, the curve's position on the dose scale is determined by the human surveillance record. Because of large differences in the behaviour of the dose-response model at low doses, the magnitude of the adjustment factor was model-dependent (Tables 2.14 and 2.15).

Table 2.14 Model-dependence of dose-response adjustment factor for intermediate-age populations.

Model	**Dose Adjustment** (Log_{10} CFU) Minimum	Maximum
Logistic	11.85	12.35
Exponential	11.85	12.35
Gompertz-Log	11.85	12.35
Probit	11.95	12.20
Multihit	11.95	12.45

SOURCE: FDA/FSIS, 2001: Table IV-5a.

Table 2.15 Model-dependence of the *Listeria monocytogenes* dose-response adjustment factor ranges for the three human populations.

Population	**Dose-Response Adjustment Factor Range** (Log_{10} CFU) Minimum	Maximum
Intermediate-Age	11.85	12.45
Neonatal (1)	7.8	8.4
Elderly	11.85	11.45

NOTE: (1) An adjustment to account for total perinatal deaths (prenatal and neonatal) is in the risk characterization section of FDA/FSIS, 2001.

SOURCE: FDA/FSIS, 2001: Table IV-5b.

After applying the virulence distribution (Table 2.13) to the normal mouse dose-response mortality curve (Table 2.10), the dose-response adjustment factor was shifted using iteration techniques, moving the curve towards the higher doses necessary for lethality estimates to agree with surveillence data. Figure 2.4 depicts the results of applying this factor to the intermediate-aged population. The distribution considered four sources of uncertainty and variability: strain virulence, host (human) susceptibility, uncertainty in the exposure, and dose-response adjustment factor.

In two subsequent dose-response curves (Figures 2.5 and 2.6), adjustments are made that reflect increased susceptibility in perinatal and elderly populations (see next section). The intermediate-age population includes higher-risk individuals not explicitly included in the perinatal and elderly groups, such as AIDS, cancer and transplant patients. These individuals probably make up a disproportionate number of the cases of serious listeriosis within this population; however, it was considered that insufficient data to further distinguish these populations were available when FDA/FSIS conducted their dose-response modelling. Because the portion of the intermediate-age population at higher risk for listeriosis is small in comparison with entire population, this leads to dose estimates at high response rates (e.g. LD_{10}, LD_{50}) that are unrealistic in terms of the number of bacteria that could be consumed by an individual. Doses greater than 10^{12} or 10^{13} CFU should be considered notional and be interpreted as indicating that a substantial segment of the population is not susceptible.

Modelling dose-response relations for perinatal and elderly populations

The FDA/FSIS LMRA adjusted the dose-response model to account for the increased susceptibility of neonates and the elderly, in order to make predicted results consistent with both the number of cases reported from surveillance data (CDC, 2001) and the range of sensitivity encountered in studies with immunocompromised mice. These models employed a susceptibility adjustment distribution for each sub-group. This included adjusting the number of servings consumed by the size of the population relative to the total population.

For neonates, the population size was adjusted for an annual birth rate corresponding to 1.8% of the total population and a distribution period for *in utero* exposure with a range of 1 to 30 days prior to birth, with an average value of 10 days. The dose-response relations were initially modelled only for neonates because the epidemiology data were for that group. Conversion to perinatal case rates (prenatal and neonatal combined) was done after dose-response simulation. Perinatal deaths were estimated at 2.5 times the neonatal deaths, based on Los Angeles County historical data. Figure 2.5 depicts the neonatal dose-response curve, based on the dose, when consumed maternally, required to produce death due to *in utero* exposure of the neonate.

For the elderly, the census estimated that 13% of the current United States of America population was aged 60 or over. Figure 2.6 depicts the elderly population dose-response curve.

The dose-response relations for mortality rate on a per-serving basis for the three population groups are summarized in Table 2.16. In general, the risk of a fatal listeriosis infection was 10 to 100 times greater for the elderly and neonate populations compared with the intermediate-age population.

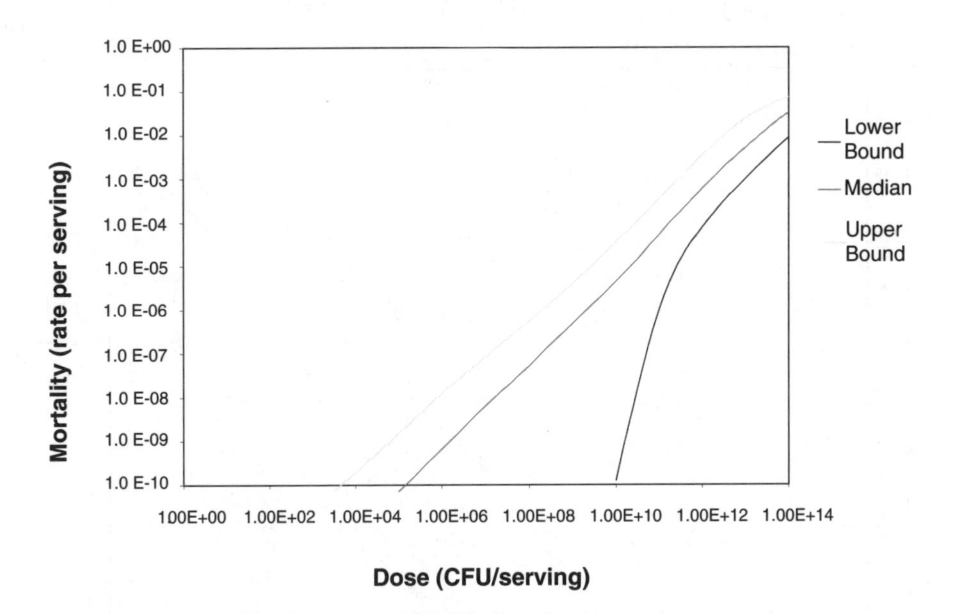

Figure 2.4 Dose response with variable strain virulence for the intermediate-age population.

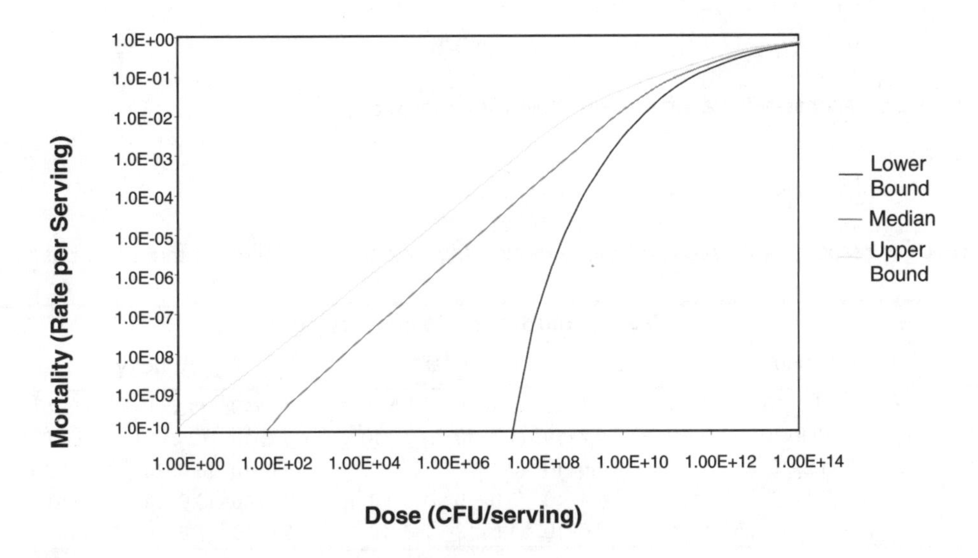

Figure 2.5 Dose response with variable strain virulence for neonates.

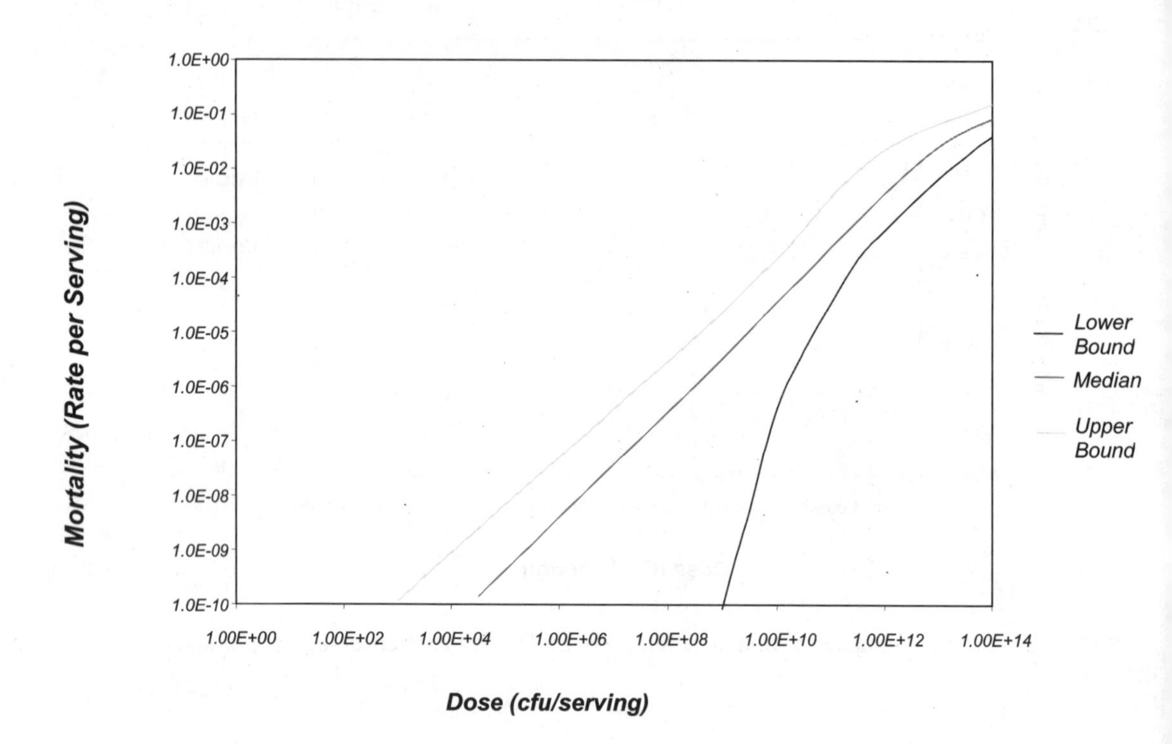

Figure 2.6 Dose response with variable strain virulence for the elderly

Table 2.16 Dose response with variable *Listeria monocytogenes* strain virulence for three age-based populations.

Dose (CFU per serving)	Median mortality rate per serving (1)		
	Intermediate-age	Neonatal(2)	Elderly
1	1.6×10^{-15} (1.9×10^{-134}, 2.7×10^{-13})	1.7×10^{-15} (7.5×10^{-84}, 1.5×10^{-10})	6.2×10^{-15} (4.3×10^{-112}, 3.9×10^{-13})
10^3	1.6×10^{-12} (3.9×10^{-83}, 7.9×10^{-11})	1.7×10^{-9} (1.3×10^{-44}, 6.2×10^{-8})	5.2×10^{-12} (2.2×10^{-65}, 1.2×10^{-10})
10^6	1.3×10^{-9} (1.2×10^{-43}, 2.9×10^{-8})	1.6×10^{-6} (2.1×10^{-18}, 2.8×10^{-5})	4.0×10^{-9} (8.6×10^{-32}, 5.0×10^{-8})
10^9	1.0×10^{-6} (7.1×10^{-18}, 1.2×10^{-5})	1.4×10^{-3} (5.0×10^{-5}, 1.3×10^{-2})	3.3×10^{-6} (3.6×10^{-11}, 2.3×10^{-5})
10^{12}	1.1×10^{-3} (3.5×10^{-5}, 5.7×10^{-3})	2.0×10^{-1} (1.5×10^{-1}, 2.6×10^{-1})	3.6×10^{-3} (7.8×10^{-4}, 2.3×10^{-2})

NOTES: (1) The 5th and 95th percentiles from the uncertainty are in parentheses.

SOURCE: FDA/FSIS, 2001: Table IV-9.

The median dose-response curves depicted in Figures 2.4, 2.5 and 2.6 approach the exponential model in shape. Using the median values for the 10^{12} CFU dose from Table 2.16, the current study used the response probability to estimate r-values for the intermediate-age, neonate and elderly population. The r-values were 8.5×10^{-16}, 5.0×10^{-14} and 8.4×10^{-15}, respectively. This allows a direct comparison of the three FDA/FSIS LMRA dose-response curves with the dose-response curves derived using only epidemiological data (Figure 2.7). All three of the FDA/FSIS LMRA models indicate a lower median probability of response at a specified dose compared with the other dose-response relations. This probably reflects the fact that the FDA/FSIS model was (1) based on mortality, not morbidity, and (2) the other models are based on strains with known high virulence, whereas the FDA/FSIS model considers the distribution of virulence that is likely to be encountered with *L. monocytogenes* isolates from foods. The predicted risk of serious listeriosis would be 5 times that for mortality.

The models include outbreaks where the primary symptoms included serious illness (including deaths) (smoked fish: Buchanan et al., 1997; smoked fish: Lindqvist and Westöö, 2000, and Farber, Ross and Harwig, 1996; butter: current study), perinatal and neonatal infections (death) (Hispanic-style cheese: current study; FDA/FSIS-neonates: FDA/FSIS, 2001), febrile gastroenteritis (sweet corn-tuna salad and chocolate milk: current study), death (general population) (FDA/FSIS, 2001), and death (elderly) (FDA/FSIS, 2001).

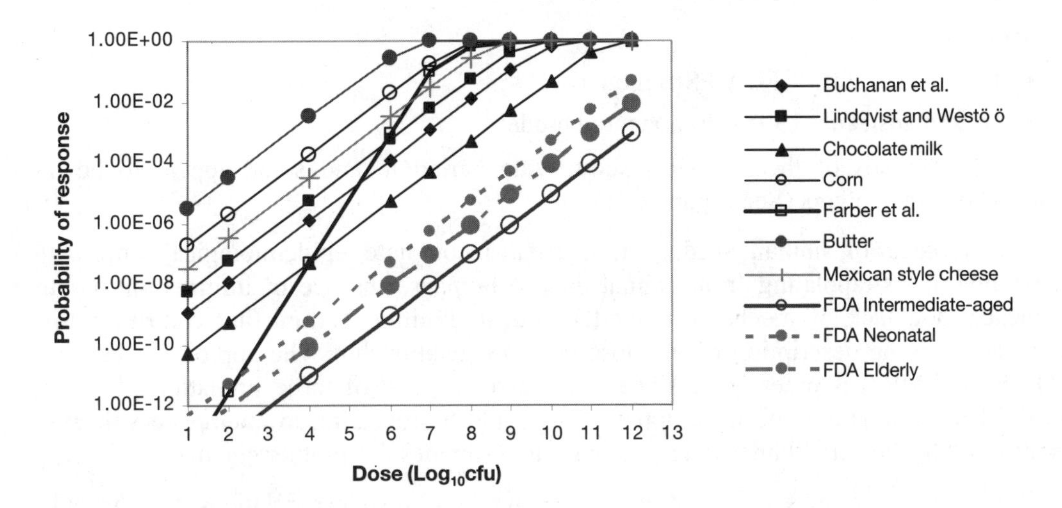

Figure 2.7 A comparison of the FDA/FSIS LMRA dose-response models for mortality with those derived earlier for morbidity based on epidemiological data or expert elicitations.

SOURCES: Buchanan et al., 1997; Lindqvist and Westöö, 2000; Farber, Ross and Harwig, 1996; FDA/FSIS, 2001.

2.3 OPTIONS FOR HAZARD CHARACTERIZATIONS TO BE USED FOR MODELLING THE PUBLIC HEALTH IMPACT OF *L. MONOCYTOGENES* IN READY-TO-EAT FOODS

Dose-response data from human volunteer studies with *L. monocytogenes* or from volunteer studies with a surrogate pathogen do not exist. Instead, dose-response relations have been developed and evaluated based on expert elicitations, epidemiological or animal data, or combinations of these. These dose-response relations, which were reviewed and summarized in the preceding sections, cover the spectrum of biological end-points, i.e. infection, morbidity and mortality, and have, to varying degrees of sophistication, been evaluated using human epidemiological data. The potential effects of the food matrix on the dose-response relation have not been considered as a variable within any of the models due to insufficient data. Available models, categorized by the end-point being modelled, include:

Infection:

- Farber, Ross and Harwig, 1996; Bemrah et al., 1998 – Weibull-Gamma model.
- Notermans et al., 1998; Haas et al., 1999 – Exponential model.
- Haas et al., 1999 – Beta-Poisson model.

Morbidity:

- Buchanan et al., 1997; Lindqvist and Westöö, 2000 – Exponential model.
- FDA/FSIS, 2001 – FDA/FSIS model.

Mortality:

- FDA/FSIS, 2001 – FDA/FSIS model.
- Notermans et al., 1998 – Exponential model.

The predictions of these models show wide variation, and some appear to be more conservative than others (See Figure 2.7).

The absence of human feeding trial data, incomplete epidemiological information, difficulties in extrapolating from animal data to humans, absence of information on strain virulence, and lack of mechanistic models are all limiting factors that contribute to the uncertainty in the description of the dose-response relationship. The approach taken in the FDA/FSIS LMRA is noteworthy since it addresses several of these limitations, but it will need further evaluation and development. It would be revealing to attempt to validate the model with health surveillance data and exposure estimates for another country.

While there are substantial differences in the dose-response relations that have been developed by different investigators (Figure 2.7), it appears that for those based on epidemiological data, a substantial part of the variability may reflect a combination of the biological endpoint being examined and the size and the characteristics of the population being considered. For example, the dose-response relation that was developed based on the outbreak of listeriosis in a hospital in Finland suggested that the LD_{50} for humans was approximately 10^6 CFU. This was based on 15 cases from a population of 234 individuals, all of who were highly immunocompromised. However, if this group were considered in relation to the entire population of individuals that had consumed the butter, this would have a great affect on the calculated dose-response. For example, if the dose-response model had been based on the entire population of Finland (5.2×10^6 individuals) having consumed the butter, but with only the hospital patients becoming seriously ill, the dose leading to a 50% serious infection rate in the population would be approximately 1×10^{11} CFU.

The calculated dose-response model will be influenced strongly by the numbers and the characteristics of the individuals included within that population. This is particularly important when epidemiological data is used to determine the dose-response relation. In part, the selection of the population to be considered will be a risk management decision related to the degree of conservatism that is to be built into the model and the degree to which even the most at-risk individuals are to be protected. However, by selecting a dose-response relation based on a specific higher-risk sub-population, any hazard characterization based on that relation would exaggerate the risk faced by the population in general.

At present there are only limited criteria on which to base the selection of the dose-response model, and better tools are needed to compare different models. Available criteria include the recommended use of non-threshold dose-response models that are linear in the low-dose region, and which have a biological basis and biologically interpretable parameters (*Hazard Characterization for Pathogens in Food and Water: Guidelines* (FAO/WHO, 2003)). However, the choice of which models to use will also depend on factors such as the purpose of the risk assessment and the level of resources and sophistication available to the risk assessors. This requires that the basis for the various dose-response relations and their impact on the overall risk assessment be adequately communicated to the risk managers who request the assessment. The use of several dose-response model relationships to frame the risk estimates is one approach to addressing the uncertainty related to current gaps in knowledge. A second approach, which has been used by at least one group of risk assessors, is the simultaneous use of several dose-response model relationships (FDA/FSIS, 2001). However, the latter choice requires a high degree of modelling sophistication, a requirement that could influence negatively the goal of providing a risk assessment that could be adapted by FAO/WHO for use internationally, where the level of risk assessment resources and sophistication varies substantially. On this basis, the risk assessment team, with the concurrence of an international panel of experts in foodborne disease, opted to develop a set of simpler dose-response models based on the use of the exponential model.

2.3.1 Exponential dose-response model used in the present risk assessment

The preceding sections discussed the various dose-response relations that have been developed and described their strengths and weaknesses. However, none of the available models were fully able to meet the needs of the current risk assessment in relation to the parameters examined and the requirement for simplicity of calculation. For these reasons, alternative approaches based on the exponential model were developed and evaluated.

The general approach was to estimate the single parameter r in the exponential model, i.e. the probability that a single cell will cause invasive listeriosis, by pairing population consumption patterns (exposure) with epidemiological data on the number of invasive listeriosis cases in the population. This was done in a manner similar to that described in Buchanan et al. (1997) and Lindqvist and Westöö (2000), but it was possible to refine their approach with the new epidemiological data and the detailed exposure assessment from the recently published draft FDA/FSIS *L. monocytogenes* risk assessment [see FDA/FSIS, 2001].

The validity of this approach is dependent on several assumptions or sources of information, or both: the percentage of individuals susceptible to severe *L. monocytogenes* infections; the appropriateness of the exponential model for describing the pathogen's dose-response relation in humans in the dose range of interest; the exposure assessment and numbers of *L. monocytogenes* consumed; and the accuracy of the statistics on the annual rate of severe listeriosis cases. The approach is based on mean population characteristics, i.e. the estimated exposure of the human population to a distribution of different strains, resulting in a number of illnesses. Consequently, variability in virulence is considered in the sense that the data, and therefore r-values, reflect the mean characteristics of many strains of *L. monocytogenes*, including frequency of occurrence and virulence. Similarly, the biological end point (response) used for the dose-response relationships is listeriosis. As indicated earlier, that term refers to "severe infection" or "invasive listeriosis" and includes those infected individuals suffering from life-threatening, systemic infections such as perinatal listeriosis, meningitis or septicaemia. Since the annual incidence of listeriosis included the entire designated population, the variability among individuals exposed to the pathogen is also inherently considered in this approach to dose-response modelling.

2.3.1.1 Overview of the estimation of parameter r in the exponential dose-response model

Specific r-values were derived for the less susceptible (healthy) and more susceptible populations as inputs to the current risk assessment, on the assumption that the overall consumption of *L. monocytogenes* was similar in these groups. The dietary consumption surveys did not indicate any major differences between population groups included in the draft FDA/FSIS (2001) risk assessment. Derivation of the r-values was achieved using the consolidated food contamination distribution from the FDA/FSIS 2001 draft exposure model in conjunction with the CDC annual estimated number of listeriosis cases (Mead et al., 1999) as a percentage of the total population of either more or less susceptible groups within the United States of America population. This provided values for P and N in the exponential model so that the r-value could be calculated by re-arranging the equation and solving.

Mathematically, the r-value is considered to be a constant parameter for a specified population. However, the accuracy of the estimate of the r-value is dependent on the size and inclusiveness of the population being considered, the accuracy of the annual disease statistics, and the reliability of data on the frequency and extent of *L. monocytogenes* contamination in foods. The uncertainty associated with the r-value included uncertainty estimates in the data used to derive the constant. Uncertainty estimates for the percentage of the population who are at increased risk range from 15 to 20% of the total population. The uncertainty estimates in the percentage of total cases in the annual disease statistics associated with the increased susceptibility population was estimated to range from 80 to 98%, and the uncertainty range in the total number of listeriosis cases in the United States of America was assumed to be from 1888 to 3148 cases (2518 cases ±25%). The derived r-values with estimated uncertainties were then determined by Monte Carlo simulation. Thus, although the r-value is mathematically a constant, due to the uncertainty in its estimation, the actual values used in the calculation of the dose-response curve were a distribution based on the estimated uncertainties.

In the FDA/FSIS 2001 draft risk assessment the total number of servings at each of five different dose levels for a number of RTE foods was estimated. The upper bound of the highest dose level, i.e. the maximum level of *L. monocytogenes* in an individual serving is uncertain and may vary for the different types of foods. Limitations in the contamination databases do not permit resolution of this issue. However, the maximum levels of *L. monocytogenes* encountered in individual servings of the different foods have a large impact on the calculated mean ingested doses. This, in turn, affects the derived r-value and the resulting dose-response curve. Consequently, this assumption was evaluated in detail. In the present study, the effect on the estimated r-value was considered by assuming different maximum contamination levels and calculating values for four point estimates of the maximum doses of 7.5, 8.5, 9.5 and 10.5 $\log_{10}$ CFU. In the studies by Buchanan et al. (1997) and Lindqvist and Westöö (2000), the maximum dose was assumed to be 5.7–6.0 $\log_{10}$ CFU, whereas the FDA/FSIS (2001) risk assessment allowed growth in a food to reach over 10 $\log_{10}$ CFU per serving. Assuming a lower maximum contamination level produces a more conservative or cautious estimation of the r-value. The lower the maximum dose assumed, the larger is the estimated r-value. The larger the r-value, the greater is the assumed virulence of the *L. monocytogenes*. In addition to using point estimates for the maximum levels of *L. monocytogenes*, r-values for the susceptible and healthy populations were also calculated using Monte Carlo simulation techniques, wherein the uncertainty in the maximum dose was addressed by combining all the previous dose levels into a discrete uniform distribution.

A schematic overview of the model used to estimate the r-value is shown in Figure 2.8. The uncertainty of the r-value due to the uncertainties in the assumed maximum dose levels in the different food categories, the size of the population of interest, and the number of cases in this population were calculated using the routine illustrated in Table 2.22 (at end of Part 2), where the input data employed and the Excel spreadsheet routine used for developing the model are indicated. From the FDA/FSIS exposure assessment, a summary table of the total number of servings at each of five different dose levels (<1; $1–10^{3}$; $10^{3}–10^{6}$; $10^{6}–10^{9}$; $>10^{9}$ CFU) was extracted (Figure 2.8 and Table 2.22, at end of Part 2). These data were fitted to an empirical cumulative frequency distribution. The equation was used to generate a new frequency distribution at closer dose intervals. The newly generated frequency distribution did not differ significantly from the original distribution (Figure 2.9). The data on the number of servings at different dose levels from the FDA/FSIS risk assessment and the number of United States of America cases of severe listeriosis were the basis for estimating the r-value for the population of interest. Finally, r-values were calculated in manners analogous to those described by Buchanan et al. (1997), for two scenarios: (1) assuming that all cases were attributable to servings from the maximum dose level only; and (2) assuming that all dose levels contributed to causing listeriosis. The first approach is based on the observation that the exponential model is generally steep, which results in the highest exposure levels having the greatest impact on the probability of disease within the dose ranges of interest.

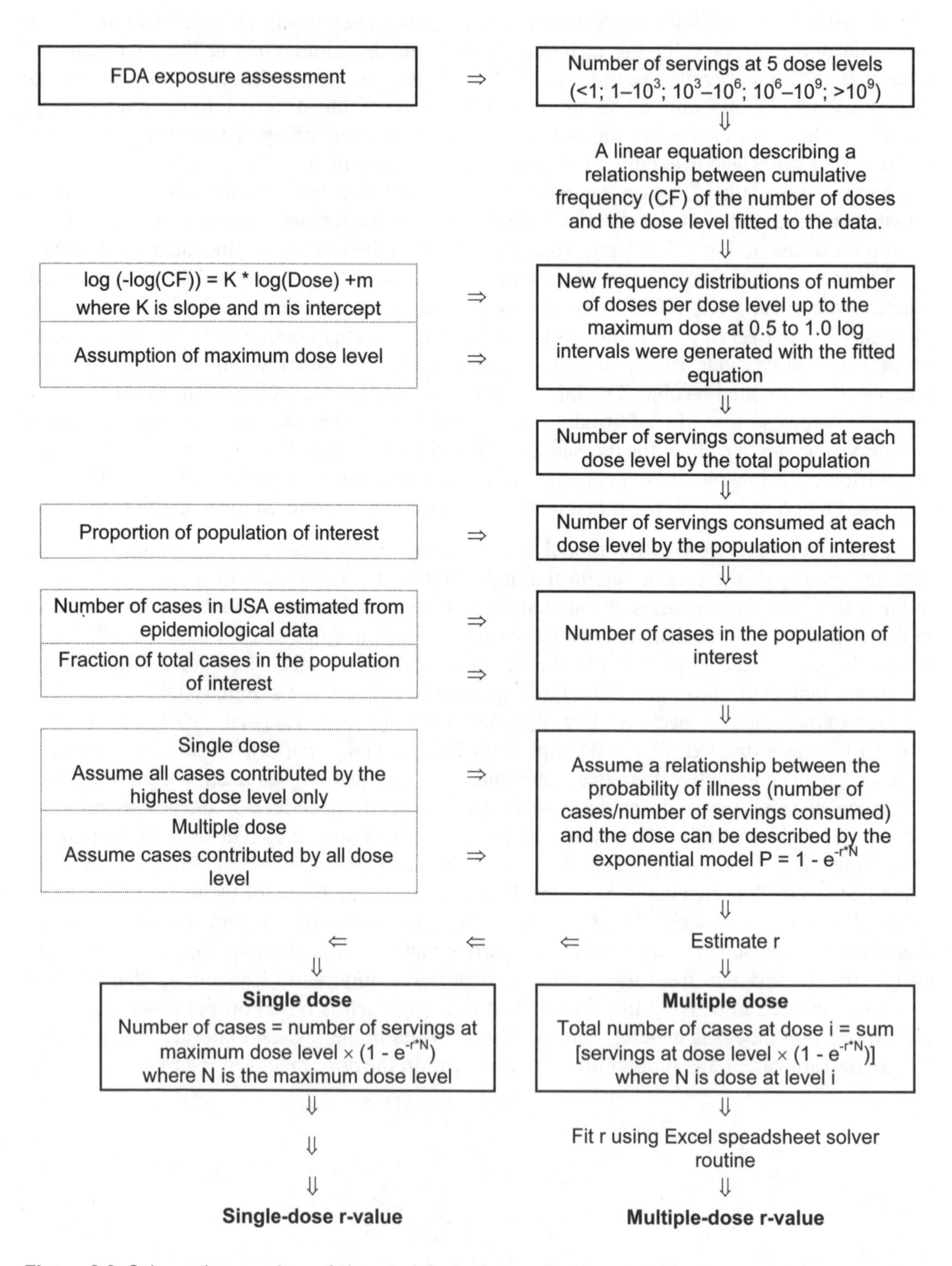

Figure 2.8 Schematic overview of the model used to estimate r in the exponential dose-response model.

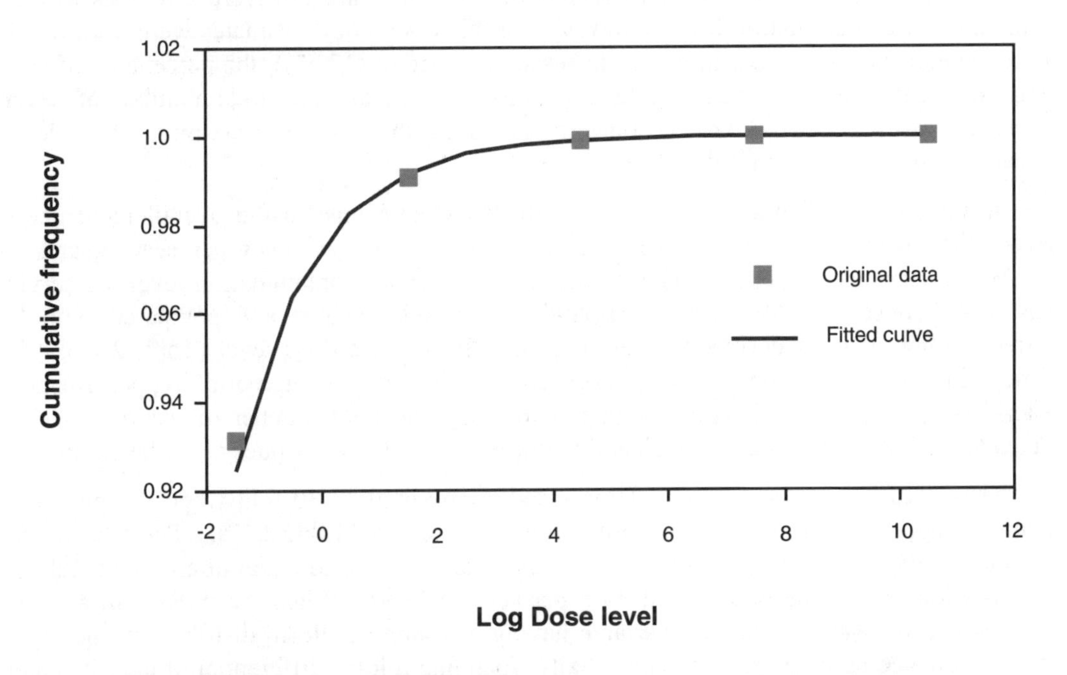

Figure 2.9 The cumulative frequency distribution of servings with different dose levels of *L. monocytogenes*. The squares represent the original data from the draft FDA/FSIS risk assessment (2001) describing the number of servings at each of five dose levels. The curve represent the resulting equation when these data were fitted to an empirical cumulative frequency distribution. This curve was used to generate a new frequency distribution at closer dose intervals.

2.3.2 Dose-response models for healthy and susceptible population

Dose-response models for the susceptible population and the healthy population were calculated in three different ways, as described below. In each case, the impact of the assumed maximum level of contamination that *L. monocytogenes* reaches in foods was considered.

The first approach was to assume that cases of severe listeriosis are due overwhelmingly to those servings of foods that have the highest level of contamination. In addition, uncertainty associated with three parameters that influence the dose-response relations was evaluated by assuming that (i) the percentage of the population with increased susceptibility to *L. monocytogenes* varied between 15% and 20%; (ii) that the percentage of cases of total severe listeriosis cases associated with this increased susceptibility population ranged from 80% to 98%; and (iii) the estimates of the total number of cases (2518) has a degree of uncertainty of 25%. The equation was then solved using Monte Carlo techniques, with 5000 iterations being run to obtain the median r-value and its 5% to 95% confidence interval for maximum assumed contamination levels per serving of 7.5, 8.5, 9.5 and 10.5 $\log_{10}$ CFU. In addition, the maximum contamination level was also considered as a variable using a discrete uniform distribution [RiskDuniform(7.5; 8; 8.5; 9; 9.5; 10; 10.5)]. The r-values obtained are presented in Table 2.17.

The second approach was to again assume that the incidence of severe listeriosis was due to the highest contamination level. However, in this case, point estimates were used for the percentage of the population in the more susceptible group (17.5%), the percentage of cases associated with the more susceptible population (83%), and the total number of severe listeriosis cases (2518). The r-values obtained at the various maximum dose levels considered are presented in Table 2.18.

The third approach was to assume that all doses contribute to the overall incidence of severe listeriosis, in accordance with the level of *L. monocytogenes* per serving and the number of servings consumed. The influence of maximum contamination level per serving was again considered for these multiple-dose derived r-values and compared with the corresponding r-values derived considering only the highest dose level (Table 2.18). The same point estimates for percentage of population with increased susceptibility, percentage of listeriosis cases associated with that population, and the total number of listeriosis events were used, so the single-dose derived and multiple-dose derived r-values could be compared.

The r-value for *L. monocytogenes* increased approximately 30-fold when the maximum assumed $\log_{10}$ dose level was decreased from 10.5 to 7.5 (Table 2.17). The 5% to 95% confidence interval for the r-value was typically small ($\log_{10}$ differential of approximately 0.2 to 0.3) within a maximum assumed dose per serving level. When the maximum assumed $\log_{10}$ dose level was treated as a variable having a discrete uniform distribution, the 5% to 95% confidence range increased substantially, spanning a $\log_{10}$ differential of approximately 1.6. This indicated that the impact of the assumed maximum level of contamination in food had a substantially greater effect than the other parameters considered.

The multiple-dose derived r-values (Table 2.18) were consistently lower than, but not very different from, the corresponding r-value based on maximum-dose derivation, i.e. the assumption that all cases were due only to servings contaminated by the maximum dose level that would be encountered. The maximum-dose assumption simplifies calculation but in reality all dose levels may contribute to the incidence of foodborne listeriosis. This is depicted as a function of the maximum assumed level of contamination per serving in Table 2.19. The estimated number of cases does not increase monotonically going from a lower to a higher dose level. This is because the number of cases contributed by food in a specific dose category depends not only on that dose, but also on the number of servings in that dose category.

Table 2.17 The effect of the assumed maximum individual dose level on the calculated r-values for *Listeria monocytogenes* for the fraction of the population with increased susceptibility. Estimations assume all cases of severe listeriosis to be due to ingestion of servings only at the highest dose level.

Maximum Log Dose per Serving	Median Maximum-Dose Derived r-value	5th Percentile r-value	95th Percentile r-value
7.5	8.61×10^{-12}	6.38×10^{-12}	1.13×10^{-11}
8.5	2.09×10^{-12}	1.53×10^{-12}	2.76×10^{-12}
9.5	5.59×10^{-13}	4.14×10^{-13}	7.47×10^{-13}
10.5	2.79×10^{-13}	2.06×10^{-13}	3.70×10^{-13}
7.5 to 10.5(2)	1.06×10^{-12}	2.47×10^{-13}	9.32×10^{-12}

NOTE: (1) Values were obtained by Monte Carlo simulation techniques assuming that (i) the percentage of the population with increased susceptibility to *L. monocytogenes* varied between 15 and 20%; (ii) the percentage of cases of total severe listeriosis cases associated with this increased susceptibility population ranged from 80 to 98%; and (iii) the uncertainty of the estimates of the total number of cases is ±25%.
(2) Uncertainty of maximum dose level described by RiskDuniform (7.5; 8.0; 8.5; 9.0; 9.5; 10.0; 10.5) distribution.

Table 2.18 The effect of the assumed maximum individual dose level on the calculated r-values for *Listeria monocytogenes* for the fraction of the population with increased susceptibility. The estimations are based on calculations using the single-value estimates of the maximum (maximum-dose derived r-values) or using the entire range of maximum dose values in deriving a single r-value (multiple-dose derived r-values). These estimations assume that all cases of severe listeriosis are due to ingestion of servings only at the highest dose level (see Note 1).

Maximum Log Dose per Serving (CFU)	Maximum-Dose Derived r-value	Multiple-Dose Derived r-value
7.5	8.05×10^{-12}	5.85×10^{-12}
8.5	1.95×10^{-12}	1.45×10^{-12}
9.5	5.24×10^{-13}	3.72×10^{-13}
10.5	2.61×10^{-13}	1.34×10^{-13}

NOTE: (1) Point estimates were used for (i) the percentage of the population in the more susceptible group (17.5%); (ii) the percentage of cases associated with the more susceptible population (83%); and (iii) the total number of severe listeriosis cases (2518).

The exponential dose-response model is a non-threshold model. Consequently, there is no dose value other than zero that results in a prediction that there is no risk of illness. As mentioned previously, the r-value in the exponential model can be viewed as the probability that a single cell of *L. monocytogenes* would cause an illness. Table 2.19 indicates that when the most conservative assumption for the numbers of *L. monocytogenes* consumed in a serving ($10^{7.5}$ maximum CFU/serving) was used, over 99% of the cases arise from the consumption of servings that contain $10^{5.5}$ CFU per serving. If the maximum number in a serving was assumed to be $10^{10.5}$, over 99% of the cases of listeriosis arise from 10^{8} CFU or more per serving. Only a very small fraction of the 3.66×10^{11} United States of America servings need to achieve these levels of *L. monocytogenes* to account for 2500 cases of severe listeriosis. The estimated number of servings for each dose category can be found in spreadsheet cells F82 to F98 in Table 2.22 (at the end of Part 2). The distribution is shown in Figure 2.9.

The second set of r-values was developed for the remainder of the population that did not have increased susceptibility for *L. monocytogenes*. The same types of calculations were performed as for the susceptible population. The parameters were (i) the portion of the population with decreased susceptibility for *L. monocytogenes* was 80% to 85% (point estimate = 83%) of the total population; and (ii) this portion of the population is associated with 2–20% of the cases (point estimate = 11%). It was again assumed that the total number of cases of severe listeriosis was 2518 ± 25%. The effect of the maximum assumed level of contamination on maximum-dose derived r-values based on Monte Carlo simulations with estimates of uncertainty for size of this population, the fraction of listeriosis cases with which it is associated, and the total number of listeriosis cases is summarized in Table 2.20. The corresponding comparison of the maximum-dose derived values and the multiple-dose derived r-values is presented in Table 2.21.

The r-values for the less susceptible portion of the population were 1 to 2 orders of magnitude smaller than the corresponding r-values for the more susceptible population. In general, the uncertainty associated with the less susceptible population (Table 2.20) was greater than that for the more susceptible population (Table 2.17). Like the more susceptible population, the multiple-dose derived r-values for the less susceptible population was consistently smaller (i.e. less conservative) than the corresponding maximum-dose derived r-values (Table 2.21). It should be noted that calculating morbidity$_{50}$ values for r-values of the magnitude observed in Table 2.21 should be considered notional, since the doses required are higher than achievable in a food serving. It is highly unlikely than any individual would ever encounter a dose in foods greater than 10^{10} or 10^{11} CFU per serving. In such instances, the dose-response curve would most appropriately be interpreted as indicating that a large portion of the population would not acquire severe listeriosis even in the presence of extremely high doses. This also indicates why most cases of listeriosis are sporadic.

Table 2.19 The effect of assumed maximum individual dose level per serving on the number of cases contributed per dose level per serving of food. The predictions are for the susceptible population and were based on the exponential dose-response model, the distribution of servings per dose level and the multiple-dose derived r-values in Table 2.18.

Log Dose	**Estimated number of cases with different presumed maximum $\log_{10}$ doses**			
	7.5	8.5	9.5	10.5
-1.5	<1(1)	<1	<1	<1
-0.5	<1	<1	<1	<1
0.5	<1	<1	<1	<1
1.5	<1	<1	<1	<1
2.5	1	<1	<1	<1
3.5	2	1	<1	<1
4.5	12	3	1	<1
5.5	56	14	3	1
6.5	265	64	16	6
7.0	235	57	14	5
7.5	1 519	123	31	11
8.0		268	68	25
8.5		1 561	149	53
9.0			323	116
9.5			1 483	252
10.0				547
10.5				1 073
Total cases(2)	2 090	2 090	2 090	2 090
Multiple-dose derived r-value used	5.85×10^{-12}	1.45×10^{-12}	3.72×10^{-13}	1.33×10^{-13}

NOTE: (1) Predicted number of cases attributed to a specific dose. Total cases based on the assumption behind the r-values in Table 2.18 that 83% of total cases (2518) are in the susceptible group; 0.83 × 2518=2090 cases

Table 2.20 The effect of assumed maximum individual dose level on the calculated r-values for *Listeria monocytogenes* for the population with decreased susceptibility. The estimations asssume that all cases of severe listeriosis are due to ingestion of servings only at the highest dose level (see Note (1)).

Maximum log dose per serving (CFU)	**Median Maximum-Dose Derived r-value**	**5th Percentile r-value**	**95th Percentile r-value**
7.5	2.23×10^{-13}	5.82×10^{-14}	4.22×10^{-13}
8.5	5.34×10^{-14}	1.42×10^{-14}	1.02×10^{-13}
9.5	1.45×10^{-14}	3.75×10^{-15}	2.74×10^{-14}
10.5	7.18×10^{-15}	1.85×10^{-15}	1.15×10^{-14}
7.5 to 10.5(2)	2.37×10^{-14}	3.55×10^{-15}	2.70×10^{-13}

NOTES: (1) Values were obtained by Monte Carlo simulation techniques, assuming that (i) the percentage of the population with increased susceptibility to *L. monocytogenes* varied between 80 and 85%, (ii) that the percentage of cases of total severe listeriosis cases associated with this increased susceptibility population ranged from 2 to 20%, and (iii) the uncertainty of the estimates of the total number of cases is ±25%. (2) Uncertainty of maximum dose level described by RiskDuniform(7.5; 8.0; 8.5; 9.0; 9.5;10.0; 10.5) distribution.

Table 2.21 The effect of assumed maximum individual dose level on the calculated r-values for *Listeria monocytogenes* for the population with decreased susceptibility: The estimations are based on calculations using the single-value estimates of the maximum dose (maximum-dose derived r-values) or using the entire range of maximum dose values in deriving a single r-value (multiple-dose derived r-values), assuming that all cases of severe listeriosis are due to ingestion of servings only at the highest dose level (see Note (1)).

Maximum log dose per serving (CFU)	Maximum-dose derived r-value	Multiple-dose derived r-value
7.5	2.25×10^{-13}	1.64×10^{-13}
8.5	5.45×10^{-14}	4.07×10^{-14}
9.5	1.47×10^{-14}	1.07×10^{-14}
10.5	7.27×10^{-15}	3.73×10^{-15}

NOTES: (1) Point estimates were used for (i) the percentage of the population in the more susceptible group (17.5%), (ii) the percentage of cases associated with the more susceptible population (83%), and (iii) the total number of severe listeriosis cases (2518).

2.3.3 Differences in susceptibility to listeriosis for different human populations.

In addition to developing dose-response models for the entire more-susceptible population, the Codex Committee for Food Hygiene also requested estimates of the relative susceptibility of different sub-populations that have specific chronic diseases. These had not been developed in previous risk assessments, so a means of fulfilling this request had to be developed. The approach taken was to estimate the relative susceptibility based on detailed epidemiological data and to estimate the dose-response relations in conjunction with the exponential dose-response model (*see* Section 5.2).

2.4 r-VALUES FOR RISK CHARACTERIZATION

As explained in the preceding sections, the available contamination and epidemiological data do not permit an unequivocal choice of the most appropriate r-values for different populations. Accordingly, the risk assessment team, in consultation with the international panel of experts, used the following r-values to illustrate various attributes associated with the risk assessment and to address the CCFH questions.

- For CCFH Question 1, on the risk from consuming different numbers of *L. monocytogenes*, an r-value of 5.85×10^{-12} was used for the susceptible population. This was the most conservative dose-response curve used in the current risk assessment and was calculated on the assumption that the maximum individual dose was 7.5 $\log_{10}$ CFU per serving (Table 2.18).
- To illustrate how to estimate r-values based on the relative risks for different susceptible sub-populations in CCFH Question 2, an r-value of 5.34×10^{-14} was selected as the reference value for the general healthy population. This r-value was derived based on an assumption of an intermediate level of maximum individual dose, 8.5 $\log_{10}$ CFU per serving, in food (Table 2.20).
- For the food examples described in the risk assessment and CCFH Question 3, the r-values used were based on the use of Monte Carlo simulation techniques in combination with a discrete uniform distribution wherein the maximum number of *L. monocytogenes* consumed varied from 7.5 to 10.5 $\log_{10}$ CFU per serving. For the

population with increased susceptibility, the median r-value used with its distribution was 1.06×10^{-12} (Table 2.17). For the healthy population, the median r-value used with its distribution was 2.37×10^{-14} (Table 2.20).

Table 2.22. Spreadsheet-based exponential *Listeria monocytogenes* dose-response model (See following pages).

Exponential Lm Dose-response model

For the population of interest, R is estimated based on the U.S. FDA/FSIS assessment of the annual exposure to different doses of Lm and the annual number of listeriosis cases. The estimated uncertainty of R due to the uncertainties in the assumed maximum dose levels in the different food categories, the size of the population of interest, and the number of cases in this population is calculated. Blue indata, Red outdata, Green results calculated and used in the spreadsheet model

Assumptions

* The same as FDA/FSSIS Exposure assessment.
* Maximum dose levels: vary between 7.5 and 10.5
* The total number of listeriosis cases 2518 +/-25%
* The fraction of cases within each of the subgroups were based on outbreak data shown in worksheet proportion susceptible (T. Ross), and estimates from the U.S. Risk assessment (98% cases belonging to the susceptible group, R. Whiting)

Input data	Input	Formula		Output	Formula
Population of interest	Susceptible				
Fraction of total population	1.75E-01	C4 = RiskUniform(0.15;0.2)	**Estimated R**	2.80E-13	G4 = E106
Total # of listeriosis cases	2.52E+03	C5 = RiskUniform(1888;3148)			
Fraction of listeriosis cases	8.90E-01	C6 = RiskUniform(0.8;0.98)			
# listeriosis cases in this population	2.24E+03	C7 = C5*C6			
Assumed maximum log dose level	10.5	C8 = Input data			

Dose (CFU) as % of serves at point of consumption

Food category	Total consumption (Servings)	< 1 g	1 to 1000	10^3 to 10^6	10^6 to 10^9	> 10^9	TOTAL %	Formula
Smoked Seafood	2.05E+08	70.64%	14.29%	11.06%	3.42%	0.20%	0.996	B12:G31 = Input data
Raw Seafood	1.82E+08	92.07%	6.66%	1.21%	0.07%	0.00%	1.000	H12:H31 = Sum(C12:G12)
Preserved Fish	1.05E+08	84.77%	10.42%	3.89%	0.49%	0.04%	0.996	
Cooked RTE Shellfish	5.52E+08	94.50%	4.01%	1.28%	0.20%	0.05%	1.000	
Vegetables	1.17E+11	91.11%	7.23%	1.54%	0.07%	0.00%	0.999	
Fruits	5.03E+10	81.37%	18.49%	0.13%	0.00%	0.00%	1.000	
Soft mold-ripened	2.44E+08	92.81%	3.21%	3.34%	0.67%	0.01%	1.000	
Goat/Sheep etc cheese	2.55E+08	92.18%	6.24%	1.48%	0.07%	0.00%	1.000	
Fresh Soft Cheese	1.34E+08	89.72%	3.20%	4.31%	2.51%	0.19%	0.999	
Heated and Processed	1.82E+10	98.20%	1.71%	0.08%	0.01%	0.00%	1.000	
Aged Cheese	1.38E+10	98.07%	1.82%	0.03%	0.00%	0.00%	0.999	
Pastuerised Milk	8.72E+10	99.20%	0.74%	0.05%	0.00%	0.00%	1.000	
Raw Milk	4.36E+08	91.87%	7.56%	0.55%	0.01%	0.00%	1.000	
Ice Cream	1.49E+10	99.08%	0.53%	0.02%	0.00%	0.00%	0.996	
Miscellaneous Dairy	2.81E+10	98.26%	1.64%	0.07%	0.00%	0.00%	1.000	
Frankfurters	6.52E+09	92.40%	6.08%	1.37%	0.21%	0.02%	1.001	
Dry/Semi-Dry	1.79E+09	90.27%	6.83%	2.40%	0.10%	0.00%	0.996	
Deli Meats	2.07E+10	90.66%	5.40%	3.29%	0.70%	0.12%	1.002	
Pâté	1.18E+08	91.52%	4.01%	2.87%	1.06%	0.22%	0.997	
Deli Salads, Non-	5.63E+09	86.30%	8.77%	3.98%	0.80%	0.03%	0.999	
Total Servings	3.66E+11	SUMMA(B12:B31)						

	A	B	C	D	E	F	G	H
34								
35			Predicted Number of Doses at indicated dose level					
36	**Food category**	< 1 g	1	10^3	10^6 to	> 10^9	Formula	
37	Smoked Seafood	1.45E+08	2.93E+07	2.27E+07	7.01E+06	4.20E+05	B37:F56 = C12 * $B12	
38	Raw Seafood	1.68E+08	1.21E+07	2.20E+06	1.24E+05	5.46E+03	B57:F57 = Sum(B37:B56)	
39	Preserved Fish	8.90E+07	1.09E+07	4.09E+06	5.19E+05	4.27E+04	G57 = Sum(B57:G57)	
40	Cooked RTE Shellfish	5.22E+08	2.21E+07	7.05E+06	1.13E+06	2.59E+05	B58:F58 = SUM(B57:B57)/SUM(B57:F57)	
41	Vegetables	1.07E+11	8.46E+09	1.80E+09	7.80E+07	0.00E+00		
42	Fruits	4.09E+10	9.30E+09	6.49E+07	1.34E+06	0.00E+00		
43	Soft mold-ripened	2.26E+08	7.83E+06	8.15E+06	1.62E+06	2.20E+04		
44	Goat/Sheep etc cheese	2.35E+08	1.59E+07	3.76E+06	1.84E+05	0.00E+00		
45	Fresh Soft Cheese	1.20E+08	4.28E+06	5.77E+06	3.36E+06	2.60E+05		
46	Heated and Processed	1.79E+10	3.12E+08	1.50E+07	1.05E+06	3.03E+05		
47	Aged Cheese	1.35E+10	2.51E+08	3.84E+06	0.00E+00	0.00E+00		
48	Pastuerised Milk	8.65E+10	6.43E+08	4.24E+07	3.63E+06	1.60E+06		
49	Raw Milk	4.01E+08	3.30E+07	2.39E+06	5.38E+04	1.02E+04		
50	Ice Cream	1.48E+10	7.83E+07	2.38E+06	7.45E+04	0.00E+00		
51	Miscellaneous Dairy	2.76E+10	4.61E+08	1.88E+07	1.40E+06	4.92E+05		
52	Frankfurters	6.02E+09	3.96E+08	8.95E+07	1.40E+07	1.30E+06		
53	Dry/Semi-Dry	1.62E+09	1.22E+08	4.30E+07	1.82E+06	5.37E+04		
54	Deli Meats	1.88E+10	1.12E+09	6.80E+08	1.45E+08	2.41E+07		
55	Pâté	1.08E+08	4.73E+06	3.39E+06	1.25E+06	2.57E+05		
56	Deli Salads, Non-	4.86E+09	4.94E+08	2.24E+08	4.51E+07	1.84E+06		
57	**Total servings at each dose level**	3.41E+11	2.18E+10	3.05E+09	3.07E+08	3.10E+07	3.662E+11	
58	**Cumulative frequency**	0.931297	0.990762	0.999077	0.999915	1.000000		
59								
60								
61	**Empirical cum freq distribution. The max log dose (A66) is variable between 8 and 10.5**						**log no**	**Formula**
62	**Log Dose levels**	**No of servings**	**Cum no servings**	**Cum Freq**	**Log Cum Freq**	**log(-log(CF))**	**servings**	
63	-1.5	3.41E+11	3.41E+11	9.313E-01	-3.091E-02	-1.509874594	11.53286444	A63:A67 = Input data
64	1.5	2.18E+10	3.63E+11	9.908E-01	-4.031E-03	-2.39464095	10.33804167	B63:B67 = B57
65	4.5	3.05E+09	3.66E+11	9.991E-01	-4.011E-04	-3.396742519	9.483611912	C63:C67 = C63 + B64
66	7.5	3.07E+08	3.66E+11	9.999E-01	-3.678E-05	-4.434399463	8.487262531	D63:D67 = C63/($B$68)
67	10.5	3.10E+07	3.66E+11	1.000E+00	0.000E+00		7.491574159	E63:E67 = Log (D63)
68	**Total**	3.662489E+11						F:63:F67 = Log (-E63)
69				4.92E+00				G63:G67 = Log (B63)
70								
71	**Fitted curve to Empirical cum freq distribution**							
72	**Log Dose level**	**Log (-log(CF))**			**Comment**	**Formula**		
73	-1.5	-1.509874594	slope	-0.32585587	Linear regression	A73:A76 = Input data		
74	1.5	-2.39464095	intercept	-1.95634676	Log(dose) vs Log (	B73:B76 = F63		
75	4.5	-3.396742519			Log(CF))	D73 = Slope(B73:B76;A73:A76)		
76	7.5	-4.434399463				D74 = Intercept(B73:B76;A73:A76)		
77								

	A	B	C	D	E	F	G	H
77								
78	**Generation of new frequency distribution data at closer intervals depending on the simulated**							
79	**maximum dose levels and using the fitted curve derived above**							
80								
81	**Log Dose level**	**log(-log(CF))**	**Cum Frequency**	**Frequency**	**From original data**	**Meals consumed in each category**	**# Meals consumed by population**	**Formula**
82	-1.5	-1.467562955	0.924538301	0.9245383	9.313E-01	3.39E+11	5.93E+10	A82:A98 = Input data
83	-0.5	-1.793418828	0.963627242	0.03908894		1.43E+10	2.51E+09	B82:B91 = $D$74+$D$73*A82
84	0.5	-2.1192747	0.982656078	0.01902884		6.97E+09	1.22E+09	B92:B98 = IF($C$8>=7.5;$D$74+$D$73*A92;MISSING())
85	1.5	-2.445130573	0.991772038	0.00911596	9.91E-01	3.34E+09	5.84E+08	C82:C98 = 10^(-(10^B82))
86	2.5	-2.770986445	0.996106122	0.00433408		1.59E+09	2.78E+08	C92:C98 = IF($C$8>=7.5;10^(-(10^B92));MISSING())
87	3.5	-3.096842318	0.998159341	0.00205322		7.52E+08	1.32E+08	D82:D98 = C83-C82
88	4.5	-3.42269819	0.999130382	0.00097104	9.99E-01	3.56E+08	6.22E+07	D92:D98 = IF($C$8=A92;$B$100-C91;IF($C$8>A92;C92-C91;M
89	5.5	-3.748554063	0.999589255	0.00045887		1.68E+08	2.94E+07	E82:E98 = Input data
90	6.5	-4.074409935	0.999806017	0.00021676		7.94E+07	1.39E+07	E92:E98 = IF($C$8=7.5;1;D66)
91	7.0	-4.237337872	0.999866694	6.0677E-05		2.22E+07	3.89E+06	F82:F98 = D82*$B$68
92	7.5	-4.400265808	0.999908393	4.1698E-05	9.99915E-01	1.53E+07	2.67E+06	F92:F98 =IF($C$8>=7.5;D92*$B$68;
93	8.0	-4.563193744	0.999937048	2.8655E-05		1.05E+07	1.84E+06	G82:G98 = F82*$C$4
94	8.5	-4.72612168	0.99995674	1.9692E-05		7.21E+06	1.26E+06	G92:G98 = IF($C$8>=7.5;F92*$C$4;"")
95	9.0	-4.889049617	0.999970272	1.3532E-05		4.96E+06	8.67E+05	F99 = SUM(F82:F98)
96	9.5	-5.051977553	0.999979572	9.2992E-06		3.41E+06	5.96E+05	G99 = SUM(G82:G98)
97	10.0	-5.214905489	0.999985962	6.3903E-06		2.34E+06	4.10E+05	
98	10.5	-5.377833425	0.999990353	3.9656E-06	1.00E+00	1.45E+06	2.54E+05	
99					**Total**	3.662452E+11	6.41E+10	
100	Fitted cum fr (10.5)	0.999989927573						
101								
102	**Collation of Annual US National Exposure at each Dose Level**							
103								
104	**Estimation of R**			**Solving for R**			**Input for calculation of attack rate**	
105	**Assumption**	**Endpoint**	**# Meals Max Dose**	**Attack rate**	**R**		**Max Log Dose**	**# Meals**
106	Cases caused by highest dose level only	Listeriosis	2.54E+05	8.82E-03	2.80E-13		7.5	
107							8.0	
108		**Formula table**					8.5	
109		H106:H112 = IF($C$8=G110;G96;"")					9.0	
110		C106 = HLOOKUP($C$8;G106:G112;H106:H112)					9.5	
111		D106 = C7/C106					10.0	
112		E106 = -(LN(1-D106))/10^$C$8					10.5	2.54E+05
113								
114								

Part 3.

Exposure Assessment

3.1 INTRODUCTION

In a quantitative microbiological risk assessment, the exposure assessment describes the pathways through which a pathogen population is introduced, distributed and altered in the production, distribution and consumption of food. The result desired from the exposure assessment is the prevalence, concentration and, if possible, virulence of the pathogen in foods at the point that they are eaten and the level of consumption of the food by the population of interest.

In many cases, data necessary to complete the exposure assessment are usually not known, in particular the frequency of contamination of foods and the total pathogen numbers ingested by consumers. An estimate can be derived, however, based on knowledge of contamination levels and prevalence at some earlier point in the farm-to-fork chain, and on models of the effect of physical processes and conditions that the food undergoes from then until the point of consumption, i.e. final pathogen numbers ingested by consumers.

This section aims to identify the data needed to assess human exposure to *L. monocytogenes* in ready-to-eat (RTE) foods; potential sources of that data; tools and techniques to overcome gaps in the data; and approaches for synthesizing data using models to enable estimation of exposure.

Conceptual and mathematical approaches that can be used in exposure assessment are also described, such as "predictive microbiology" models that can help provide necessary information and fill some of the data gaps. Such models need to be validated in products of similar microbial ecology to the product of interest. Existing data concerning current understanding of the microbial ecology of *L. monocytogenes* in foods is presented to assist in assessment of predictive microbiology models for use in exposure assessment.

Thus, an assessment of foodborne exposure to *L. monocytogenes* typically requires acquisition of data that:

- describe the prevalence of *L. monocytogenes* in ingredients, or specific finished products of interest, or both;
- describe the concentration of *L. monocytogenes* in ingredients, or specific finished products of interest, or both;
- describe the amount of the product eaten at each meal or serving and the frequency of eating, and, if possible, the consumption characteristics of sub-groups of the population that are particularly susceptible to listeriosis;

- enable the prevalence and concentration at one point in the food chain to be determined from an earlier point in the chain, e.g. storage times and temperatures, and from the microbial ecology, e.g. growth potential in the food; and
- determine the simplifying assumptions and process model that the exposure assessment will include. It is impossible to include in a model all of the situations that a food may experience.

Many of these data are typically derived from studies intended for other purposes and are not ideally suited for the objectives of exposure assessment. Often, they are published in the scientific literature, or appear in reports from regulatory authorities performing routine monitoring. Other sources for these data are import and export control services for quarantine purposes; outbreak investigation reports; and industry files. Unpublished reports from government or industry are not always accessible because of confidentiality concerns. Ideally, the studies used for exposures assessment should be comprehensive national surveys of the specific foods in question, with information on the extent of contamination (prevalence) and level of *L. monocytogenes* contamination in the product (concentration). These are rarely available, and smaller surveys within several countries often have to be used to estimate the contamination of RTE foods by *L. monocytogenes*.

In such studies, information about concentration is often lacking. Under the zero-tolerance regulatory approach adopted by many authorities towards *L. monocytogenes* in RTE foods, concentration is not of particular interest to the requestor and supervisor of the surveys, particularly when faced with the fact that concentration data are more time consuming and costly to acquire. Zero tolerance implies regulations that require that the hazard not be detectable in a test sample of specified size. Many countries specify the absence of *L. monocytogenes* in a 25-g test sample in RTE foods as the tolerable limit.

Data about consumption of RTE foods are also limited. These are usually available only from government sources, usually through national or regional nutrition surveys. The surveys often capture covariate information about those consumers and non-consumers. Those data help, for example, to estimate consumption patterns separately for age and gender classes, enabling inferences to be drawn about consumption by at-risk groups. Some surveys, though, do not have the level of detail to identify a specific RTE food, the "foods eaten" tending to be grouped into broader categories based on nutritional composition, but which may not be related to the risk of listeriosis. More specific consumption data can be derived from the individual records of each consumer surveyed. These data are kept by some survey authorities, and are available under some circumstances, but are not publicly released for reasons of confidentiality. If available, those data can also be used to better determine the consumption patterns of at-risk groups. For example, the Australian National Nutrition Survey (ABS, 1995) included a health status survey, but few of the health-related questions addressed known susceptibility factors for listeriosis.

Another source of data, complementary to that of the consumption surveys, is the inventory databases of food retailers, which provides complete and specific data on the number of units of every product type sold. Most stores and chains can provide estimates of their market share and "wastage" (i.e. product not sold but discarded because of spoilage, damage or other loss), and, from this, estimates of specific consumption levels from national to local levels can be derived. Commercial confidentiality and consumer privacy are a potential issues in collecting and accessing these data. Information is also available commercially from market research companies that specialize in determining consumer preferences and volume of products purchased. These reports are used by industry for marketing, but risk assessors may purchase some data from them.

These data help to get close to consumption characteristics like the ones listed above. The statistical agencies of many countries publish aggregate disappearance data – production, import and export – for some raw and processed foodstuffs. Some of the data are detailed enough for purposes of exposure assessments.

3.2 EXPOSURE DATA

3.2.1 Introduction

In a quantitative risk assessment, the key desired output of the exposure assessment is prevalence, concentration and, if possible, physiological state of *L. monocytogenes* in foods at the point of consumption. In the case of *L monocytogenes*, although the final numbers ingested by consumers are usually not known, an estimate can be derived based on models of the effect of physical processes and conditions that the food undergoes through the farm-to-fork chain. Such estimates are based on predictive microbiology models, and these are discussed in Appendix 3, including their limitations and methods for assessing their reliability. A strength of the risk assessment approach is that it can assess contributions to risk from all points along a food's journey from the point of harvest or slaughter to when it is consumed, enabling prioritization of risk management actions. While much attention has been paid to modelling risk from farm to fork, it is not always necessary to include the entire food chain to answer the risk management question, as in the case of the questions addressed in the current risk assessment (see Part 5 of this report).

The models are parameterized by data from studies carried out on products or their ingredients at different stages in the production-to-consumption chain. Information on what is in a serving requires information on the extent (prevalence) and level (concentration) of *L. monocytogenes* in a single package of the food, i.e. individual consumer units. Even if this is known at the point of manufacture, an estimate of the extent of growth or die-off during retail and consumer storage and handling has to be made. The only practical means of doing this is through modelling different components in the production-to-consumption chain. Mathematical models have been developed for growth, survival and inactivation of *L. monocytogenes* in laboratory broth media and some foods. The most reliable of these models are developed from systematic studies under carefully controlled conditions known to exert a major influence on *L. monocytogenes* growth, namely temperature, water activity (a_w) or NaCl concentration, pH and levels of preservatives, including organic acids and nitrite, etc. These models may have to be modified for specific foods and their full complement of ingredients. A last step estimates the meal sizes for the RTE foods and frequency of eating.

Process models are sometimes developed to examine how prevalence and concentration changes at points along the food chain. Models for microbial growth, survival or inactivation are developed for each step (unit operation – production, processing and handling, transportation, storage, and consumer preparation) in the progression from production up to preparation prior to consumption. The concentration at the conclusion of one step is the initial concentration for the next.

3.2.2 Prevalence

Recorded prevalence of *L. monocytogenes* in RTE foods varies with the product type and the stage in the production-to-consumption chain at which it is measured. The degree of *L. monocytogenes* contamination in ingredients differs substantially, depending on whether they are derived from farm animals, fish or shellfish, or produce. *L. monocytogenes* occurs in both uncultivated and cultivated soils and in silage and manure piles. It is less frequent in water or fish. Some geographical differences in prevalence may occur. For example, the prevalence of *L. monocytogenes* is considered to be much lower in fish products harvested from tropical waters than those derived from temperate waters (FAO, 1999). Prevalence on raw ingredients can be affected by various factors such as climate or health status of workers. Although *L. monocytogenes* in RTE foods is primarily reported in industrialized countries, it has been detected in foods produced in developing countries (Kovacs-Domjan, 1991; Salamah, 1993; Arumugaswamy, Ali and Hamid, 1994; Gohil et al., 1995; Luisjuanmorales et al., 1995; Warke et al., 2000; Xiumei Liu, pers. comm., 2000; A.S. Anandavally, pers. comm., 2000; Carlos, Oscar and Irma, 2001; Eleftheriadou et al., 2002; Dhanshree et al., 2003) and its occurrence in these countries may be more frequent than the literature suggests.

Contamination of foods by *L. monocytogenes* appears to occur most often at the processing level. *L. monocytogenes* may be present on processing equipment and facilities (walls, floors, drains, etc.), and contaminate food via water droplets, splashing, dust particles from the ceiling, and contact surfaces, including transfer by workers hands (Grau, 1993). Some RTE products may not undergo thermal or other processing sufficient to inactivate *L. monocytogenes*. In those products receiving a listericidal treatment, the presence of the pathogen is generally associated with recontamination from environmental sources prior to final packaging. Other RTE foods may be contaminated at the point of sale, for example, due to slicing of processed meats. Within the home, opened packages may be contaminated from *L. monocytogenes* present within the refrigerator or in other refrigerated foods, from the kitchen environment or from family members. Surveys of *L. monocytogenes* prevalence, conducted for purposes other than risk assessments are usually available for at least some of the RTE foods.

Section A2.8.2 in Appendix 2 describes the beta-binomial model for combining prevalence estimates from disparate sources.

3.3 MODELLING EXPOSURE: APPROACHES

3.3.1 Introduction

Microbial food safety risk assessment is a relatively new development. For developing and structuring a risk assessment, there is no one standard accepted at international or even national levels. Primarily, the exposure assessments in risk assessments conducted to date have been conducted beginning from either production stages or retail stages. Some have modelled prevalence and concentration at the time of consumption by allowing for the effects of time and temperature on growth and survival of *L. monocytogenes* from an earlier point in the chain. If necessary to meet the purpose of the risk assessment, a few have started the exposure assessments as far back in the food chain as the farm, or the water for fisheries or aquaculture products. However, lack of data about the impact of various environmental sources of contamination means that knowledge of the significance of early production stages is limited, at best. To date, they have not been used to any great extent in published exposure assessments. In risk assessments, and therefore in exposure assessments, there is a gradation of approaches – from descriptive, through qualitative to fully quantitative – for characterizing the variable of interest, whether risk or exposure. These include:

- qualitative expressions, e.g. high, average, low, more than, less than;
- an estimate relative to some known or existing level of exposure;
- a single numerical estimate for the end result based upon a series of point estimates, e.g. the average, or the worst case;
- a set of estimates that describes the range of possible outcomes as well as the one considered most likely, e.g. an average, worst-case and conservative estimate based on series of average, worst-case and conservative estimates for each variable in the assessment affecting exposure; and
- an estimate derived by combining the frequency distribution of variables in the assessment, characterized by a frequency distribution of possible outcomes. This approach gives as complete a representation as possible of the range of possible outcomes and the probability of each, providing all the information that the other methods do, and considerably more. This approach requires the greatest amount of information and the use of mathematical modelling techniques.

Van Gerwen et al. (1997) presented a three-step plan for hazard identification in the context of risk assessment, aimed at discerning those perceived hazards that represented the greatest risk, and which warranted more detailed study. Their plan involved "rough", "detailed" and "comprehensive" hazard identification. "Rough" hazard identification selects pathogens that have been implicated in foodborne outbreaks in the food of interest. The "detailed" hazard identification selects pathogens that have been reported as being *present* in the ingredients of the food of interest. The "comprehensive" procedure considers all pathogens, and even those less likely to arise in a specific food are included in the assessment. By including those hazards currently considered to be unlikely to be present, it should be possible to create an estimate of potential problems and to deal with them proactively. That philosophy can be extended to the performance of exposure assessments. The effort expended to undertake an exposure assessment must be commensurate with the magnitude of the risk. Pre-screening of the magnitude of exposure, using simple methods, can aid decisions about the value of investing in fully quantitative assessment methods. The approach can also show where greater detail should be built into the risk assessment model and where higher quality data will be required. If a risk assessment, for example, is intended to evaluate various options in a food process, details about on-farm contamination are unnecessary and modelling the consumer handling of the food can be simplified.

Microbial hazards in foods can arise at any stage in the food chain, and be affected by subsequent processing and handling steps. Thus, the system under analysis is a continuum, often from the point of production (farm, sea) to the point of consumption, and the risks presented by hazards at one point in the chain cannot be considered in isolation from the system as a whole.

To assess exposure it is necessary to understand both:

- the amount of food consumed and by whom, and
- where in that system the hazards arise, and all factors that affect the prevalence and concentration of the hazard in the food at the time of consumption.

This section provides an overview of methods used to estimate exposure. The ideas introduced here will be discussed further when reviewing existing exposure assessments.

3.3.2 Prevalence and concentration

Prevalence and concentration of *L. monocytogenes* in foods can change as a result of:

- initial and subsequent contamination;
- physical processes, e.g. dilution by mixing with uncontaminated ingredients, or division of batches into smaller units for distribution and sale; and
- growth or inactivation in the product.

To date, despite some assessors (Bemrah, et al., 1998; FDA/FSIS, 2001) noting that *L. monocytogenes* is probably heterogeneously distributed in some foods, all published exposure assessments have assumed that pathogens are distributed homogenously within a food. Multiple sampling of a food would presumably show a normal distribution of the $\log_{10}$ CFU/g of the microorganisms. This is a clearly a simplification. A consequence of the assumption of homogeneity is that in exposure assessments prevalence and concentration of *L. monocytogenes* in foods are often considered to be related properties, particularly at very low concentrations. The observed prevalence will depend on the sample size and the extent of contamination of the batch. If the batch is contaminated at a level of >1 CFU/g, there is high probability that each 25-g sample would test positive for *L. monocytogenes*. If, however, the sample size were only 1 g, some samples would test negative. If the contamination level were 1/100 g, we would expect only 1 in 4 samples of 25 g would test positive and it would be more typical to describe this concentration as "25% prevalence".

The distribution of bacteria in a homogeneous sample is likely to follow a Poisson distribution. In that case, if the mean concentration is X per gram, and there are Y grams per sample the count of *L. monocytogenes* per *sample* is Poisson distributed, with mean $X*Y$. More importantly, the probability of a positive result for a sample of Y grams then becomes $1 - \exp(-X*Y)$. Therefore, for large amounts of product, the prevalence and concentration are related and the estimate of the prevalence depends on the level of contamination and sample size. This is explicitly considered in a recent risk assessment (FDA/FSIS, 2001), although sample data for RTE foods were in some cases aggregated without regard to sample size. Thus, when incorporating data from many sources into an exposure assessment, it is important to consider the sampling methodology and test protocols, because sample sizes may differ and test methodology may differ in sensitivity Furthermore, some methods offer better sensitivity for specific types of foods than do other methods.

Similarly, products that permit the growth of *L. monocytogenes* may exhibit a low prevalence of contamination at the point of production and a higher prevalence at the point of consumption. This is not necessarily due to re-contamination, but may arise because the product was initially contaminated at a very low level. Subsequent growth in the product increases the probability of detection of that contamination. It is important, then, to recognize prevalence as "detected" prevalence. Also, the use of a more sensitive analytical method will find a higher prevalence of contaminated samples than will a less sensitive method. The estimated prevalences of the studies carry introduced uncertainty from the test methods and protocols used.

Qualitative risk assessments may be undertaken, for example, using the process of "expert elicitation". Synthesizing the knowledge of experts and describing some uncertainties permits at least a ranking of relative risks, or separation into risk categories. No true qualitative risk assessment has been conducted, however, in the area of microbial food safety. As assessors understand how qualitative risk assessments are done, they may become effective tools for risk managers because they can be conducted quickly and used to address specific questions or to demonstrate that extensive, fully quantitative exposure, and risk, assessment is not required. While there is no universally agreed methodology for qualitative exposure assessment, a useful discussion is presented in FAO/WHO [2004], which also includes a detailed example.

Many assessments of exposure of human populations to foodborne *L. monocytogenes* have been undertaken (Peeler and Bunning, 1994; Farber, Ross and Harwig, 1996; Hitchins, 1996; Lindqvist and Westöö, 2000; Buchanan et al., 1997; Bemrah et al., 1998; FAO, 1999). Most have included numerical epidemiological and prevalence data and some included concentration data for *L. monocytogenes* in specific RTE foods or classes of RTE foods. Nonetheless, in some cases the resulting assessments are descriptive or have simply ranked exposure or risk relative to some other, unquantified, level of risk (FAO, 1999; Ross and Sanderson, 2000; FDA/FSIS, 2001). Few have quantified exposure in terms of probability and magnitude of exposure, and fewer still (FDA/FSIS, 2001) have reported rigorously on the sources and magnitude of uncertainty in the estimates.

Methods for modelling growth and inactivation are discussed in detail in Appendix 2.

3.3.3 Conceptual model

The food production and distribution system being assessed can be described in a number of ways, but it is often easiest to start the process using diagrams, such as flow charts, to show the origin of hazards and the relationships and operations that can change the level and prevalence of the hazard in the food. An example of a flow chart, describing a very generic model for microbial food safety exposure assessment, is shown in Figure 3.1. That qualitative description of the factors that affect exposure (or more generally the risk), and the relationships among them, is described as a "conceptual model".

Semi-quantitative assessments can be developed using descriptors for each variable such as {high, low, normal}, or {better, worse, same}, or {+, -, 0}, or by applying a weighting system, or a combination. These methods rely implicitly on some known reference value, and have not been widely used in food safety risk assessments. Such approaches are often found in decision trees, such as that shown in Figure 3.2.

3.3.4 Mathematical models

A refinement of the conceptual model is to construct a mathematical model of the relationships. In principle, the entire system and the relationships between all variables could be explicitly defined by expressing the relationships mathematically, i.e. using algebraic notations and equations. By substituting data or values based on expert opinion for the variables in the model, the equations describing the origin and amount of *L. monocytogenes* in the food and the factors that impinge upon it can, in principle, be solved to yield a numerical estimate of exposure. Mathematical expertise is required to accurately describe the system, but it is now possible to model very complex systems relatively easily using the so-called Monte Carlo techniques and "spreadsheet models" written using computer spreadsheet software. Frequently the conceptual model can be very complex, and the solution of the corresponding mathematical model is also made easier using spreadsheet models. While it is easy to develop spreadsheet models, it is also easy to make mathematical and logical errors in the construction of the model. It is therefore very important to verify both the accuracy of the mathematical model as a description of the system being assessed and its mathematical reliability (Starfield, Smith and Bleloch, 1990; Morgan, 1993; Vose, 1996). Texts that teach modelling skills are available (e.g. Starfield, Smith and Bleloch, 1990).

3.3.5 Point estimates

When solving exposure assessment models, a decision has to be made regarding the value of the variables to be used in the model. Typically, the factors in a system that affect exposure do not have single, fixed values but are characterized by a range of possible values. The most obvious method is to characterize the variable quantity by its central tendency value (e.g. mean, median). Thus, the mathematical model would produce an estimate of the risk characterized by the most commonly occurring scenario.

Point in Food Continuum	Variables Affecting Dose		
	Consumption	Concentration in contaminated units	Prevalence of contaminated units
Raw Ingredients		environmental sources affecting concentration in ingredients	season, harvest area, fodder and feeding regimes, irrigation water, etc.
⇓			
Processing		*volumetric changes:* mixing with other ingredients, changes due to dilution or concentration steps (e.g. evaporation, removal of whey) *growth or inactivation changes* brining, heating steps, holding times and temperatures,	cross-contamination, mixing with other bulk ingredients, splitting into smaller units for retail or food service
⇓			
Transport and Storage		time, temperature, product composition	
⇓			
Retail Sale		time, temperature, product composition, breakdown to smaller units	packaging and cross-contamination, portioning, breakdown to smaller units
⇓			
Home/food service		time, temperature, product composition	cross-contamination, combination with other foods
⇓			
Consumption	frequency and amount consumed affected by: season, wealth, age, sex, culture/region, etc.	heating; mixing with other components (e.g. vinegar in salads); breakdown to smaller units	breakdown to smaller units/serving portions

Figure 3.1 A generic exposure assessment model for pathogens in foods.

However, this ignores important risk characteristics, as will be discussed in more detail in the risk characterization, as the highest risk is associated with the small percentage with the highest levels of *L. monocytogenes*. An alternative approach is to use worse-case scenarios, based, for example, on the 90th or 95th percentiles. One problem with this approach, particularly when dealing with a multiple step conceptual model, is the "compounding conservatism" (Cassin et al., 1996). If conservative or worst-case values are taken for each variable, the resulting risk estimate is characterized by an extremely improbable event. It should also be noted that point estimates based on measures of central tendencies, e.g. average, or modes will not necessarily lead to an answer that represents the most likely outcome and can lead to large errors (Cassin et al., 1996).

The use of point estimates of parameters determining the probability of an adverse event has severe limitations in relation to providing an "accurate" assessment of risk (Buchanan and Whiting, 1997), and, increasingly, stochastic modelling techniques are being employed for hazard characterizations, exposure assessments and risk characterizations.

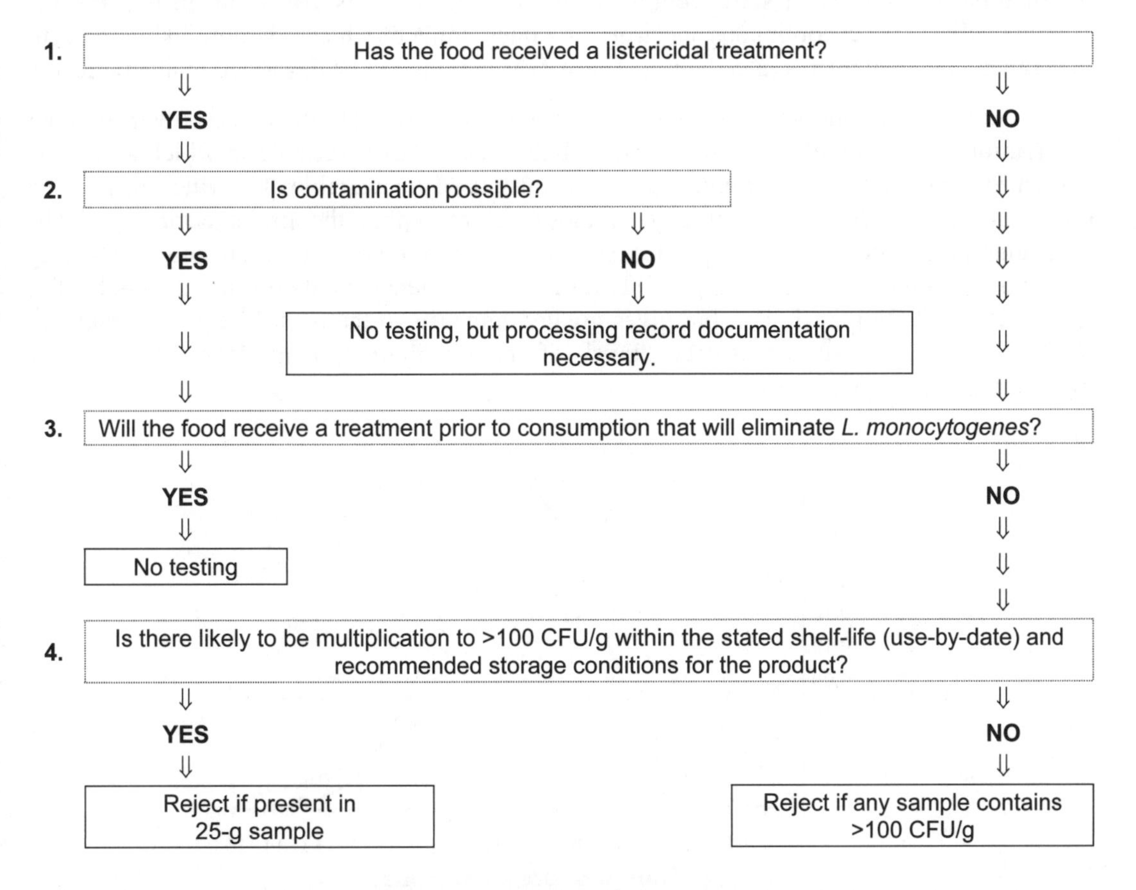

Figure 3.2 A decision tree to aid the management of the hazard of listeriosis from foods showing qualitative risk assessment decisions within a risk management scheme.
SOURCE: Reproduced from CCFH, 1999.

While increasing the potential accuracy of microbial risk assessments, two potential disadvantages are associated with stochastic modelling methods. The first is that the time it takes to develop such models may delay risk-management decisions. The second is that the complexity of the model increases the bounds of uncertainty and variability, which may become so wide as to lead to questions on the part of the risk manager regarding the reliability of the information. However, this must be put in context, namely that, in most instances, food safety decisions will be reached with or without the availability of a risk assessment.

3.3.6 Distributions and stochastic approaches

Point estimates are useful to provide a quick estimate of the magnitude of risk. To support critical decisions, however, a more accurate estimate conveys an understanding of the complete range and probability of all possible outcomes, and their consequences.

The *range* of possible values can be characterized by a minimum and maximum. More information is conveyed if some central, or *most-likely,* value is also used. In general, the possible values form a continuous spectrum of values, some of which are more likely to occur than others, i.e. they form a *distribution*. These distributions can be described mathematically.

The normal distribution is well known, but many data sets are better described by other distributions. For example, the uniform distribution describes a variable in which a value is known to vary between two limits. It is frequently used for variables for which there is no knowledge of the probability of any of those values within the limits occurring. The triangular distribution is the simplest description of minimum, maximum and most-likely values and is used to represent a possible range when extensive data are not available. The Beta-Pert distribution is similar, but gives greater emphasis to the most-likely value and less to the upper and lower limits (the "tails" of the distribution) than does the triangular distribution (see Figure 3.3).

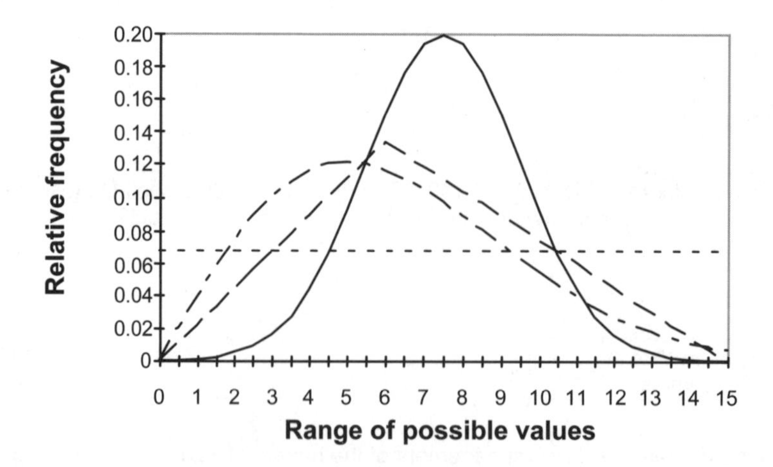

Figure 3.3 Some types of distributions used to describe ranges of values for observations. Normal (————); triangular (—— —— ——); uniform (- - - -); Beta-Pert (— - — - — -).

In certain instances, naturally occurring phenomena can be described by a mathematical equation (e.g. decay of a radionuclide) instead of the fixed variable values in the conceptual model. Lognormal, beta, gamma and Weibull distributions, for example, are used frequently to describe data. This results in the solution of the model being a distribution of possible values, based on all the possible combinations of scenario sets. The results are the range of possible outcomes. The answer obtained by this method is called an *explicit* solution. The explicit solution offers a complete representation of the range and probability of possible outcomes of a process, and provides much more insight than does a calculation based on average values. In most cases, however, the calculations and resulting equations for an explicit solution become so complicated so quickly that they cannot be solved for anything but the simplest models.

A third approach to describing risks is through simulation modelling. This is based on the Theory of Large Numbers, which effectively states that an accurate answer to a complex model can be deduced if the model is solved repeatedly using the various distributional inputs in accordance with the likelihood of occurrence.

3.3.7 Simulation modelling

Computer simulation modelling software (e.g. @Risk, Crystal Ball, Analytica,) offers a means to calculate the results for complex systems or processes for which explicit mathematical models do not exist or are difficult, if not impossible, to solve analytically. After the model is constructed, the software calculates all of the possible combinations of factors by calculating the answer many times, each cycle of which is called an iteration. At each iteration a value is selected from each variable range, at random according to the probability distribution describing that variable. The outcome is then calculated for that specific set of circumstances, i.e. that iteration. All of those values are collated to generate a distribution of possible outcomes. Because some or all of the independent variables in the model are characterized by a range of possible values, there is a range of outcomes, some of which will occur more often than others.

The distributions used to describe both the inputs and outputs of a model are composed of two components: variability and uncertainty. It is important to be able to differentiate between uncertainty and variability. Variability describes diversity that is inherent in any population. Uncertainty refers to the situation where assumptions have to be made about the ranges of values and their probabilities of occurrence. The degree of uncertainty will be reduced by the acquisition of new data or knowledge, whereas additional data will not decrease variability.

The results of an exposure assessment that employs simulation modelling techniques depend on the model, the data ranges and distributions that are used, and on the assumptions made in setting up the model. Detailed consideration of the potential pitfalls in simulation modelling are available in general references and guidelines for the use of simulation modelling in risk assessment (e.g. Vose, 1996; Morgan, 1993; Burmaster and Anderson, 1994; EPA, 1997).

3.3.8 Uncertainty and variability

Acceptance of a degree of uncertainty and variability is fundamental to an estimation of exposure in any model. Uncertainty refers to information that is required for completion of the assessment but that is not available and has to be assumed or inferred. The basis of uncertainty is twofold: information uncertainty and model uncertainty. The information on which the exposure estimates are made is often limited. Population characteristics must be inferred from observations made on a sample drawn from the population at a specific point in time, and observed phenomena must be extrapolated to the situation under study. The assumptions on which the exposure estimates are based introduce uncertainty: simplification of complex processes into mathematical models for physical processes, inactivation and growth introduce uncertainty; small sets of scenarios are generalized to all scenarios of importance; and assumptions are made about how recognizable components of processes operate. In addition, the limitations in testing methods for *L. monocytogenes* also introduce uncertainty in the levels of the pathogen in the food supply. Many surveys test only for presence per 25 g of product.

Variability is an inherent property of all physical, chemical and biological systems. There is natural variability (heterogeneity) among the constituents of a population. In the case of the current risk assessment there are multiple factors influencing risk that each have inherent variability. The prevalence and concentration of *L. monocytogenes* in RTE foods vary, and the composition of the foods, serving sizes and frequencies, the virulence of *L. monocytogenes* isolates and the susceptibility of infected individuals were among the long list of variable parameters encountered in the risk assessment.

3.4 MODELLING THE PRODUCTION-TO-CONSUMPTION CHAIN

3.4.1 Environmental niche

Sources of *L. monocytogenes* in the environment were described in Section 1.2.

3.4.2 Preharvest

A complete exposure assessment starts at the earliest stages in the production of a food so that it can include the effect of the environment. Green vegetables or berry crops might be affected by contamination from soil, manure, irrigation, silage and the pathogens in them, for example. Insects may also play a role in the spread of organisms to crops. Pathogens may survive in manure or soil for long periods (Dowe et al., 1997); inside protozoa (Barker and Brown, 1994); and some may also penetrate the vasculature of leafy plants like lettuce, and alfalfa or mung bean seeds. *L. monocytogenes* does not occur naturally in oceans. Some aquatic environments may become contaminated with *L. monocytogenes* from human or animal sewage or from soil from cultivated and uncultivated fields carried in rainwater runoff. In such cases *L. monocytogenes* might contaminate fish and shellfish.

3.4.3 Production

After harvest, preliminary washing or cleaning of the product may remove some of the initial contamination. Transport may introduce additional or new pathogens. At each of the succeeding stages of production, changes in prevalence and concentration are likely to occur. However, unless actual measurements are taken at each these stages, they must be modelled based on the knowledge that already exists.

3.4.4 Processing and packaging

Subsequent production steps include holding, mixing and aggregation, fermentation, heating, pasteurization, brining, smoking and pickling. Some of these steps increase, but most decrease, the prevalence and concentration of pathogens. Much of *L. monocytogenes* contamination arises from environmental contamination in the processing plant. For example, aerosols from cleaning water and dirty equipment may be sources. Cooked products, e.g. processed RTE meats, should be free of *L. monocytogenes,* but may become recontaminated during subsequent handling and contact with equipment before final packaging. Slicing operations appear to be common sources of re-contamination of cooked products. Sources and routes of contamination of food with *L. monocytogenes* in food processing facilities are extensively reviewed in Ryser and Marth (1999). More recent studies include those by Norton et al. (2001) and Chasseignaux et al. (2002).

3.4.5 Transportation

Changes in the frequency of *L. monocytogenes* contamination can occur after final packaging for products that remained sealed until consumption. The number of *L. monocytogenes* can increase if the food and the storage conditions support the growth of the microorganism. This can lead to an apparent increase in the frequency of contamination if the product was initially contaminated at a level below the limit of detection of the method used to enumerate *L. monocytogenes* (see Table 3.1).

3.4.6 Retail

Changes to populations of the microorganisms can take place during storage and display. The prevalence and levels of a pathogen may change through recontamination from portioning of the opened packaged products through slicing, chopping and then repackaging. Other packages or other RTE foods then may be cross-contaminated by the same process. Ambient temperatures can permit the growth of the pathogen on contaminated slicing equipment, cutting boards, etc., and could increase the level of hazard.

Table 3.1 Ranges of environmental factors that permit growth of *Listeria monocytogenes* when all other factors are optimal.

Environmental Factor	Limits	
	Lower Limit	Upper Limit
Temperature (°C)	-2 to +4	~ 45
Salt (% water phase NaCl)	<0.5	13 – 16
(and corresponding a_w)	(0.91–0.93)	(> 0.997)
pH (HCl as acidulant)	4.2–4.3	9.4 – 9.5
Lactic acid (water phase)	0	3.8–4.6 mM, MIC(1) of undissociated acid(2) (800–1000 mM, MIC of sodium lactate(3))
Acetic acid	0	~20 mM (MIC of undissociated acid)
Citric acid	0	~3 mM (MIC of undissociated acid)
Sodium nitrite	0	8.4 – 14.4 µM (undissociated)

NOTES: (1) MIC = minimum inhibitory concentration, i.e. the minimum concentration that prevents growth. (2) From Tienungoon, 1998. (3) From Houtsma, de Wit and Rombouts, 1993.

SOURCES: The overall ranges are summarized from Ryser and Marth, 1991; ICMSF, 1996; and Augustin and Carlier, 2000a.

3.4.7 Home and foodservice

For foods that support growth of *L. monocytogenes*, time and temperature of storage are the most critical parts of this stage since RTE products may be kept refrigerated for long periods. In addition, cross-contamination to opened RTE food packages may occur in the refrigerator from other foods with *L. monocytogenes*. For some RTE foods that do not support its growth, such as dry fermented sausages, levels of *L. monocytogenes* are expected to diminish during storage, and probably at a faster rate if held at ambient temperature than if refrigerated. If there is no final heating step prior to eating, as is the usual case for RTE foods, the concentration of *L. monocytogenes* at the end of the storage period in the home or foodservice establishment will be the concentration when the food is eaten.

3.5 MICROBIAL ECOLOGY OF *LISTERIA MONOCYTOGENES* IN FOODS

3.5.1 Introduction

The dose ingested, and hence the risk of listeriosis, is dependent on the mass of food consumed and the level and frequency of contamination. However, surveys of the level of *L. monocytogenes* in foods are not conducted; instead, the dose must be inferred from exposure data acquired earlier in the food chain. In the case of the current risk assessment, retail data were employed in conjunction with predictive microbiology models and data on storage times and temperatures to predict the levels ingested. The need for this modelling reflects that when *L. monocytogenes* is present in food its numbers may increase, decrease or remain constant as a result of growth, death (or inactivation) or stasis, respectively. The degree to which growth and inactivation occur is governed by the composition of the food, the conditions under which the food is stored or subject, and the time during which those different conditions apply.

While the distributions of serving sizes of RTE foods generally only differ by a 5–10-fold range (e.g. 10–100 g), the concentration of *L. monocytogenes* within the serving can range over many orders of magnitude. Given sufficient time, *L. monocytogenes* can reach concentrations of 10^6 to 10^9 CFU/g in many RTE foods that support microbial growth. Conversely, heat treatments can effectively eliminate the microorganism in a matter of minutes. Typically, microbial populations increase or decrease exponentially over time. Consequently, if growth is possible in the product, the predicted risk resulting from that growth generally changes exponentially with time. The same is true of pathogen inactivation.

Since predictive microbiology plays such an important role in the current microbiological risk assessment, it is important that the application of predictive microbiology methods and its limitations are well understood by risk assessors, stakeholders and risk managers. A review of predictive microbiology concepts and limitations, methods of assessing predictive model performance, and techniques for the application of predictive models in risk assessment is given is Appendix 3, including a compendium of published predictive models for *L. monocytogenes* relevant to foods.

The current section presents patterns of microbial behaviour in foods and food processing, and identifies unifying principles to aid understanding of the factors that affect the ecology of *L. monocytogenes* in foods. Reviews of the ecology and physiology of *L. monocytogenes* in food products in general (Lou and Yousef, 1999) and in specific food products (Ryser, 1999a,b; Farber and Peterkin, 1999; Cox, Bailey and Ryser, 1999; Jinneman, Wekell and

Eklund, 1999; Brackett, 1999) have recently been presented. Many relevant data are collated and tabulated in ICMSF (1996) and Augustin and Carlier (2000a). The following material is based on Ross, Baranyi and McMeekin (1999) and Ross, Dalgaard and Tienungoon (2000) who reviewed the microbial ecology of *L. monocytogenes* in relation to the risk assessment of RTE seafood.

3.5.2 Growth limits

The ranges of environmental factors that permit growth of *L. monocytogenes* are discussed in detail in a number of reviews (Lou and Yousef, 1999; ICMSF, 1996; Augustin and Carlier, 2000a), as summarized in Table 3.1. These limits are not absolute, however, as discussed below, but represent the widest range of that factor when all other factors are optimal for growth. When several factors are suboptimal for growth, the ranges of each that will permit growth of *L. monocytogenes* are restricted. This is the basis of the Hurdle Concept, or "multiple barrier methods" in food preservation. There are exceptions to this behaviour. While slightly elevated salt concentration may inhibit growth rate, it has also been reported to increase the high-temperature tolerance of many bacterial species, though the effect is not universal (Gould, 1989).

For several foodborne pathogens, including *L. monocytogenes,* greatest tolerance to sub-optimal conditions is exhibited at conditions optimal for growth yield[1] (George, Richardson and Peck, 1996; Presser, Ross and Ratkowsky, 1998; Tienungoon, 1998). Conditions that maximize the growth *rate* of *L. monocytogenes* are not necessarily the same as those that maximize growth *yield.* For *L. monocytogenes,* yield is maximal when temperature is in the range of 20° to 25°C, while the growth rate is fastest at ~37°C. It is often important in growth modelling of *L. monocytogenes* to calculate the growth yields at temperatures in the 0° to 7°C range. At temperatures above or below 20–25°C, the water activity or pH growth limits of *L. monocytogenes* will not be as wide as the extreme values listed in Table 3.1. Similarly, recovery of *L. monocytogenes* from injury is most rapid at 20–25°C (Mackey et al., 1994; see also Figure 3.4).

3.5.3 Growth: rate, lag and maximum population density

Where the interaction of factors permits growth, the amount of growth that occurs in a specified time will be governed by:

- the growth rate;
- whether there is a lag time before growth is initiated; and
- the total concentration of bacteria that the food will support.

These three topics are considered individually below.

3.5.3.1 Growth rate

Growth rate is affected by factors that include:

- temperature;
- storage atmosphere;

1. In this context, yield is taken to represent the maximum cell biomass produced in a given (batch) environment. An analogous measure is maximum population density.

- salt or sugar content (often expressed as water activity);
- pH and presence of organic acids;
- preservatives such as nitrite, sorbate, etc.; and
- the presence of high levels of other microorganisms of other strains or species.

Many of these factors act independently and can be understood in terms of the relative inhibition of growth rate due to each factor. Under completely optimal conditions, each microbial strain has a unique maximum growth rate. For *L. monocytogenes,* the fastest doubling time is in the range of 35 to 40 minutes, and occurs at temperature of ~37°C, when pH is neutral, and in a rich medium that contains sufficient nutrients and has a water activity in the range 0.990 to 0.995 (1±0.5% NaCl). As any environmental factor becomes less optimal, the growth rate declines in a predictable manner. The cumulative effect of many factors at suboptimal levels can be estimated by multiplying the relative inhibitory effect of each factor. The relative inhibitory effect can be determined from the "distance" between the optimal level of the factor and the minimum (or maximum) level that completely inhibits growth. This concept is embodied in the structure of a number of the square-root type models (Ratkowsksy et al., 1982, 1983; Presser, Ross and Ratkowsky, 1998), "gamma" models (Zwietering, De Wit and Notermans, 1996) and "cardinal parameter" models (Rosso et al., 1995) derived from them.

Interactions can occur between some factors used to preserve foods. The activity of many preservatives is pH dependent. The effect is best described for organic acids. The inhibitory effect of organic acid is almost completely determined by the concentration of the undissociated form of the acid. The concentration of undissociated form can be readily calculated from the total concentration of the organic acid and the pH. If the inhibitory activity of organic acids is described in terms of the undissociated form of the acid the simple multiplicative rule (as described above) works well, as illustrated by Presser, Ross and Ratkowsky (1998) and by Tienungoon (1998) for *L. monocytogenes*. Nitrite activity is also reported to be pH dependent (Woods, Wood and Gibbs, 1989) and the results of studies by the USDA Agricultural Research Service Eastern Regional Research Centre in Philadelphia (embodied in the Pathogen Modelling Program[2]) also show a pH dependence of nitrite on the growth rate of *L. monocytogenes,* particularly at levels >125 ppm in broth. The relative inhibition of a specific concentration of nitrite is equivalent at all experimental conditions of pH, temperature and water activity. That inhibition is approximately linearly related to the total nitrite concentration.

In general, the growth of *L. monocytogenes* is reported to be little affected by anaerobic, or oxygen reduced, atmospheres (Buchanan and Phillips, 1990; Pelroy et al., 1994; Buchanan and Golden, 1995; ICMSF, 1996). However, growth is reduced by CO_2 when it used in modified atmosphere packaging (Davies, 1997; Bell, Penney and Moorhead, 1995; Ingham, Escude and McCown, 1990; Szabo and Cahill, 1998; Nilsson, Huss and Gram, 1997).

Growth rate may also be affected by the presence of high levels of other microorganisms, in a phenomenon described as the "Jameson effect" by Stephens et al. (1997). Jameson (1962), in studies concerning the growth of *Salmonella*, reported the suppression of growth of

2. Pathogen Modelling Program. Available free of charge from USDA. Download from: http://www.arserrc.gov/mfs/pathogen.htm

all microorganisms on the food when the total microbial population achieved the maximum population density (MPD) characteristic of the food. The same effect has been reported for *Staphylococcus aureus* in seafood (Ross and McMeekin, 1991), *L. monocytogenes* in meat products (Grau and Vanderlinde, 1992), in fresh-cut spinach (Babic, Watada and Buta, 1997), co-cultures of *L. monocytogenes* and *Carnobacterium* spp. in laboratory broth, fish juice and seafood (Buchanan and Bagi, 1997; Duffes et al., 1999; Nilsson, Gram and Huss, 1999), and was discussed by Peeler and Bunning (1994) in relation to their predictions of the growth of *L. monocytogenes* in raw milk.

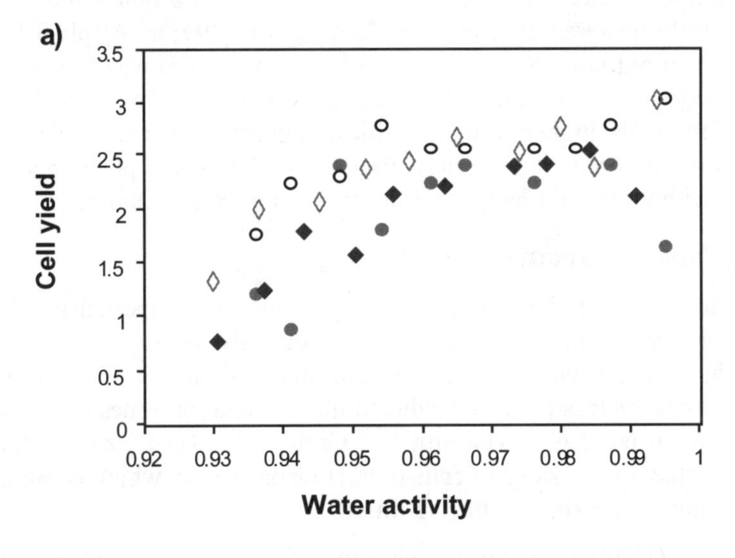

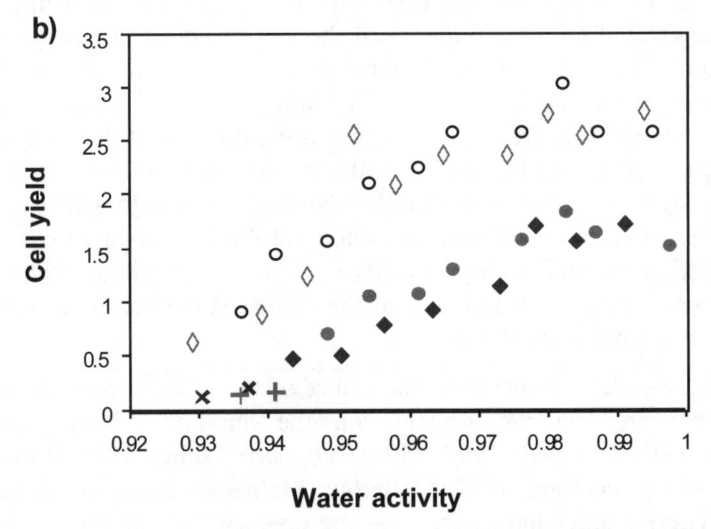

Figure 3.4 The observed cell yield of *Listeria monocytogenes* "corrected" for the non-linearity of the Optical Density (OD)-concentration relationship using the function of Dalgaard et al. (1994) and plotted against water activity (NaCl as humectant), demonstrating the influence of lactic acid, and pH; a) pH ≈5.7, and b) pH ≈5.4. Strain Scott A; growth in the absence of lactic acid (◊), and growth (◆) and no growth (×) in the presence of 50 mM lactic acid. Strain L5; growth in the absence of lactic acid (○), and growth (●) and no growth (+) in the presence of 50 mM lactic acid.

SOURCE: Reproduced from Tienungoon, 1998.

3.5.3.2 Maximum concentration

A corollary of the Jameson effect is that there is an upper concentration limit to the growth of *L. monocytogenes* and other bacteria in foods. Under optimal conditions, this level is of the order of 10^9 CFU/g or CFU/ml. However, the conditions of growth may limit the maximum concentration of *L. monocytogenes* that can occur. This phenomenon was reviewed by FDA/FSIS (2001) and incorporated in that exposure assessment. Specifically, at lower temperatures, the maximum growth predicted to occur was limited to levels up to 1000-fold lower than at temperatures above 8°C. Similar behaviour as a function of water activity, pH and lactic acid in broths was described by Tienungoon (1998). At pH 6.1, decline in final population numbers did not occur unless water activity (NaCl) was less than 0.935. As pH decreased, or lactic acid concentration increased, or both, the final cell density began to be reduced at progressively higher water activities, suggesting that multiple hurdles to growth reduce the maximum population density. Figure 3.4 shows this phenomenon at pH 5.4 and 5.7 and with or without 50 mM lactic acid for two strains of *L. monocytogenes*.

3.5.3.3 Lag phases or recovery from injury

Upon transfer to a new environment, microorganisms may experience a lag phase before growth begins or recommences. The effect is to reduce the amount of growth predicted. Lag time duration has often been considered erratic and evaluations of predictive models have shown that lag times are less reliably predicted than generation times (Walls and Scott, 1997; Dalgaard and Jørgensen, 1998; Augustin and Carlier, 2000a,b). This variability has often been attributed to the prior history of cells (e.g. Hudson, 1993), which is usually ill-defined or unknown, affecting the duration of the lag time.

Robinson et al. (1998) formalized a concept of the lag time as being dictated by two elements: (i) the amount of work required of the cell to adjust to a new environment or to repair injury due to the shift to the new environment, or both; and (ii) the rate at which those repairs and adjustments can be made. The latter rate is presumed to respond to the environment in the same way, relatively, as generation time, i.e. if the environment causes the generation time to double, the lag time will also double, and so forth. In recognition of this, the ratio of the lag time : generation time has been introduced to enable comparison of lag times measured in different environments (Mellefont, McMeekin and Ross, 2003) This ratio can be considered as the relative lag time (RLT). The RLT can be considered as the amount of work (whether adjustment or repair) that the cell must perform in a new environment or after injury before growth can recommence.

Systematic studies have considered the effect of the prior history of the cell, including prior temperature and osmotic stresses, on the duration of lag time and RLT of *L. monocytogenes* (Bréand et al., 1997, 1999; Delignette-Muller, 1998; Robinson et al., 1998; Ross, 1999; Whiting and Bagi, 2002, Mellefont, McMeekin and Ross, 2003; Mellefont and Ross, 2003). These studies have supported the concept that the RLT is greater, i.e. more work is required, when there is a larger shift in environmental conditions. Generally, the effect is more pronounced when cells are shifted away from optimal conditions rather than towards conditions more optimal for growth.

Ross (1999) undertook a review of published lag time data for *L. monocytogenes*, expressing the results as RLTs. The distribution of reported RLTs has a sharp peak in the range 3 to 6. Augustin and Carlier (2000a) presented similar information expressed as ln(RLT). Both analyses are highly consistent. These distributions of RLT can be exploited for "exposure assessment", either as point values taken from the cumulative distribution, or by providing a distribution of lag times from which to sample in Monte Carlo simulations (Ross and McMeekin, 2003).

It has also been proposed that lag times may be a function of the concentration of cells present, with fewer cells leading to longer lag times (Zhao, Montville and Schaffner, 2000; Robinson et al., 2001). This may reflect the probability of a cell being ready to grow; with more cells present, it is more likely that at least one cell will have a short lag.

The integration into a conceptual model of factors that may affect the rate and amount of growth of *L. monocytogenes* is shown in Figure 3.5.

3.5.4 Death or inactivation

3.5.4.1 Death rates

When conditions are outside the ranges that permit growth, microorganisms will either survive or be inactivated. Inactivation has traditionally been considered to follow log-linear kinetics, characterized by D and *z*-values (see next section), although the actual kinetics may be complex and involve several distinct phases, each with its own log-linear rate (Cerf, 1977; Augustin, Carlier and Rozier, 1998; Humpheson et al., 1998; Peleg and Cole, 1998). Until recently, D and *z* values were the primary methods of modelling thermal inactivation of microorganisms.

Recent reports indicate that log-linear models are inadequate to describe the death kinetics of *L. monocytogenes*, and that more complex (e.g. sigmoidal) functions are needed. Augustin, Carlier and Rozier (1998) used the concept of heat resistance *distributions* to develop models. The issue of variability in responses between strains, or due to uncontrolled variables, is currently a major theme in predictive microbiology.

The use of temperatures above the biokinetic range to inactivate microorganisms may be termed "thermal" processes, while the use of other growth preventing conditions, e.g. high salt or low pH, that result in inactivation have been called "non-thermal inactivation".

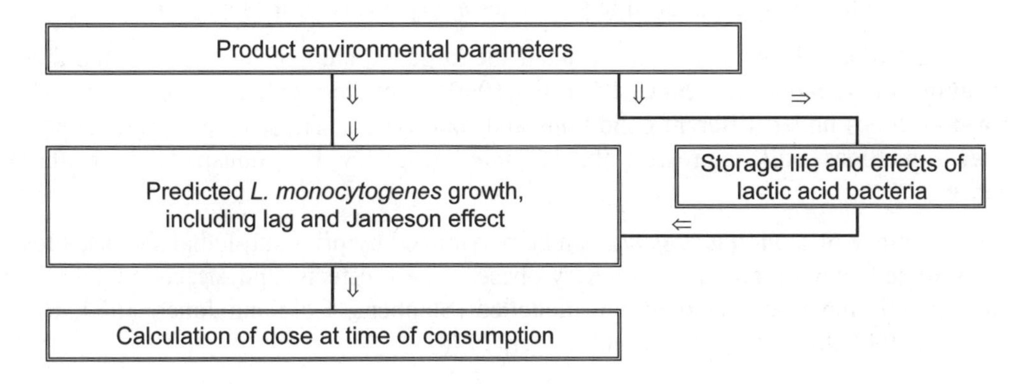

Figure 3.5 Overall model structure for the conceptual model and influence diagram for the interaction of factors governing the extent of growth of *Listeria monocytogenes* in ready-to-eat foods. Each of the boxes represents a "module" of calculations. Details of the predictive growth module are shown in Figure 3.6.

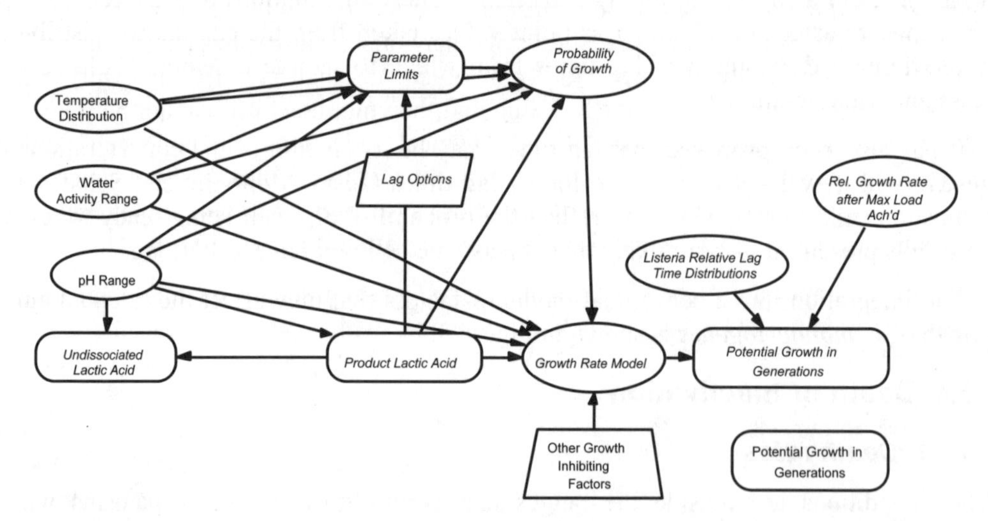

Figure 3.6 Detail of "predicted growth ..." module in the conceptual model (Figure 3.5) and influence diagram of the interaction of factors governing the extent of growth of *Listeria monocytogenes* in ready-to-eat foods. Each of the boxes represents a "module" of calculations.

3.5.4.2 Thermal inactivation

A number of measures are used to describe the effect of heat, expressed as temperature, on the rate of death of microorganisms. The first of these is the D value, which is the time required for a ten-fold reduction, i.e. 90% of the population killed, at some specified condition of temperature and other factors. Implicit in the use of the D value to describe death rate is the assumption that death follows log-linear kinetics. Temperature is very effective in killing microorganisms and the rate of killing by temperatures even slightly above the upper limit for growth of vegetative bacteria is many times faster than death due to other factors (Shadbolt, Ross and McMeekin, 1999). Once temperature exceeds the physiological range for the organisms, small increases in temperature cause large increases in mortality. The increase in lethality due to heat is described by the z value, which is the temperature increase required to increase the lethality by a factor of ten. z values are typically of the order of 5–15°C.

Lou and Yousef (1999) reviewed in detail the large expanse of literature on the thermal inactivation of *L. monocytogenes*. ICMSF (1996) provides extensive lists of thermal inactivation times under different conditions and food types. Those data do not support the opinion sometimes still expressed that *L. monocytogenes* has unusually high thermal tolerance.

Heat tolerance of *L. monocytogenes* can be maximized by prior sub-lethal shocks, stress or applying to cells having reached stationary phase. These effects, and effects on subsequent lag and growth, have been studied and modelled (Stephens, Cole and Jones, 1994; Bréand et al., 1997, 1999; Augustin, Carlier and Rozier, 1998).

3.5.4.3 Freezing

Freezing damages and kills some microorganisms, mainly due to the increasing osmotic potential, i.e. as the water in the food freezes it increases the effective concentration of solutes in the remaining liquid water, causing osmotic stress to those organisms suspended in that water. As water freezes, ice crystals may also cause physical disruption of cell membranes, further reducing the viability of organisms that have frozen. During thawing, further damage to cells can occur. Freezing and thawing, however, cannot be relied upon to eliminate contaminating microorganisms. Typical reductions in viable cell numbers on freezing and thawing for foodborne microorganisms of public health significance are of the order of a 10 to 100-fold reduction in the most susceptible types of organisms. Multiple freeze-thaw cycles are more lethal than a single freeze-thaw cycle.

The most important factor influencing the effect of freezing on microbial cells is the suspending medium. Certain compounds enhance, while others diminish, the effects of freezing. Glycerin, saccharose, gelatin and proteins in general act as cryoprotectants. Common salt (NaCl) increases the effect of freezing, due to depression of the freezing point of the water in the system, which has the effect of prolonging the cell's exposure to damaging high osmotic stress. The rate of freezing and thawing will also affect the lethality of these processes, with more rapid rates of both being less lethal. During frozen storage there will be a gradual loss of viability, the rate being slower at colder temperatures below freezing. Fluctuations in temperature during frozen storage will increase the rate of loss of viability. Cells are also re-exposed to damage through osmotic stress during thawing. A review of the studies concerning freezing and thawing effects on foodborne microorganisms is given in Lou and Yousef (1999) and Singhal and Kalkarni (2000).

3.5.4.4 Non-thermal inactivation

Conditions that prevent growth of microorganisms ultimately lead to their inactivation. Low temperature seems to be an exception to the general rule that more extreme conditions accelerate rates of microbial inactivation. Lower temperatures reduce the rate of death when other factors prohibit growth: very low temperature is routinely used as a method of culture preservation.

Non-thermal inactivation may be very slow. Seeliger (1961) reported that *L. monocytogenes* can survive for up to a year in 16% NaCl ($a_w = 0.883$). The mechanisms of non-thermal inactivation are currently poorly understood but have recently been reviewed (Mackey, 1999).

Buchanan and his colleagues have provided much of the published non-thermal inactivation data for *L. monocytogenes* (Buchanan and Golden, 1994, 1995; Golden, Buchanan and Whiting, 1995; Buchanan, Golden and Phillips, 1997). In most of those studies, organic acid was considered the main factor causing inactivation. A single predictive model encompassing much of the USDA data was presented in Buchanan, Golden and Phillips (1997). The inactivation kinetics were not log-linear. The model predicts the time required to reduce the original population by 99.99%, a time termed t_{4D}. It should be pointed out that because inactivation rates are not log-linear, the model cannot be used reliably to predict inactivation times beyond a 4 D kill, i.e. an 8 D kill will not necessarily occur after two t_{4D}s.

Data for rates of radiation inactivation are summarized in ICSMF (1996). The lethality of irradiation depends on the medium in which the cell is suspended, including factors such as temperature, water activity and pH.

3.6 SUMMARY

This section has attempted to identify the data needed to assess human exposure to *L. monocytogenes* in RTE foods, as well as tools and techniques to overcome missing data, and approaches for synthesizing data through models to enable estimation of exposure. Those data include the incidence of contamination; level of contamination; point of contamination; time and temperature history between contamination and consumption; volume of food per meal; and total consumption of the food in the community of interest.

Incidence and prevalence data at the point of consumption will rarely be available and quantitative or semi-quantitative assessment of exposure will probably have to rely on predictive microbiology models. Those models will have to have been successfully validated in products of similar microbial ecology to the product of interest. Some models have been shown to be too "fail-safe" and to produce unrealistically high estimates of exposure. Any exposure assessment should explicitly recognize the limitations of existing data, our understanding of the microbial ecology of *L. monocytogenes* and of the current generation of predictive microbiology models, so that the risk assessment process remains transparent.

Subsequent sections will provide specific examples of exposure assessments of *L. monocytogenes* in RTE foods, demonstrating the above principles.

Part 4.

Example Risk Assessments

4.1 OVERVIEW

4.1.1 Introduction

This section presents four risk assessments of *L. monocytogenes* in specific RTE foods: pasteurized milk; ice cream; fermented meats; and cold-smoked, vacuum-packed fish. All were modelled from production or retail to the point of consumption.

The four commodities were selected to exemplify estimation of the difference in risk associated with foods that do support and those that do not support the growth of *L. monocytogenes*, and foods with different contamination rates, shelf-life and levels of consumption. They also serve to answer one of the CCFH questions on the risk from *L. monocytogenes* in foods that support growth and foods that do not support growth under specific storage and shelf-life conditions. Various microbiological and mathematical and statistical considerations are also discussed and illustrated using different modelling approaches. Two examples attempt to estimate risk to a consumer in a specific nation, and two examples attempt to assess an average risk for any consumer in the world. The former approach is limited in its applicability to other nations, while the latter "ignores" the effect on risk of differences between nations, e.g. in consumption.

The dose-response relationship elaborated in Part 2 was linked with an exposure assessment developed for each commodity in order to generate an estimate of the risk of acquiring listeriosis. The risk estimates were expressed per 100 000 population and per 1 million servings, to illustrate the importance of the risk metric in understanding comparative risk. "Cases per 1 million servings" illustrates the risk to an individual consumer of that food, whereas "cases per 100 000 population" includes the effect of number of servings per year in a country, and reflects the comparative risk to a population originating in different foods.

Risk estimates ranged from 1 case per 20 million servings for smoked fish to 4 cases per 100 000 million servings for fermented meats, or from 9 cases per 10 million consumers per year for pasteurized milk to 7 cases per 100 000 million consumers per year for fermented meats.

4.1.2 Approaches taken

The risk characterization begins with the prevalence and concentration of *L. monocytogenes*, nominally at the point of completion of production or retail, in packages or containers of the selected RTE foods. Changes are followed in the pathogen population in contaminated product through to the point where the consumer eats a portion but the risk assessment does not consider cross-contamination. The aim is to simulate the prevalence and levels of *L. monocytogenes* in those portions. Ancillary information is simulated for the frequency of consumption and the annual number of meals in a large population of susceptible and non-susceptible people.

To characterize the risk to consumers, the variables that have to be considered in these exposure assessment examples include: *L. monocytogenes* prevalence and concentration in finished products; product formulation; growth and inactivation rates; period and temperature of storage; and national and regional consumption patterns. The aim of these examples is to illustrate the effect of (i) potential for *L. monocytogenes* growth; (ii) low contamination levels in products that do not permit growth of *L. monocytogenes*; (iii) long-term storage on *L. monocytogenes* concentration; (iv) consumption patterns and volumes on dose eaten; and (v) low prevalence, or low concentration, of contamination of product on the risk of listeriosis from RTE foods.

The foods were selected based on various criteria, to exemplify various issues and effects of factors such as: different food commodities; potential for growth or not during long-term storage; cold-chain integrity; inactivation processes (e.g. pasteurization); post-process contamination; expected high contamination load of final RTE foods; high consumption rates; and products in international trade.

In addition to being used to estimate the risk of listeriosis from various RTE foods, and to contribute to providing answers to the questions posed by CCFH, the examples chosen are used to illustrate approaches to estimation of the risk of foodborne microbial illness. Examples 1 and 2 illustrate, in detail, appropriate statistical approaches to modelling the risk of microbial foodborne illness, including a description of prevalence and concentration of contaminants, while examples 3 and 4 emphasize modelling of the microbial ecology of *L. monocytogenes* in foods. Both of these topics – the statistical aspects of modelling and the microbial ecology of *L. monocytogenes* in foods – have been discussed in detail earlier in this report.

4.1.3 Choice of example risk assessments

4.1.3.1 Example 1: Fluid milk

The criteria for choosing milk were that it is widely consumed and the source is from many local suppliers. The variables were the prevalence and concentration of contamination with *L. monocytogenes*, post-processing contamination and growth during consumer refrigeration, and consumption patterns. The aim is to illustrate the interactive effects on risk deriving from consumption levels, contamination levels, shelf-life, contamination rates per package, and effects of times between exposures from a single contaminated unit.

4.1.3.2 Example 2: Ice cream

The criteria for selecting ice cream were the fact that no growth should occur during storage life and that the product is eaten worldwide, with a high consumption rate, particularly for some immunocompromised persons. The variables were contamination levels and national and regional consumption rates. The aim is to illustrate the relative risk of low contamination in a non-growth-permissive product, i.e. to estimate whether ice cream represents an important potential source of risk of listeriosis.

4.1.3.3 Example 3: Semi–dry fermented meats

The criteria for selection of fermented meats were that they are frequently contaminated but do not support growth. These products are widely consumed around the world, with many different varieties. The purpose in this risk assessment is to illustrate the effects of product formulation on potential for growth and the subsequent risk, and to attempt to contrast this with the risk from RTE foods that do allow the growth of *L. monocytogenes*.

4.1.3.4 Example 4: Cold-smoked fish

The criteria for selecting cold-smoked, vacuum-packed fish were that it is frequently contaminated; its formulation, storage conditions and long shelf-life suggest potential for extensive *L. monocytogenes* growth; and there is extensive international trade in the product. Variables modelled include formulation of the product, contamination levels, time and temperature of storage, national and regional consumption data and, in particular, the complex microbial ecology of the product, including the effect of lactic acid bacteria on product shelf-life and potential for growth of *L. monocytogenes.* The aim of the assessment is to illustrate the effects of the interaction of patterns and volumes of consumption with contamination frequency and potential for growth in a long-shelf-life product.

4.1.4 Common elements used in risk assessments

4.1.4.1 Definition of risks that were calculated

Key elements of the exposure assessment are the probability of consuming the food and the levels of pathogen consumed on each eating occasion. The latter reflects the hazard identification, namely the risk arises from the acute hazard attributable to exposure to individual meals, rather than a chronic hazard from repeated exposure. Two measures are used to characterize the risk: the number of illnesses per 100 000 population per year, and the number of illnesses per 1 000 000 servings of the food.

In examples 1 and 2, the risk to "susceptible" populations and normal consumers was estimated separately, as described below (see Section 4.1.4.5 – Dose-response modelling). In examples 3 and 4, data to enable differentiation of consumption by these groups was not available. Development of the dose-response relationships for these two groups relied on epidemiological data that indicate that the susceptible population ranges from 15 to 20% of the total population, and that individuals within the susceptible population account for 80–98% of all cases of listeriosis. As such, calculation of the risk to each sub-population would only reflect the assumptions concerning their relative susceptibility (defined by the r-value used) and the proportion of the population that each group represents (also defined in the modelling as between 15 and 20% of the total population) and would not provide additional insight. Thus, in examples 3 and 4, the risk to the total population alone was estimated.

It is probable, though not certain, however, that a more precise estimate of risk is generated by calculating separately the risk outcome for the susceptible and non-susceptible populations, and then combining the estimates to obtain the final total population outcome.

4.1.4.2 Simulation modelling

Simulated results are, themselves, subject to uncertainty introduced by the modelling algorithms used to perform the computations. The extremely low probabilities associated with acquiring listeriosis as a result of consuming any single serving of food means the estimates from risk characterization are very sensitive to extreme values from input distributions (the right-hand tails of the distributions). Those values are infrequently sampled but, when sampled, greatly increase the risk estimate. To overcome this problem, models were simplified to reduce processing time so that more iterations could be performed, and more replicates of each simulation model run. Summary statistics of replicated runs of the models using different random seeds to initialize the software were used to describe some notion of that variability (A. Fazil, pers. comm., 2001; G. Paoli, pers. comm., 2001).

Simulations were done using Analytica™ 1.1.1, Analytica™ 2.0.1 or Analytica™ 2.0.5, software using Median Latin Hypercube sampling, generating random numbers using the Minimal Standard method (multiplicative congruential), with various random seeds. The seeds 203132, 6821, 113307, 651757, 201246, 421952, 323512, 71796, 311868, 300896, 197545, 496893, 692118, 726146, 242899 and 959784 were selected at random from a Uniform(0, 1 000 000). For examples 1 and 2, the simulations were run on a personal computer with a Pentium®III processor. For examples 3 and 4, the simulations were run using Analytica™ 1.1.1 on a Macintosh Powerbook G4 computer. Unless otherwise noted, each simulation involved 32 000 iterations.

4.1.4.3 Estimation of consumption

Two approaches were taken to estimate consumption. In examples 1 and 2, Canadian consumption data were used and enabled the differentiation of consumption patterns by age and gender for adults in that population. In examples 3 and 4, the approach taken was to attempt to estimate the risk to a consumer from any nation. Estimates of annual per capita consumption were derived from national consumption and national population estimates for five nations. This approach resulted in very coarse estimates of consumption, and did not allow differentiation of consumption by age or gender.

Relatively few countries collect information on consumption that is useful for risk assessment purposes, i.e. on a daily or per-serving basis; most databases are cumulative over a year for nutritional purposes. The Canadian Nutrition Surveys (CFPNS, 1992–1995) for pasteurized milk and ice cream were used because the exposure assessment working team members were more familiar with this set of data than others, and there was enough information to have distributions based on daily meal portions by gender and age. For both these products, however, the consumption by young children and teenagers, as well as those >74 years old, were not considered in the survey, despite those in these age ranges possibly being high consumers. The exposure assessment, therefore, is most meaningful for a Canadian adult situation, although many other countries probably have similar consumption patterns. This differs from the consumption data generated for cold-smoked fish and semi-dry fermented meats, where survey data from several countries were combined. These scenarios show two approaches to generating information on eating practices, one at a national level and one with a more global focus.

4.1.4.4 Temperature data

Several studies reporting temperatures of distribution, retail display and commercial or home storage are available (Willocx, Hendrickx and Tobback, 1993; Notermans et al., 1997; Sergeledis et al., 1997; O'Brien, 1997; Johnson et al., 1998; MLA, 1999). For simplicity, in the example studies reported here, all product temperature data were derived from Audits International (2000) survey of home refrigerators in the United States of America.

4.1.4.5 Dose-response modelling

The functional form for the dose-response relationship is Pr{illness|dose} = $1\text{-}e^{-r.dose}$. Two distributions for the r-value of the exponential model were used, for consumers of increased susceptibility and for healthy consumers, respectively. Uncertainty about the appropriate parameterization and variability across the population of interest in the response to the same *L. monocytogenes* dose (e.g. due to variability in individual consumers health status at any given time, the type of meal and factors that could affect the survival of *L. monocytogenes* during passage through the stomach, variability in virulence of strains of *L. monocytogenes*, etc.), a distribution of r-values for each subpopulation was generated from 5000 iterations of the dose-response model, following the procedure described in Sections 2.3 and 2.4. The dose-response distribution for individuals from the susceptible population is stochastically smaller than the dose-response distribution for individuals from the non-susceptible population.

In each iteration of the model, the calculated dose is combined with an estimate of the r-value from the distribution outlined above for either a susceptible or normal consumer. For examples 3 and 4, the models were constructed so that in 15–20% of iterations, an r-value was drawn from the distribution of r-values for a susceptible consumer, but in all other cases a value was selected from the r-value distribution for "normal" consumers. The dose-response model is then combined with serving size data, and the modelled contamination level data, to predict probability of illness from the serving in that iteration.

4.2 EXAMPLE 1. PASTEURIZED MILK

4.2.1 Statement of purpose

This pasteurized milk assessment begins with the prevalence and concentration of *L. monocytogenes*, nominally at retail, in packages or containers of this RTE product and traces growth of the pathogen population in contaminated product through to the point where the consumer drinks a portion. The aim is to simulate the prevalence and levels of *L. monocytogenes* in those portions that, along with serving sizes, determine the size of the dose of *L. monocytogenes* that a consumer might ingest. Ancillary information is simulated for the frequency of consumption and the annual number of servings in a large population of susceptible adults and non-susceptible adults. Among those annual servings are some contaminated milk portions, which might lead to illness, according to the hazard characterization. The situation modelled is based upon Canadian data and practices.

4.2.2 Hazard identification

L. monocytogenes is found throughout the farm environment and can be transmitted to cows through consumption of silage and hay (Farber and Peterkin, 2000; Ryser, 1999a). The pathogen can also cause mastitis that allows the organism to be continually excreted into milk. It has frequently been isolated from milking barns and parlours and from dairy processing equipment. It is therefore not surprising that it has been found in raw milk around the world. It has been implicated in one outbreak of listeriosis attributed to pasteurized milk and another attributed to chocolate milk. In 1983, in Massachusetts, 49 people suffered from listeriosis after consuming one brand of 2% fat pasteurized milk (Fleming et al., 1985). The milk came from several farms, one of which had animals with bovine listeriosis at the time of the outbreak. The milk was apparently properly pasteurized, which indicates there was such a high level of contamination in the milk that some organisms survived the pasteurization or, more likely, post-process contamination occurred in the plant. In 1994, in the midwest United States of America, 54 people at a summer picnic developed gastroenteritis following consumption of chocolate milk in cartons that were later found to contain up to 10^9 *L. monocytogenes* CFU/ml (Dalton et al., 1997; Ryser, 1999a). Again, post-process contamination and storage for at least 2 hours at ambient temperatures was the most likely scenario.

4.2.3 Exposure assessment results

4.2.3.1 Prevalence of L. monocytogenes *at retail in pasteurized milk*

Prevalence at retail is based on 10 separate prevalence estimates for *L. monocytogenes* in pasteurized bovine milk produced in various countries, from retail or distribution, in packaged amounts. *L. monocytogenes* prevalence ranged from 0 to 1.1% of samples, in studies reporting from 14 to 1039 samples, 2157 samples in total (Table 4.1). Considered, but not included in the results, is the information from the study that reports only prevalence without noting also a sample size (in Baek et al., 2000). Also considered, but not included in the results, is the information from the studies that reported prevalence in samples drawn from bulk tanks of pasteurized milk. Samples drawn from bulk amounts were considered to represent prevalence of *L. monocytogenes* in pasteurized milk, but at a stage in the production-to-consumption chain earlier than the starting point used here. Among samples drawn from bulk tanks, prevalence of contamination was also generally very low, except for one study that found contamination in 21.4% of samples (Fleming et al., 1985; Fernandez-Garayzabal et al., 1986; Venables, 1989; Destro, Serrano and Kabuki, 1991; Harvey and Gilmour, 1992; Moura, Destro and Franco, 1993; Pitt, Harden and Hull, 1999). The stochastic structure of the collection of studies presented in Table 4.1 is represented by attributing binomial variability to the within-study estimates to account for their individual precision, and attributing a Beta distribution to the between-study variability of the true study prevalences, π_i, from data y_i of n_i samples positive for *L. monocytogenes* in the ith study, a two-stage hierarchical model $Y_i|n_i,\pi_i \sim$ Binomial(n_i,π_i), i=1, …, 9 and $\pi_i \sim$Beta(α,β). This leads to the inference that average prevalence is 3.50×10^{-3} [4.39×10^{-4}, 3.87×10^{-3}] at the 95% confidence interval when maximum likelihood estimates are $\hat{\alpha}$=0.55 and $\hat{\beta}$=155.47 (Figure 4.1).

4.2.3.2 Concentration of L. monocytogenes *in contaminated milk at retail*

No studies that described *L. monocytogenes* concentrations in pasteurized milk samples were found. This assessment relies on information summarized in FDA/FSIS (2001) (Table 4.2a), assumes that these are concentrations as if measured at retail, and constructs a distribution with estimated minimum and maximum concentrations (Table 4.2b). Minimum concentration in positive samples was assumed to be 0.04 CFU/ml and maximum concentration was assumed to be 250 CFU/ml, based on the authors' judgment. Variability in *L. monocytogenes* concentrations in contaminated pasteurized milk, at retail, is constructed by simulating concentrations in [0.04, 250] CFU/ml, assuming that concentrations are block Uniform between the $\log_{10}$ quantiles in Table 4.2b.

Table 4.1 Data sets used to estimate prevalence of *Listeria monocytogenes* in pasteurized milk, at retail.

Food	Stage	Country of study	Positive	Samples	Proportion +ve	Source
Pasteurized milk	Retail	Brazil	0	20	0	[1]
Pasteurized milk	Retail	Canada	0	14	0	[2]
Pasteurized milk	NA	Germany	0	651	0	[3]
Pasteurized milk	Retail	Korea	0	26	0	[4]
Pasteurized milk	Retail	Poland	0	73	0	[5]
Cow cream pasteurized	Retail	UK	0	40	0	[6]
Pasteurized milk	Retail	UK	11	1 039	0.011	[6]
2% low-fat milk	Retail or distribution	USA	0	125	0	[7]
Whole milk	Retail or distribution	USA	1	169	0.006	[7]
Pasteurized milk	Retail	Japan	NA	NA	0.009	[6] [8]

NOTES: NA = not available

SOURCES: [1] Casarotti, Gallo and Camargo, 1994. [2] Farber, Sanders and Johnston, 1989. [3] Hartung, 2000. [4] Baek et al., 2000. [5] Rola et al., 1994. [6] Greenwood, Roberts and Burden, 1991. [7] US FDA, 1987, (cited in Hitchins, 1996). [8] MacGowan et al., 1994 (cited in Baek et al., 2000).

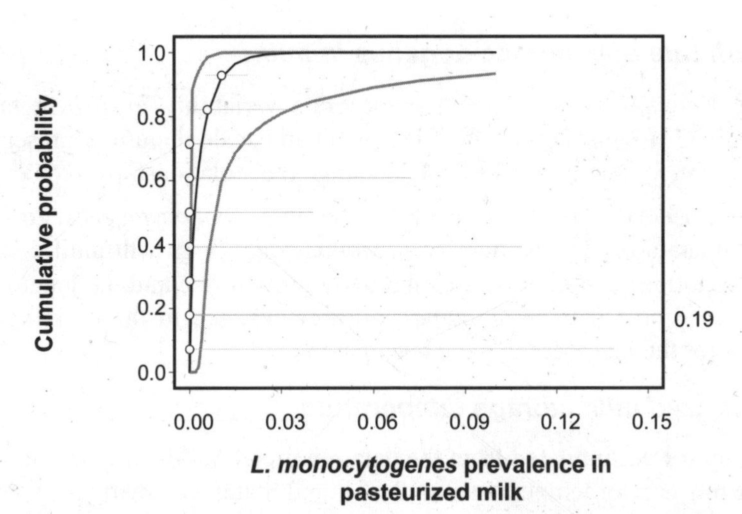

Figure 4.1 Empirical cumulative distribution function (open circles with 95% confidence intervals) for Table 4.1's individual studies' prevalence estimates, and fitted Beta distribution (solid line with shaded lines for 95% confidence limits) show the plot of the 2-stage hierarchical model that combined individual studies' estimates into a single estimate for the distribution of the prevalence of *Listeria monocytogenes* in pasteurized milk.

Table 4.2 *Listeria monocytogenes* concentration in contaminated pasteurized milk.

4.2a Data set used to estimate *L. monocytogenes* concentration in contaminated pasteurized milk.

Data set	1 CFU/ml	$\leq 10^2$ CFU/ml	Samples
FDA/FSIS, 2001	39	2	41

4.2b Assumed distribution function for *L. monocytogenes* concentration in contaminated pasteurized milk, with estimated minimum (0.04 CFU/ml) and maximum (250 CFU/ml) concentration.

CFU/ml	$\log_{10}$ CFU/ml	Cumulative probability [95% confidence interval]
0.04	-1.4	0
1	0	0.935 [0.835, 0.994]
100	2	0.984 [0.914, 1]
250	2.4	1

4.2.3.3 Growth of L. monocytogenes *in milk*

L. monocytogenes in pasteurized milk can grow at refrigerator temperatures, increasing the concentration in milk at the point of consumption from the concentration that is observed at retail. Although other conditions also explicitly define boundaries between growth and no growth and parameterize the growth rate, this exposure assessment has accounted for only the effect of storage temperature and the length of time that the product is stored before consumption.

4.2.3.4 Growth rate of L. monocytogenes *in milk*

Simulations of the exposure assessment incorporate variability in growth rates at 5°C as Uniform(0.092, 0.434) $\log_{10}$/day (FDA/FSIS, 2001) and scale them to represent growth rates at the storage temperatures in Table 4.3 using the relationship $\sqrt{\mu_T} = \sqrt{\mu_5}(T - T_{\min})(5 - T_{\min})^{-1}$ (McMeekin et al., 1993) to incorporate variability in *L. monocytogenes* growth associated with storage temperature. The amount of *L. monocytogenes* growth until consumption of a pasteurized milk portion is the product of the daily growth rate and the length of the storage time. It is assumed also that the *L. monocytogenes* detected in the milk were in the milk sufficiently long for the lag phase to have been passed.

4.2.3.5 Pasteurized milk storage temperature

Storage temperature was simulated from the data reported (Audits International, 2000) from a survey of home refrigerator temperatures in the United States of America (Table 4.3).

Table 4.3 Selected quantiles (1%, 5%, 50%, 95% and 99% points) from simulated distributions of characteristics controlling *Listeria monocytogenes* growth.

Storage temperature (Audits International, 2000)		**Storage time** (FDA/FSIS, 2001) truncated to respect pasteurized milk storage life (Neumeyer, Ross and McMeekin, 1997; Neumeyer et al., 1997)	
Cumulative probability	Storage temperature quantile (°C)	Cumulative probability	Storage time quantile (days)
0.01	0.06	0.01	1.58 (0.0002 s.e.)
0.05	0.53	0.05	2.31 (0.0002 s.e.)
0.50	3.41	0.50	5.29 (0.0003 s.e.)
0.95	6.88	0.95	10.47 (0.0016 s.e.)
0.99	8.61	0.99	12.68 (0.0026 s.e.)

4.2.3.6 Pasteurized milk storage time

The number of days that the consumer stores pasteurized milk before consumption is described by a Triangular(1, 5, 12) distribution, nominally, allowing the most likely value to vary as Uniform(4, 6) and the maximum value to vary as Uniform(6, 18) (FDA/FSIS, 2001), but restricting storage time to be shorter than storage life. Storage life for pasteurized milk depends on the growth of spoilage bacteria, which is assumed to be 12 days at 4°C, with storage life at other temperatures determined by the relationship $Life(T) = 12 \times \left[\frac{4+7.7}{T+7.7}\right]$ (Neumeyer, Ross and McMeekin, 1997; Neumeyer et al., 1997). Quantities vary among 16 simulations, each involving 32 000 iterations from the input distributions (Table 4.3).

4.2.3.7 Concentration of L. monocytogenes *in contaminated milk at consumption*

Concentrations in contaminated milk at retail (Table 4.2) increase due to growth of *L. monocytogenes* during the storage time and under the temperature conditions modelled (Table 4.3), leading to a simulated distribution of *L. monocytogenes* concentrations at the point of consumption (Table 4.4). Growth was assumed to occur when simulated storage temperatures exceeded a minimum temperature that varied from iteration to iteration (Uniform(-2°C, -1°C)). Maximum population densities are modelled to depend on temperature, after FDA/FSIS (2001): <5°C, 10^7 CFU/g; 5°–7°C, $10^{7.5}$ CFU/g; >7°C, 10^8 CFU/g. Initial concentrations were low enough and growth rates were low enough that limits imposed by maximum population densities were seldom invoked in the simulations that gave the results in Table 4.4. Quantities in Table 4.4 vary among 16 simulations, each involving 32 000 iterations.

Table 4.4 Selected quantiles from simulated distributions of *L. monocytogenes* concentration in contaminated milk at point of consumption.

Quantile ($\log_{10}$ CFU/ml)	**Cumulative probability**
-1	0.011 (3.39×10^{-5} s.e.)
0	0.374 (8.38×10^{-5} s.e.)
1	0.771 (1.13×10^{-4} s.e.)
2	0.914 (5.19×10^{-5} s.e.)
3	0.970 (5.52×10^{-5} s.e.)
4	0.991 (2.46×10^{-5} s.e.)
5	0.9977 (2.05×10^{-5} s.e.)
6	0.9996 (6.67×10^{-6} s.e.)
7	0.99998 (1.82×10^{-6} s.e.)
8	1

Only servings from contaminated milk will contain any *L. monocytogenes* organisms. Furthermore, only some servings from a multiple-serving container of contaminated milk that contains very low levels of contamination will contain any of the pathogen. Assumptions about homogeneity or heterogeneity of the organisms in a contaminated foodstuff can have a great effect on the simulated results. Clustering of colonies of pathogens would introduce extra variability into the results (Haas, Rose and Gerber, 1999). Here, it is assumed that the organism is distributed homogeneously through the product in a way that counts of organisms in samples from the product would follow a Poisson distribution, but will ignore the small variations in the number of *L. monocytogenes* organisms present in servings drawn from a packaged product with the same average concentration. It is assumed also that all organisms present would be in a part of the milk that would be consumed.

4.2.3.8 Consumption characteristics for milk

Defining milk consumption

Selection of foods from Canadian Federal-Provincial Nutrition Surveys' (CFPNS, 1991–1995) databases reflects both consumption frequency and amount of milk consumed on eating occasions. Results are based on the reported consumption practices of the 12 089 consumers who were respondents to the Nutrition Surveys, among whom 8365 consumed milk. Milk consumption, except when the eating episode involved preparation like cooking, were aggregated from all of an individual's eating occasions on the same day for this representation of milk consumption, giving, for a day at random, the estimated fraction of the population who are milk consumers and the daily amount of milk consumed. When milk forms an ingredient in a serving, an appropriate fraction of the food to represent the amount of milk included was derived or estimated (Table 4.5). Preparations using powdered milk and foods that included milk as an ingredient but that were processed before reaching the consumer were specifically excluded.

Table 4.5 Food commodities used to describe milk consumption frequency and amount consumed.

Food code	Food name	Milk (%)	Eating occasions	Average serving (g) per occasion
432	Instant breakfast, made with whole milk	90%	1	327.9
546	Milk, whole, fluid, producer, 3.7% B.F.	100%	196	123.8
547	Milk, fluid, partly skimmed, 2% B.F.	100%	10225	122.2
548	Milk, fluid, partly skimmed with added milk solids, 2% B.F.	100%	2	38.4
549	Milk, fluid, partly skimmed, 1% B.F.	100%	2292	151.2
550	Milk, fluid, partly skimmed with added milk solids, 1% B.F.	100%	1	230.6
551	Milk, fluid, skim with added milk solids	100%	15	122.4
552	Milk, fluid, buttermilk, cultured	100%	34	222.5
558	Milk, fluid, chocolate, whole	100%	4	412.6
559	Milk, fluid, chocolate, partly skimmed, 2% B.F.	100%	226	370.3
563	Milk shake, chocolate, thick	50%	1	105.7
593	Milk, fluid, skim	100%	2388	151.9
600	Milk, fluid, whole, pasteurized, homogenized, 3.3% B.F.	100%	3808	98.9
2918	Chocolate syrup, unenriched, + whole milk	90%	1	432.3
4001	Milk; cow, chocolate drink, fluid, commercial, lowfat, 1% fat	100%	8	372.8
11742	Potatoes, mashed, home-prepared, +whole milk +butter	10%	18	13.8
11899	Milk, fluid, homogenized, triple-milk	100%	27	30.5

NOTE: B.F. = butterfat

The simulated distributions constructed for annual meals and daily consumption amounts with respect to the Age × Gender groups' contributions to non-susceptible and susceptible populations were defined similarly to Miller, Whiting and Smith (1997). For the fraction of individuals that possess the same age and gender characteristics among Canadian adults 18–74 years of age, 15% (3.3 million) would fall into the susceptible group and 85% (18.7 million) would fall into the non-susceptible group.

Annual milk servings

Uncertainty about the point estimates for the estimated fraction of the population who are milk consumers is described by attributing a beta distribution to the proportion of the sample respondents that would consume pasteurized milk on a random day.

The simulated distribution for the number of days per year with milk consumption can be attributed to the gender and age groups that make up those populations. It is assumed that the daily consumption probability is the same on every day of the year for individuals in the same Gender × Age group, whether the individuals are in the non-susceptible population or the susceptible population, that days are independent, and that binomial sampling can be used to represent day-to-day variability (Table 4.6).

Amounts of milk consumed

The distribution in Table 4.7 was constructed by sampling from the Nutrition Survey data within the Age × Gender groups defined, collecting results into simulated milk consumption amounts distributions for non-susceptible and susceptible populations. The distribution that was constructed represents the proportion of the population groups that constitute the non-susceptible and susceptible populations.

Table 4.6 Selected quantiles (1%, 5%, 10%, 25%, 50%, 75%, 90%, 95% and 99% points) from simulated distribution of annual days with milk consumption among all individuals in non-susceptible and susceptible adult populations in Canada.

Population	Cumulative probability								
	.01	.05	.10	.25	.50	.75	.90	.95	.99
Non-susceptible	2.3×10^9	2.8×10^9	3.1×10^9	3.5×10^9	4.0×10^9	4.5×10^9	4.9×10^9	5.2×10^9	5.6×10^9
Susceptible	3.5×10^8	4.5×10^8	5.1×10^8	6.1×10^8	7.2×10^8	8.2×10^8	9.0×10^8	9.4×10^8	1.0×10^9

Table 4.7 Selected quantiles (1%, 5%, 10%, 25%, 50%, 75%, 90%, 95% and 99% points) from simulated distribution of daily amount (g) of milk consumption among milk consuming individuals in non-susceptible and susceptible populations.

Population	Cumulative probability								
	0.01	0.05	0.10	0.25	0.50	0.75	0.90	0.95	0.99
Non-susceptible	5.3 g	15.4 g	20.7 g	61.8 g	185.0 g	365.9 g	671.1 g	889.2 g	1 363 g
Susceptible	5.3 g	15.5 g	30.9 g	62.0 g	182.7 g	335.4 g	519.5 g	686.7 g	1 011 g

4.2.3.9 Simulated L. monocytogenes *in contaminated pasteurized milk at consumption*

The simulated distribution for the number of *L. monocytogenes* in a contaminated pasteurized milk serving (Table 4.8) is constructed from the serving size (Table 4.7) and from the distribution of concentrations at point of consumption (Table 4.4). Quantities vary among 16 simulations, each involving 32 000 iterations.

Table 4.8 Selected quantiles from simulated distributions of $\log_{10}$ number of *Listeria monocytogenes* organisms in contaminated milk servings at point of consumption.

$\log_{10}$ CFU in serving	**Cumulative probability**	
	Non-susceptible population	Susceptible population
0	0.004 (2.06×10^{-5} s.e.)	0.003 (1.26×10^{-5} s.e.)
1	0.055 (5.74×10^{-5} s.e.)	0.046 (5.83×10^{-5} s.e.)
2	0.298 (1.07×10^{-4} s.e.)	0.304 (1.28×10^{-4} s.e.)
3	0.686 (8.31×10^{-5} s.e.)	0.701 (1.06×10^{-4} s.e.)
4	0.884 (7.10×10^{-5} s.e.)	0.890 (8.36×10^{-5} s.e.)
5	0.957 (5.94×10^{-5} s.e.)	0.960 (5.62×10^{-5} s.e.)
6	0.987 (3.55×10^{-5} s.e.)	0.988 (4.12×10^{-5} s.e.)
7	0.996 (1.87×10^{-5} s.e.)	0.997 (2.40×10^{-5} s.e.)
8	0.9992 (8.95×10^{-6} s.e.)	0.9993 (9.32×10^{-6} s.e.)
9	0.99987 (3.06×10^{-6} s.e.)	0.99988 (3.31×10^{-6} s.e.)
10	0.999994 (7.81×10^{-7} s.e.)	0.999996 (6.62×10^{-7} s.e.)
11	1	1

NOTE: s.e. = standard error of the mean.

4.2.4 Risk characterization

4.2.4.1 Annual illnesses per 100 000 population

The simulated distribution for the number of illnesses per year per 100 000 population (Table 4.9) is developed using the probability of illness from consuming a contaminated serving and the number of contaminated servings per year as intermediate calculations. The distribution of annual contaminated milk servings accounts for variability and uncertainty associated with the average prevalence of contaminated servings and the distribution for the number of annual milk servings (Table 4.6). Critical to the development of risk characterization measures is the mean value of that simulated distribution, for individuals from the non-susceptible population and for individuals from the susceptible population (G. Paoli, pers. comm., 2001). The distribution for the probability of illness from consuming a contaminated milk serving is constructed from the distribution for the number of *L. monocytogenes* organisms in a contaminated serving (Table 4.8) and the dose-response function described in Section 4.1.4, an output of the hazard characterization.

Results are reported separately for a susceptible and a non-susceptible adult population, and for a mixed (entire) adult population that consists of approximately 85% non-susceptible adults and 15% susceptible adults. Summary statistics for the simulated distribution of annual illnesses per 100 000 population vary as shown among 16 simulations, each involving 32 000 iterations from the input distributions.

Table 4.9 Selected quantiles (1%, 5%, 50%, 95% and 99% points) and distribution mean from simulated distributions for annual illnesses per 100 000.

Cumulative probability	Annual illnesses per 100 000 population		
	Non-susceptible population	Susceptible population	Mixed population
.01	0.000	0.002 (0.0005 s.e.)	0.001 (0.0002 s.e.)
.05	0.000	0.027 (0.0018 s.e.)	0.007 (0.0004 s.e.)
.50	0.01 (0.0003 s.e.)	0.22 (0.009 s.e.)	0.04 (0.0015 s.e.)
.95	0.05 (0.002 s.e.)	1.37 (0.055 s.e.)	0.27 (0.011 s.e.)
.99	0.17 (0.005 s.e.)	4.93 (0.222 s.e.)	1.25 (0.063 s.e.)
Mean	0.016 (0.0005 s.e.)	0.519 (0.0312 s.e.)	0.091 (0.0047 s.e.)

NOTE: s.e. = standard error of the mean.

Table 4.10 Mean values from simulated distributions for number of illnesses per 1 000 000 servings.

	Illnesses per 1 000 000 servings		
	Non-susceptible population	Susceptible population	Mixed population
Mean	0.001 (0.0001 s.e.)	0.022 (0.0009 s.e.)	0.005 (0.0002 s.e.)

NOTE: s.e. = standard error of the mean.

4.2.4.2 Illnesses per 1 000 000 servings

The simulated distribution for the number of illnesses per 1 000 000 servings (Table 4.10) is developed using the prevalence of contaminated servings and the probability of illness from consuming a contaminated serving for individuals from non-susceptible and susceptible populations as intermediate calculations. The resulting distribution is concentrated at less than one illnesses per 1 000 000 servings, sometimes beyond the 99th percentile. Only mean values for the distributions are quoted for the results. Values vary as shown among 16 simulations, each involving 32 000 iterations from the input distributions.

4.2.5 Uncertainty and variability

A last step in this assessment for *L. monocytogenes* in milk examines the simulated results to consider how much the various inputs affect the outputs. As they are based on a simulation model, the risk characterization results are subject to uncertainty associated with a modelled representation of reality, involving assumed simple relationships among prevalence, concentration, consumption characteristics and adverse response to consumption of some number of *L. monocytogenes* organisms.

4.2.5.1 Effects of hazard characterization's dose-response

There is uncertainty in the hazard characterization's dose-response associated both with the form of the dose-response function used and with the parameterization. Describing distributions for the parameters captures how the response varies among individuals in a sub-population to the same pathogen dose. However, there is also uncertainty associated with the values assumed for the parameters.

4.2.5.2 Effects of estimated consumption frequency

Simulated milk consumption frequency is sensitive to the survey estimates of consumption frequency. Sample sizes, though, are large enough for the amount of uncertainty associated with the point estimate to have only a minor influence. Among individuals from the non-susceptible population, defined to include only males and females less than 65 years of age, consumption frequency differences are small. So, simulated consumption frequency for the non-susceptible population is not sensitive to allocation of individuals based on gender and age. Consumption frequency in the susceptible population is sensitive to changes to the gender and age composition. Individuals from the 65–74-year-old age group dominate the characteristics of the susceptible population, but estimates of milk consumption probabilities are less precise than in other age groups. Therefore the uncertainty that would be associated with those estimates plays a more significant, although still minor, role. There is uncertainty associated with extrapolation of daily consumption characteristics to annual consumption for populations of individuals. There is also uncertainty associated with extrapolation of survey results from 1991–1995 to the present day.

4.2.5.3 Effects of estimated consumption amounts

Simulated distributions for milk consumption amounts are less sensitive to composition of non-susceptible and susceptible populations than for other parameters. The age, more so than the gender, of individuals contributes more to variability in the non-susceptible population. Gender, more so than age, of individuals contributes to variability in simulated consumption amounts for the susceptible population. Estimates of milk consumption have uncertainties, including errors associated with under- and over-reporting, estimation methods for the amount of milk consumed, the use of several food codes, and the derivation or estimation of an appropriate amount of milk to include when the milk was an ingredient in the meal. All of a respondent's identified milk amounts within a day were aggregated into a daily amount. That practice loses the distinction that one might wish to make among different eating occasions within the day, whether the milk was consumed alone or as part of a meal, and whether the milk was consumed at home or away from home. However, the practice does retain the variability in milk amount consumed among individuals in the population. As with consumption frequency there is uncertainty associated with extrapolation of daily consumption characteristics to annual consumption for populations of individuals as well as uncertainty associated with extrapolation of survey results from 1991–1995 to the present day.

4.2.5.4 Effects of L. monocytogenes prevalence on risks of listeriosis

Simulated distributions for *L. monocytogenes* prevalence in the milk portions that consumers eat depend on estimates of prevalence of the pathogen in packages of milk, here assumed to have been measured at retail, from studies reported in the literature and on inferences from those data about the variability of prevalence. Sensitivity to prevalence of *L. monocytogenes* in pasteurized milk at retail is nearly multiplicative. If the mean prevalence is reduced by a factor of 10, then simulated annual illnesses per 100 000 population and simulated illnesses per 1 000 000 milk servings are also reduced by approximately a factor of 10. Risk characterization results (Tables 4.9 and 4.10) assume that *L. monocytogenes* prevalence estimates are appropriately pooled using a beta mixing distribution, yielding an inference that the average prevalence is 3.50×10^{-3} [4.39×10^{-4}, 3.87×10^{-3}] 95% confidence interval. Alternatively, one can proceed under the assumption that all prevalence studies, regardless of source, have sampled the same phenomenon, namely a single, fixed prevalence, estimated to be 5.56×10^{-3} [2.88×10^{-3}, 9.70×10^{-3}] 95% confidence interval (12 samples positive for *L. monocytogenes* in 2157 samples). In either case, the inference describes uncertainty about the prevalence of *L. monocytogenes* contamination in milk in a large number of servings. In the second case, there is less uncertainty about the mean prevalence, leading to a simulated distribution for risk characterization measures like the number of annual illnesses per 100 000 population that is less dispersed about the distribution mean. Mean values of the risk characterization results, though, are not sensitive to the different inferences.

4.2.5.5 Effects of L. monocytogenes *concentration at retail*

Simulated distributions for *L. monocytogenes* concentration in contaminated milk at the point of consumption (Table 4.4) depend very little on initial *L. monocytogenes* concentration (Table 4.2) in contaminated milk or at retail purchase, but are greatly dependent on the estimated maximum concentration at retail.

Simulated distributions for *L. monocytogenes* concentration in contaminated pasteurized milk at consumption showed levels that exceed 10^2 CFU/g in a small fraction of cases (Table 4.2). A set of simulations were done, setting maximum concentrations of *L. monocytogenes* in contaminated product at retail to levels $<10^2$ CFU/g at retail and to levels up to 10^3 CFU/g, still subject to growth under the same storage time and temperature (Table 4.3), to compare risk characterization measures (Table 4.11). Values vary as shown among 16 simulations, each involving 32 000 iterations from the input distributions.

Table 4.11 Comparison of simulated mean annual illnesses per 100 000 population associated with *Listeria monocytogenes* concentrations in contaminated pasteurized milk at retail from distributions with different assumed truncation points.

4.11a Concentrations at retail in contaminated pasteurized milk truncated to be <100 CFU/ml.

	Annual illnesses per 100 000 population		
	Non-susceptible population	Susceptible population	Mixed population
Mean	0.006 (0.0003 s.e.)	0.153 (0.0065 s.e.)	0.028 (0.0011 s.e.)

4.11b Baseline case, concentrations at retail in contaminated pasteurized milk modelled to be [100, 250] CFU/ml in approximately 1.6% of cases.

	Annual illnesses per 100 000 population		
	Non-susceptible population	Susceptible population	Mixed population
Mean	0.016 (0.0005 s.e.)	0.519 (0.0312 s.e.)	0.091 (0.0047 s.e.)

4.11c Concentrations at retail in contaminated pasteurized milk modelled to be [100, 1000] CFU/ml in approximately 1.6% of cases.

	Annual illnesses per 100 000 population		
	Non-susceptible population	Susceptible population	Mixed population
Mean	0.023 (0.0012 s.e.)	0.681 (0.0218 s.e.)	0.121 (0.0035 s.e.)

NOTE: s.e. = standard error of the mean.

4.2.5.6 Effects of higher storage temperatures

Simulated distributions for *L. monocytogenes* concentration in contaminated pasteurized milk are subject to pathogen growth that is modelled to depend on storage time and temperature (Table 4.3). To examine the effect of storage conditions on contamination levels, sets of simulations were done where storage temperatures were increased and where storage times were increased.

Storage temperature was simulated from the data that Johnson et al. (1998) reported from a survey of home refrigerator temperatures in the United Kingdom (Table 4.12a). Storage times were defined as a nominally Triangular(1, Uniform(4,6), Uniform(6,18)) distribution and then truncated to represent the effects of spoilage of milk held at temperatures described in Johnson et al. (1998). Daily growth was defined as Uniform(0.092, 0.434) (log_{10}/day) at 5°C and adjusted to storage temperature. Total growth was constrained to respect maximum population densities at the storage temperatures, as explained earlier. Summary statistics for storage conditions and mean values (Table 4.12b) of simulated distributions for annual illnesses per 100 000 population vary as shown among 16 simulations, each involving 32 000 iterations.

Table 4.12 Comparison of mean annual illnesses per 100 000 population for pasteurized milk held at refrigerator storage temperatures simulated from different assumed distributions.

4.12a Selected quantiles (1%, 5%, 50%, 95% and 99% points) from simulated distributions of storage temperature and storage time.

Storage temperature (from Johnson et al., 1998)		**Storage time** (FDA/FSIS, 2001) truncated to respect pasteurized milk storage life	
Cumulative probability	Storage temperature quantile (°C)	Cumulative probability	Storage time quantile (days)
0.01	-0.1	0.01	1.54 (0.0004 s.e.)
0.05	1.7	0.05	2.21 (0.0004 s.e.)
0.50	6.2	0.50	4.91 (0.0004 s.e.)
0.95	8.5	0.95	8.57 (0.0018 s.e.)
0.99	10.3	0.99	11.07 (0.0056 s.e.)

4.12b Mean values from simulated distribution for annual illnesses per 100 000 population, with scenario of warmer storage temperatures compared to baseline case.

1. Storage temperatures from Johnson et al., (1998)

	Annual illnesses per 100 000 population		
	Non-susceptible population	Susceptible population	Mixed population
Mean	0.23 (0.012 s.e.)	6.41 (0.252 s.e.)	1.15 (0.045 s.e.)

2. Baseline case, storage temperatures from Audits International (2000)

	Annual illnesses per 100 000 population		
	Non-susceptible population	Susceptible population	Mixed population
Mean	0.016 (0.0005 s.e.)	0.519 (0.0312 s.e.)	0.091 (0.0047 s.e.)

NOTE: s.e. = standard error of the mean.

4.2.5.7 Effects of longer storage times

Storage temperature was simulated from the data that were reported (Audits International, 2000) from a survey of home refrigerator temperatures in the United States of America (Table 4.13a). Storage times defined as a nominally Triangular(1, Uniform(4, 6), Uniform(6, 18)) distribution were lengthened by 1 day and truncated to represent the effects of spoilage. Daily growth was defined as a Uniform(0.092, 0.434) distribution (log_{10}/day) at 5°C and adjusted to storage temperature to complete the specification of growth conditions, and total growth was constrained to respect maximum population densities at the storage temperatures. Summary statistics for storage conditions and mean values (Table 4.13b) of simulated distributions for annual illnesses per 100 000 population vary as shown among 16 simulations, each involving 32 000 iterations.

Table 4.13 Effects of changes to storage time distribution on risk characterization measures.

4.13a Selected quantiles (1%, 5%, 50%, 95% and 99% points) from simulated distributions of storage temperature distribution and storage time distribution.

Storage temperature (from Audits International, 2000)		**Storage time** (FDA/FSIS, 2001) truncated to respect pasteurized milk storage life lengthened by 1 day	
Cumulative probability	Storage temperature quantile (°C)	Cumulative probability	Storage time quantile (days)
0.01	0.06	0.01	3.07 (0.0006 s.e.)
0.05	0.53	0.05	3.78 (0.0008 s.e.)
0.50	3.41	0.50	6.68 (0.0010 s.e.)
0.95	6.88	0.95	11.66 (0.0039 s.e.)
0.99	8.59	0.99	13.78 (0.0063 s.e.)

4.13b Mean values from simulated distribution for annual illnesses per 100 000 population, with scenario of longer storage time distribution compared to baseline case.

1. Storage time lengthened by 1 day

	Annual illnesses per 100 000 population		
	Non-susceptible population	Susceptible population	Mixed population
Mean	0.073 (0.0073 s.e.)	0.950 (0.0573 s.e.)	0.204 (0.0115 s.e.)

2. Baseline case

	Annual illnesses per 100 000 population		
	Non-susceptible population	Susceptible population	Mixed population
Mean	0.016 (0.0005 s.e.)	0.519 (0.0312 s.e.)	0.091 (0.0047 s.e.)

NOTE: s.e. = standard error of the mean.

4.2.5.8 Effects of growth

Risk characterization measures depend markedly on the amount of growth of the pathogen populations before consumption. Estimated amount of growth is modelled simply as the product of the daily growth rate at the storage temperature and the number of days of storage. That amount of growth is constrained by the maximum population density, which is modelled as a deterministic function of the storage temperature, but only seldom invoked, within the conditions modelled here.

If held under conditions under which no growth of *L. monocytogenes* occurs, with the same prevalence and level of contamination at retail (Table 4.2) and with the same consumption characteristics (Tables 4.6 and 4.7) as in the other cases examined, the simulated annual illnesses per 100 000 population and the illnesses per 1 000 000 servings decrease (Table 4.14).

Table 4.14 Risk characterization results for the pasteurized milk example, assuming no growth of *Listeria monocytogenes* in contaminated product.

4.14a Annual illnesses per 100 000 population.

	Non-susceptible population	Susceptible population	Mixed population
Mean	1.26×10^{-5} (6.72×10^{-8} s.e.)	3.76×10^{-4} (1.58×10^{-6} s.e.)	6.68×10^{-5} (2.42×10^{-7} s.e.)

4.14b Illnesses per 1 000 000 servings.

	Non-susceptible population	Susceptible population	Mixed population
Mean	5.87×10^{-7} (3.14×10^{-9} s.e.)	1.72×10^{-5} (7.50×10^{-8} s.e.)	3.64×10^{-6} (1.41×10^{-8} s.e.)

NOTE: s.e. = standard error of the mean.

4.3 EXAMPLE 2. ICE CREAM

4.3.1 Statement of purpose

The ice cream assessment begins with an estimation of the prevalence and concentration of *L. monocytogenes*, nominally at retail, in packages or containers of this RTE product, thus simulating the prevalence and levels of *L. monocytogenes* in consumed portions. Growth of the pathogen population in contaminated ice cream does not occur. Ancillary information is simulated for the frequency of consumption and the annual number of servings consumed by a large population of susceptible adults and non-susceptible adults. Among those annual servings are some contaminated ice cream portions, as estimated in the exposure assessment phase, which might lead to illness, as defined in the hazard characterization.

4.3.2 Hazard identification

The raw ingredients of ice cream and the processing environment may contain *L. monocytogenes*, which has been found in frozen dairy products. In the United States of America, there have been many recalls of ice cream, ice milk, sherbet and ice cream novelties by the Food and Drug Administration in implementing its zero tolerance policy, at a cost of many millions of dollars. However, no illness has been conclusively linked with these types of products in that country (Ryser, 1999a). In 1986, the mother of an infected newborn had eaten ice cream sandwiches 3 days before delivery. In 1987, a cluster of 31 cases seemed to be epidemiologically linked to consumption of ice cream (Schwartz et al., 1989). In neither of these scenarios were any strains isolated from the implemented products. However, one case in an immunocompromised man was caused by *L. monocytogenes* serotype 4b infection arising from consumption of a commercially prepared ice cream in Belgium (Andre et al., 1990). The ice cream was found to contain 10^4 CFU/g, which probably arose because of post-pasteurization recontamination. The epidemiological and laboratory evidence indicates that contamination of ice cream occurs, but, with no opportunity for growth after production, levels are typically very low.

4.3.3 Exposure assessment results

4.3.3.1 Prevalence of L. monocytogenes *at retail*

For prevalence and concentration data, studies were selected based on the types of products – ice cream, ice cream mix and ice cream novelties – sampled from retail outlets, distribution centres or processing facilities. It is assumed that *L. monocytogenes* survives but does not grow at the temperatures appropriate for storing ice cream. So, any source of prevalence information, after final packaging of the product, should be appropriate for this exposure assessment. Thirteen studies contributed 24 separate prevalence estimates for *L. monocytogenes* contamination in ice cream. Extensive data are available from North America and Europe, but fewer studies have reported data collected from ice cream obtained in countries in Asia, Australia and South America. Prevalence estimates ranged from 0 to 8.3%, in studies involving from 5 to 48 520 samples; there were 191 461 samples in total (Table 4.15). Considered, but not included in the results, is the information from one study or data set, which reported prevalence but without stating a sample size (in Pitt, Harden and Hull, 1999). The study by Pitt, Harden and Hull (1999) gave a prevalence estimate of 0.139, which is the highest reported prevalence found in the literature (no other prevalence estimate exceeded 0.083). However, without knowing the sample size, it is difficult to know how much weight to give that individual point when determining an appropriate description of the variability in prevalence.

The stochastic structure of the collection of studies in Table 4.15 is represented by attributing binomial variability to the within-study estimates to account for their individual precision and attributing a Beta distribution to the between-study variability of the true study prevalences, π_i, from data y_i of n_i samples positive for *L. monocytogenes* in the ith study, giving a two-stage hierarchical model $\boldsymbol{Y}_i|n_i,\pi_i \sim$ Binomial(n_i,π_i), $i = 1, \ldots, 24$ and $\pi_i \sim$Beta(α,β). This leads to the inference that average prevalence is 1.75×10^{-2} [8.31×10^{-3}, 0.042] at the 95% confidence interval when maximum likelihood estimates are $\hat{\alpha} = 0.42$ and $\hat{\beta} = 23.86$ (Figure 4.2).

Table 4.15 Data sets used to estimate prevalence of *Listeria monocytogenes* in ice cream.

Food	Stage	Country of study	Positive	Samples	Fraction	Ref.
Ice cream	NA	Austria	0	5	0	[1]
Ice cream	Retail	Canada	1	394	0.003	[2]
Ice cream mix	Retail	Canada	0	85	0	
Ice cream novelties	Retail	Canada	1	51	0.020	
Ice cream	Processing	Finland	4	603	0.007	[3]
Ice cream	Processing	Finland	0	188	0	
Ice cream	Processing	Finland	2	264	0.008	
Ice cream	Processing	Finland	0	74	0	
Ice cream	NA	Germany	1	2490	4.02×10^{-4}	[4]
Ice cream	NA	Germany	1	43	0.023	[5]
Ice cream, parfait	Retail or consumption	Hungary	1	15	0.067	[6]
Ice cream	Distribution	Korea	8	132	0.061	[7]
Ice cream (18 ewe milk; 1 goat milk; 131 cow milk)	Retail	UK	3	150	0.020	[8]
Ice cream	Retail	USA	23	659	0.035	[9]
Ice cream novelties	Retail	USA	29	351	0.083	
Ice milk	Retail	USA	0	42	0	
Ice cream	NA	USA	6	231	0.026	[10
Ice cream novelties	NA	USA	10	145	0.069	
Ice cream	Processing	various	48	48520	0.001	[11]
Ice cream	Processing	various	33	36661	0.001	
Ice cream	Processing	various	10	32078	3.12×10^{-4}	
Ice cream	Processing	various	11	36873	2.98×10^{-4}	
Ice cream	Processing	various	13	31407	4.19×10^{-4}	
Chocolate ice cream	NA	Australia	NA	NA	0.139	[12]

NOTE: NA = not available.

SOURCES: [1] From data submitted to FAO/WHO by the Austrian authorities, March 2000. [2] Farber, Sanders and Johnston, 1989. [3] Miettinen, Bjorkroth and Korkeala, 1999. [4] Hartung, 2000. [5] Steinmeyer and Terplan, 1990, cited in Klein, 1999. [6] Kiss et al., 1996. [7] Baek et al., 2000. [8] Greenwood, Roberts and Burden, 1991. [9] Kozak et al., 1996, citing unpublished 1987 data of Kozak. [10] US FDA, 1987, cited in Hitchins, 1996. [11] ICD, 2000. [12] [author not given] cited in Pitt, Harden and Hull, 1999.

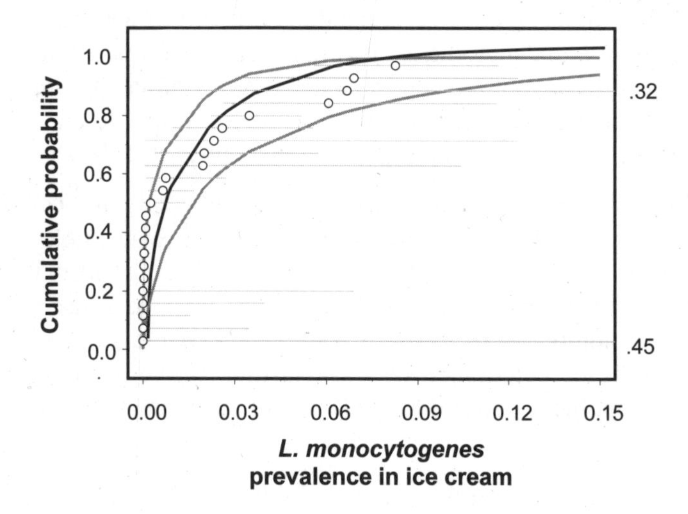

Figure 4.2 Empirical cumulative distribution function (open circles with 95% confidence intervals) for individual study prevalence estimates in Table 4.15, and fitted Beta distribution (solid line with shaded lines for 95% confidence limits) describe 2-stage hierarchical model for combining individual studies' estimates into an estimate for *Listeria monocytogenes* prevalence in ice cream.

4.3.3.2 *Concentration of* L. monocytogenes *at retail*

Kozak (1996, citing unpublished 1987 data from Kozak) provides the only information found to describe the concentration[3] of *L. monocytogenes* in contaminated ice cream (Table 4.16a). These data were used to construct a distribution with estimated minimum and maximum concentrations. Minimum concentration in positive samples was assumed to be 0.04 CFU/g and maximum concentration was assumed to be 100 CFU/g, based on the authors' judgment (Table 4.16b). Variability in *L. monocytogenes* concentrations in contaminated ice cream, at retail, was constructed by simulating concentrations in [0.04, 100] CFU/g, assuming that concentrations are block Uniform between the $\log_{10}$ quantiles in Table 4.16b.

4.3.3.3 *Growth of* L. monocytogenes *in ice cream*

No growth or die-off is modelled for *L. monocytogenes* in ice cream (FDA/FSIS, 2001).

4.3.3.4 *Consumption characteristics for ice cream*

Defining ice cream consumption

Selection of foods from Canadian Federal-Provincial Nutrition Surveys (CFPNS, 1992–1995) databases was intended to reflect both consumption frequency and the amount of ice cream consumed on eating occasions. Results are based on the reported consumption practices of

---

3. A comment on a late draft of this example exposure assessment pointed to further information in Stainer and Maillot (1996), which has not been incorporated here.

the 12 089 consumers who were respondents to the Nutrition Surveys, among whom 1409 consumed ice cream. Ice cream consumptions were aggregated from all an individual's eating occasions on the same day to give, for a single day at random, the estimated fraction of the population who consume ice cream and the daily amount of ice cream consumed. When ice cream was reported as an ingredient in a meal, an appropriate fraction of the food to represent the amount of ice cream included was derived or estimated. Some foods were used as surrogates for the amount of ice cream consumed in a serving, to enrich the database, but were not used to estimate frequency of ice cream consumption (Table 4.17).

The simulated distributions constructed for annual meals and daily consumption amounts respect the Age × Gender groups' contributions to a non-susceptible and a susceptible population defined as in Miller, Whiting and Smith (1997), attributing a fraction of individuals that possess the same age and gender characteristics to a susceptible population so that, among Canadian adults 18–74 years of age (71.5% of the population), for whom the consumption characteristics determined by the Nutrition Surveys apply, 15% (3.3 million) would fit into the susceptible group and 85% (18.7 million) would fit into the non-susceptible group.

Table 4.16 *Listeria monocytogenes* concentration in contaminated ice cream.

4.16a. Data set used to estimate *L. monocytogenes* concentration.

	<5 CFU/g	<15 CFU/g	Samples
Kozak, 1996	1	1	2

4.16b. Assumed cumulative distribution function for *L. monocytogenes* concentration in contaminated ice cream, with estimated minimum (0.04 CFU/g) and maximum (100 CFU/g) concentration.

CFU/g	$\log_{10}$ CFU/g	Cumulative probability [95% confidence interval]
0.04	-1.4	0
5	0.7	0.286 [0.013, 0.987]
15	1.18	0.714 [0.158, 1]
100	2	1

Table 4.17 Food commodities used to describe ice cream consumption frequency and amount consumed.

Food code	Food name	Respondent eating occasions	Average serving (g) per occasion
536	Ice cream, vanilla, regular, hardened, 10% B.F.[(1)]	1 184	79.1
537	Ice cream, vanilla, rich, hardened, 16% B.F.	83	81.1
538	Ice milk, vanilla, hardened or soft serve	143	118.3
539	Sherbet, orange	24	96.2
563	Milk shake, chocolate, thick [(2)]	1	105.7
633	Yoghurt, frozen	56	111.1
11847	Light ice cream, vanilla, hardened, 7% B.F.	17	72.5
11848	Light ice cream product, vanilla, hardened, 1% B.F.	13	109.2

NOTE: (1) B.F. = butter fat. (2) Assumes 50% ice cream.

Annual ice cream servings

Uncertainty about the point estimates for the estimated fraction of the population who consume ice cream is described by attributing a beta distribution to the proportion of the sample respondents that would consume ice cream on a random day.

The simulated distribution for the number of days per year with ice cream consumption adds up the days per year with ice cream consumption in the gender and age groups that make up those populations. It is assumed that the daily consumption probability is the same on every day of the year for individuals in the same Gender × Age group, whether the individuals are in the non-susceptible population or the susceptible population, that days are independent, and that binomially sampling can be used to represent day-to-day variability (Table 4.18).

Amounts of ice cream consumed

The distribution in Table 4.19 was constructed by sampling from the Nutrition Survey data for Age × Gender groups defined, and collecting results into simulated ice cream consumption amounts distributions for non-susceptible and susceptible populations. The simulated distribution respects the gender and age proportions that make up the non-susceptible and susceptible populations.

Table 4.18 Selected quantiles (1%, 5%, 10%, 25%, 50%, 75%, 90%, 95% and 99% points) from simulated distribution of annual days with ice cream consumption among all individuals in non-susceptible and susceptible adult populations in Canada.

Population	Cumulative probability								
	0.01	0.05	0.10	0.25	0.50	0.75	0.90	0.95	0.99
Non-susceptible	7.4×10^8	1.0×10^9	1.2×10^9	1.5×10^9	1.9×10^9	2.2×10^9	2.6×10^9	2.9×10^9	3.3×10^9
Susceptible	1.2×10^8	1.7×10^8	2.0×10^8	2.7×10^8	3.5×10^8	4.5×10^8	5.3×10^8	5.9×10^8	6.9×10^8

Table 4.19 Selected quantiles (1%, 5%, 10%, 25%, 50%, 75%, 90%, 95% and 99% points) from simulated distribution of daily amount (g) of ice cream consumption among individuals in non-susceptible and susceptible adult populations in Canada.

Population	Cumulative probability								
	0.01	0.05	0.10	0.25	0.50	0.75	0.90	0.95	0.99
Non-susceptible	8.5 g	19.1 g	33.2 g	46.9 g	75.4 g	130.3 g	168.8 g	210.3 g	335.8 g
Susceptible	8.4 g	16.9 g	28.1 g	38.7 g	66.5 g	102.8 g	133.0 g	152.4 g	266.1 g

L. monocytogenes *in contaminated ice cream serving*

The simulated distribution for the number of *L. monocytogenes* organisms in a contaminated ice cream serving (Table 4.20) is constructed from the concentration (Table 4.16) and serving size (Table 4.19) distributions. Quantiles vary as shown among 16 simulations, each involving 32 000 iterations from the input distributions.

Only servings from contaminated ice cream will contain any *L. monocytogenes* organisms. Only a fraction of servings that contain very low levels of contamination will contain any of the pathogen. Assumptions about homogeneity or heterogeneity of the organisms in a contaminated foodstuff can have a great effect on the simulated results. Clustering of colonies of pathogens would introduce extra variability into the results (Haas, Rose and Gerber, 1999). Here, it is assumed that the organism is distributed homogeneously throughout the product in a Poisson distribution, but small variations in the number of *L. monocytogenes* organisms present in servings drawn from a packaged product with the same average concentration are ignored. It is assumed also that all organisms present would be in a part of the ice cream that would be consumed.

4.3.4 Risk characterization

4.3.4.1 Annual illnesses per 100 000 population

The simulated distribution for the number of illnesses per year per 100 000 population (Table 4.21a) is developed using the distribution for the probability of illness from consuming a contaminated serving and the distribution for the number of contaminated servings per year as intermediate calculations. The distribution of annual contaminated ice cream servings includes variability and uncertainty associated with the distribution for the average prevalence of contaminated servings and the distribution for the number of annual ice cream servings (Table 4.18). Critical to the development of risk characterization measures is the mean value of that simulated distribution for individuals from the non-susceptible population and individuals from the susceptible population (G. Paoli, pers. comm., 2001).

Table 4.20 Selected quantiles from simulated distributions of $\log_{10}$ number of *Listeria monocytogenes* organisms in contaminated ice cream servings at point of consumption.

Quantile ($\log_{10}$ CFU in serving)	Cumulative probability	
	Non-susceptible population	Susceptible population
0	7.23×10^{-4} (1.04×10^{-5} s.e.)	8.36×10^{-4} (7.99×10^{-6} s.e.)
1	0.019 (3.40×10^{-5} s.e.)	0.022 (4.22×10^{-5} s.e.)
2	0.147 (6.73×10^{-5} s.e.)	0.163 (7.93×10^{-5} s.e.)
3	0.630 (1.04×10^{-4} s.e.)	0.683 (7.98×10^{-5} s.e.)
4	0.993 (2.68×10^{-5} s.e.)	0.996 (2.27×10^{-5} s.e.)
5	1	1

NOTE: s.e. = standard error of the mean.

The distribution of the risk characterization result is concentrated at nil illnesses per 100 000 population, up to beyond the 99th percentile. Mean values for the distributions are quoted for the results. Results in Table 4.21a are reported separately for a susceptible and a non-susceptible adult population, and for a mixed (total) adult population that consists of approximately 85% non-susceptible adults and 15% susceptible adults. Summary statistics for the distributions vary as shown among 16 simulations, each involving 32 000 iterations from the input distributions.

4.3.4.2 Illnesses per 1 000 000 servings

The simulated distribution for the number of illnesses per 1 000 000 servings (Table 4.23b) is developed from the average prevalence of contaminated servings and the probability of illness from consuming a contaminated serving. The distribution of the risk characterization result is concentrated at less than one illness per 1 000 000 servings, up to beyond the 99th percentile. Mean values for the distributions are quoted for the results. Results for a non-susceptible and for a susceptible population vary as shown among 16 simulations, each involving 32 000 iterations from the input distributions.

Table 4.21 Risk characterization for ice cream.

4.21a Annual illnesses per 100 000 population

	Non-susceptible population	Susceptible population	Mixed population
Mean	2.10×10^{-5} (1.70×10^{-8} s.e.)	6.73×10^{-4} (4.24×10^{-7} s.e.)	1.18×10^{-4} (6.91×10^{-8} s.e.)

4.21b Illnesses per 1 000 000 servings.

	Non-susceptible population	Susceptible population
Mean	2.09×10^{-6} (1.19×10^{-9} s.e.)	6.08×10^{-5} (4.21×10^{-8} s.e.)

NOTE: s.e. = standard error of the mean

4.3.5 Uncertainty and variability

A last step in this assessment for *L. monocytogenes* in ice cream examines the simulation model to consider how much the various inputs affect the outputs. Based as they are on a simulation model, the risk characterization results are subject to uncertainty associated with a modelled representation of reality, involving assumed simple relationships among prevalence, concentration, consumption characteristics and adverse response to consumption of some number of *L. monocytogenes* organisms.

4.3.5.1 Effects of hazard characterization's dose-response

There is uncertainty in the hazard characterization's dose-response relationship used to relate the simulated distributions of the number of *L. monocytogenes* organisms in a serving to the measures that have been used to characterize the risk. There is uncertainty associated with the form of the dose-response function used and with the parameterization. Describing distributions for the parameters captures variability in the response to the same pathogen dose among individuals in a subpopulation. However, there is uncertainty associated with the distributions assumed for the parameters.

4.3.5.2 Effects of estimated consumption frequency

Simulated ice cream consumption frequency for non-susceptible and susceptible populations (Table 4.18) is sensitive to the survey estimates of consumption frequency. Sample sizes are large enough that the amount of uncertainty associated with the point estimate has only a minor influence. There is uncertainty due to extrapolation of those results to the present day. Further, consumption characteristics were derived for non-susceptible and susceptible individuals by imputing characteristics associated with age and gender, a source of uncertainty. There is uncertainty and variability associated with extrapolation of daily consumption characteristics to annual consumption for populations of individuals. There is uncertainty associated with extrapolation of survey results from 1991–1995 to the present day.

4.3.5.3 Effects of estimated consumption amounts

Simulated distributions for ice cream consumption amounts are less sensitive to how the composition of the non-susceptible and susceptible populations is defined. Generally, the gender and age of individuals in the non-susceptible and susceptible populations have only minor influence on the simulated distribution for the amounts of ice cream consumed. Ice cream consumption amounts have uncertainty, including errors associated with under- and over-reporting, estimation methods for the amount of ice cream consumed, the representation of ice cream consumption using several food codes and the derivation or estimation of an appropriate amount of ice cream to include when the ice cream was an ingredient in the meal. All of a respondent's identified ice cream amounts within a day were aggregated into a daily amount for the respondent. That practice loses the distinction that one might wish to make among different eating occasions within the day, whether the ice cream was consumed alone or as part of a meal and whether the ice cream was consumed at home or away from home. There is uncertainty and variability associated with extrapolation of daily consumption characteristics to annual consumption for populations of individuals. There is uncertainty associated with extrapolation of survey results from 1991–1995 to the present day.

4.3.5.4 Effects of L. monocytogenes *prevalence*

Simulated numbers of ice cream servings with any *L. monocytogenes* contamination are influenced by the number of servings that the population consumes and the prevalence of *L. monocytogenes* in packages of ice cream. Prevalence is sensitive to correct inclusion and exclusion of data sets from literature, government surveillance reports and industry (Table 4.15). Sensitivity to prevalence of *L. monocytogenes* in ice cream at retail is nearly multiplicative. If the prevalence is reduced by a factor of 10, then simulated annual illnesses per 100 000 population and simulated illnesses per 1 000 000 servings are also reduced by approximately a factor of 10.

In point of fact, risk characterization results are sensitive to the nature of the inference that one makes from Table 4.15's data concerning the prevalence in a large number of servings. Risk characterization results (Table 4.21) are based on the assumption that true *L. monocytogenes* prevalence estimates in individual studies or data sets follow a beta distribution, yielding an inference that average prevalence is 1.75×10^{-2} [8.31×10^{-3}, 0.042] at the 95% confidence interval. Alternatively, if one assumed that all prevalence studies have sampled the same phenomenon and pooled the studies' samples to provide an estimate for a single, fixed prevalence, then the inference about mean prevalence becomes 1.07×10^{-3} [9.29×10^{-4}, 1.23×10^{-3}] at the 95% confidence interval (205 samples positive for *L. monocytogenes* in 191 461 samples). Based on that inference about prevalence, risk characterization results, such as the number of annual illnesses per 100 000 population, are approximately 8% of the results in Table 4.21 (Table 4.22).

4.3.5.5 Effects of L. monocytogenes *concentration at retail*

Concentration of *L. monocytogenes* at consumption (Table 4.16) influences the simulated number of organisms in contaminated consumer servings (Table 4.20). Concentrations in consumer portions are simulated to be very low, but are based on data attributed to a single reference and are little influenced by departures from the estimated maximum concentration of 100 CFU/g in contaminated ice cream (Table 4.23).

Table 4.22 Comparison of simulated annual illnesses per 100 000 population under different inferences about the prevalence of *Listeria monocytogenes* in ice cream at retail.

4.22a Two-stage hierarchical model assuming Beta distribution as mixing distribution for prevalence estimates in the individual studies in Table 4.15.

	Annual illnesses per 100 000 population		
	Non-susceptible population	Susceptible population	Mixed population
Mean	2.10×10^{-5} (1.70×10^{-8} s.e.)	6.73×10^{-4} (4.24×10^{-7} s.e.)	1.18×10^{-4} (6.91×10^{-8} s.e.)

4.22b Pooled all studies' samples to estimate a single, fixed prevalence for all data sets in Table 4.15.

	Annual illnesses per 100 000 population		
	Non-susceptible population	Susceptible population	Mixed population
Mean	1.68×10^{-6}	5.54×10^{-5} (8.14×10^{-8} s.e.)	9.68×10^{-6} (1.21×10^{-8} s.e.)

NOTE: s.e. = standard error of the mean

Table 4.23 Annual illnesses per 100 000 population for ice cream under different inferences about the maximum concentration of *Listeria monocytogenes* in contaminated ice cream at retail.

4.23a Baseline case, assumed maximum *L. monocytogenes* concentration at retail = 100 CFU/g.

	Annual illnesses per 100 000 population		
	Non-susceptible population	Susceptible population	Mixed population
Mean	2.10×10^{-5} (1.70×10^{-8} s.e.)	6.73×10^{-4} (4.24×10^{-7} s.e.)	1.18×10^{-4} (6.91×10^{-8} s.e.)

4.23b Assumed maximum *L. monocytogenes* concentration at retail = 250 CFU/g.

	Annual illnesses per 100 000 population		
	Non-susceptible population	Susceptible population	Mixed population
Mean	2.67×10^{-5} (2.57×10^{-8} s.e.)	8.53×10^{-4} (8.79×10^{-7} s.e.)	1.50×10^{-4} (1.27×10^{-7} s.e.)

4.23c Assumed maximum *L. monocytogenes* concentration at retail = 1000 CFU/g.

	Annual illnesses per 100 000 population		
	Non-susceptible population	Susceptible population	Mixed population
Mean	4.19×10^{-5} (6.57×10^{-8} s.e.)	1.33×10^{-3} (1.93×10^{-6} s.e.)	2.34×10^{-4} (2.90×10^{-7} s.e.)

NOTE: s.e. = standard error of the mean

4.4 EXAMPLE 3. FERMENTED MEAT

4.4.1 Statement of purpose

This assessment aims to estimate the risk of listeriosis from fermented meat products (FMPs). In this assessment, fermented meats are taken to include those meat products in which reduction of water activity through addition of salt and drying and acidification due to the metabolic activity of microorganisms on added sugars are used to extend the shelf-life of meat. It does not consider risk in a specific nation, or specific regions within a nation, because consumption patterns vary widely, but it does attempt to calculate generic estimates of the risk of listeriosis per serving for any consumer of FMPs anywhere in the world. It should be noted that the data are representative of Western-style fermented meat products. The assessment begins with *L. monocytogenes* in fermented meat products after production.

4.4.2 Hazard identification

L. monocytogenes is widely distributed in raw meats. It has also often been detected in commercially produced FMPs (See Table A4.1 in Appendix 4). Investigation of an outbreak of listeriosis in Philadelphia in 1986/87 suggested that either ice cream or fermented meats were involved, based on the consumption records of victims (Schwartz et al., 1988, 1989). However, no documented cases of listeriosis have been directly attributed to FMPs (Lücke, 1995). Farber and Peterkin (1999) have reviewed the importance of *L. monocytogenes* in processed meats, including FMPs.

The acid tolerance of *L. monocytogenes* and its ability to grow at low water activity levels could allow survival or growth of the pathogen in fermented meat products. Many fermented meat products, however, do not support growth of *L. monocytogenes* in their final product form, although they can support its growth during the early stages of production.

4.4.3 Exposure assessment

4.4.3.1 Production and Consumption

Although this assessment is primarily concerned with estimating the *per serving* risk because of the paucity of national consumption data, available data describing national consumption of FMPs is presented in Table 4.24 below. The basis for the estimates is presented in Section A4.7 in Appendix 4.

Table 4.24 National population and national fermented meats consumption data used in the assessment.

Country	Population (million)[1]	Consumption (kg/person/year)	Number of 50-g servings per year
USA	271	0.295	6
Australia	19	0.4–1.68	8–34
Canada	31	0.912	18
Germany	81	0.723	14.5
Finland	5.2	3.1	62

NOTE: (1) Derived from NGS, 1999.

These data were modelled by a Triangular(6, 25, 62) distribution, empirically based on data in Table 4.24, to reflect the variability in mean national per capita consumption. While some reports indicate differences in frequency of consumption of FMPs by age and gender (e.g. CFPNS, 1992–1995; ABS, 1995), there is a lack of corresponding information on serving size. For that reason, and the inability to relate differences in susceptibility to listeriosis and the age-gender categories considered in those nutrition surveys, no attempt was made to incorporate these differences in this risk assessment.

4.4.3.2 Modelling exposure

Initial contamination

Initial contamination at production or retail was modelled using a discrete distribution (Analytica "Chancedist") based on the data presented in Section A4.1 of Appendix 4. The total number of positive results across all surveys is 13.65%. Conversely, 86.35% of samples were interpreted to have less than 1 CFU *L. monocytogenes* per 25 g, or, equivalently, <0.04 CFU *L. monocytogenes* per 1 g. Eight of the data sets included quantitative data that were used to estimate the proportion of all positive samples in three concentration ranges: 0.04 < X < 10 CFU/g; 10 < X < 100 CFU/g; and 100 < X < 10 000 CFU/g.

Other concentrations reported (see Section A4.1 of Appendix 4) had too few data to be used, or did not add any information because only two levels were specified. In these cases, the data were ascribed to the next highest concentration level with which they were consistent.

The distribution of final contamination levels at the point of production that was used in the model is shown in Table 4.25. In all cases, all samples in each range were presumed to be present at the highest level in that range – an inherently conservative decision. A non-conservative assumption, however, is also incorporated indirectly into the model because the contamination level data included data based on surveys conducted at retail. From the foregoing, it is anticipated that contamination levels will decline between the time of production and sampling at retail. The effects of this assumption are discussed in Section 4.4.5.

Table 4.25 Distribution of reported contamination levels of fermented meat products at retail.

% of samples in range[1]	Concentration of *L. monocytogenes*
86.35	<0.04 CFU/g
6.94	<10 CFU/g
6.10	<100 CFU/g
0.60	<10 000 CFU/g

NOTE: (1) Where several percentage data were available for a specific range, an average was calculated. Thus, the sum of the percentages ascribed to each range of "positive" results slightly exceeded the predicted 13.65%. Accordingly, each percentage value was adjusted in equal proportion so that the sum of the positive results was 13.65%.

Potential for Growth of L. monocytogenes *in Fermented Meat Products*

Product manufacture and composition

A description of fermented meats and their characteristics and methods of production is presented in Section A4.2 of Appendix 4.

FMPs have long shelf lives due to the combination of acidification (through fermentation of sugar added to the meat or to addition of an acidulant such as glucono-∂-lactone or encapsulated citric acid), removal of oxygen, and addition of compounds that favour the growth of desirable microbes while retarding the growth of others. Available water is typically limited through the addition of salt and the removal of water over an extended "maturation" period. These factors combine to produce products that are shelf stable and resistant to spoilage by bacteria.

Variables in the production of FMPs include:

- type of meat;
- amount of fat added;
- starter culture used (if used), and whether it produces bacteriocins;
- curing mix composition and concentration – nitrite or nitrate levels, salt concentration, spices, etc.;
- fermentation time and temperature;
- heating time and temperature (if applied);
- maturation time and temperature;
- sausage diameter;
- final pH;
- final water activity; and
- recommended storage temperatures.

The relevance of each of these variables to the microbiological safety of the product is discussed in Section A4.2 of Appendix 4.

Temperatures above 65°C are considered cooking temperatures. Such temperatures are listericidal, with D-values of a few minutes. A post-fermentation cooking step may be included with some FMPs; however, recontamination of the exterior of the product can occur.

Ecology of *L. monocytogenes* in uncooked fermented meat products.

Published literature sources indicate that *L. monocytogenes* does not grow in most FMPs once the fermentation is well underway and pH has fallen, nor does it grow during subsequent maturation (e.g. Schillinger, Kaya and Lücke, 1991; Campanini et al., 1993; Farber et al., 1993; Rödel, Stiebing and Kröckel, 1993; Samelis et al., 1998; Encinas et al., 1999; Laukova et al., 1999). Once fermentation has been established, *L. monocytogenes* is usually slowly inactivated as a result of the conditions present in shelf-stable FMPs. Some growth of *L. monocytogenes* may occur in the raw ingredients or during the initial phases of the fermentation, particularly if products are not inoculated with a starter culture (Campanini et al., 1993). Similarly, if fermentable carbohydrates are not added to the raw ingredients, the decline in pH and increase in salt concentration may be delayed. This could extend the time during which the product composition might allow growth of *L. monocytogenes*.

The extent of inactivation will depend on the time of storage, temperature and the characteristics of the FMPs, such as pH, organic acid concentration, salt concentration and presence of preservative compounds. When microorganisms cannot grow in an environment, they die at a rate governed by environmental factors, of which temperature appears to be most important (Buchanan et al., 1997; Ross and Shadbolt, 2001). The environmental limits to growth of *L. monocytogenes* are detailed in Table 3.1, and models describing them in Table A3.1. The final composition of an FMP dictates the survival of *L. monocytogenes* in the product during maturation and subsequent marketing. Composition and processing parameters for a variety of common FMPs types are shown in Appendix 4 (see Section A4.3). Each of those variables is regarded as contributing to the overall character of the product.

Change in contamination level

Gradual inactivation of *L. monocytogenes* is expected under conditions characteristic of mature FMPs, leading to an expected decline in *L. monocytogenes* levels during distribution and storage. The model of Tienungoon et al. (2000) for *L. monocytogenes* growth limits supports the belief that growth would not be expected in any FMPs that falls within the accepted pH and a_w specifications for stable FMPs (see Appendix 4, Section A4.3).

It was therefore assumed that growth of *L. monocytogenes* does not occur in the finished product. Conversely, it was assumed that inactivation would occur over time, therefore, inactivation of *L. monocytogenes* was modelled for the period that the product was held at retail and the period that the product was held in the consumer's home before consumption. It was also assumed that contamination data related to product sampled during retail storage.

Non-thermal inactivation model

The Buchanan, Golden and Phillips. (1997) model describing non-thermal inactivation of *L. monocytogenes* under reduced oxygen conditions and in response to temperature, water activity, pH, nitrite and salt concentration was selected. That model includes most factors considered relevant to inactivation of *L. monocytogenes* in FMPs, i.e. temperature (4–42°C), pH (3–7), lactic acid (0–2%), NaCl (0.5–19%) and sodium nitrate (0–200 µg/ml).

That model, however, predicts the time required under a given set of environmental conditions for a 10 000-fold reduction (t_{4D}) in *L. monocytogenes.* For the purpose of the current risk assessment, the t_{4D} value predicted by the model from the product composition data in each scenario was divided by four to generate the D-value (time for a 10-fold reduction). The storage times at retail and in the consumer's home were then divided by the modelled D-value to predict the $\log_{10}$ reduction in *L. monocytogenes* in the product at the point of consumption.

This simplification of the inactivation model of Buchanan, Golden and Phillips (1997) could be criticised because it could lead to an overprediction of inactivation. Buchanan, Golden and Phillips (1997) modelled t_{4D} rather than D-values because the inactivation kinetics observed in their experimental broth system were not always log-linear. However, Buchanan, Golden and Phillips (1997) compared the predictions of their model with published values for the inactivation of *L. monocytogenes* under analogous condition in foods. Model predictions typically over-estimated the observed t_{4D} by factors of 2–3, i.e. the predictions were inherently conservative, and it is probable that factors other than those included in the model are important in determining the rate of non-thermal inactivation. Evaluation of these values indicated that the simplification of the Buchanan, Golden and

Phillips (1997) model employed in the current assessment provides a reasonable estimate of the inactivation of *L. monocytogenes* in commercial FMPs.

Product composition data

The model of Buchanan, Golden and Phillips (1997) predicts the effect of salt concentration, pH, temperature, nitrite and undissociated lactic acid concentration on the inactivation of *L. monocytogenes*. Absolute ranges of those factors reported in FMPs were determined from the data presented in Appendix 4 (see Section A4.2), and approximations to the most probable values were made. The distributions used and their behaviour are shown in Table 4.26.

Table 4.26 Characteristics of distributions used to describe physico-chemical properties of fermented meat products that affect *Listeria monocytogenes* inactivation.

Characteristic of Variable Description		Product Parameter		
		pH	water activity[1]	nitrite (ppm)
Beta distribution		2.5, 6, 4.3, 6.6	15, 5, 0.73, 1.00	10, 90, 0, 200
Minimum estimate		4.31	0.811	3.5
Mean estimate		4.98	0.933	20
Maximum estimate		6.36	0.995	30.7
	5	4.49	0.887	11.1
Percentiles (estimated)	50	4.94	0.935	19.5
	95	5.6	0.97	53.8

NOTES: (1) a_w was converted to salt concentration using the following equation, obtained by fitting an empirical polynomial equation to a calibration curve of water activity and salt concentrations:

Salt concentration (% w/w) =√(((1000 × (1-Product water activity)+56.3904195) -7.50935546)/0.349337086)

Lactic acid concentrations were estimated from product pH data, as described in Appendix 4 (see Section A4.4).

Temperature and time of storage

Times and temperature of storage were divided into retail storage and home storage. It was assumed that the shelf-life of the product ranged between 1 and 180 days (FSIS, 1995; Ross et al., in press) using a Triangular(1, 60, 180) distribution. Some FMPs, e.g. soft spreadable types such as mettwurst, teewurst, or Braunschweiger, are considered to have a refrigerated shelf-life of only a few days. Other products, e.g. hard or dry sausages, have an indefinite unrefrigerated shelf-life. In the allocation of shelf-life, the relationship between product composition and shelf-life was not modelled.

It is assumed that the product is sold at any point in this shelf-life but, from the data shown in Appendix 4 (see Section A4.5), the probability is greatest that the sale will occur mid-way through the product's life, and no product was found available for retail sale with more than 66% of its shelf-life remaining. It is assumed that this period represents the time taken for product to reach point of sale. The data in Appendix 4 (Section A4.5) are modelled by a Triangular(33, 58, 100) distribution, representing the *percentage* of the entire shelf-life that the product is held at retail prior to sale.

The remaining shelf-life of the product is calculated from the difference between the total shelf-life and the shelf-life elapsed at retail. Of the shelf-life remaining available to the consumer, it was also considered more likely that the consumer would consume the product relatively quickly. Thus, the time of consumption was modelled by multiplying the remaining shelf-life by a Triangular(0.0, 0.1, 1.0) distribution.

The model traces the flow of an individual serving from retail to consumption. The resulting distribution is considered to be representative of a serving, not necessarily a package of servings.

Fermented meat products may be stored under refrigeration or, more traditionally, at ambient temperature. In the United States of America, many FMPs are stored, distributed and displayed at refrigeration temperature (for marketing and product quality reasons), but in Europe it is more usual for these products to be held at ambient temperature (B. Tompkin, pers. comm., 2001). It was assumed that storage at retail would be at ambient temperature in 30% of cases, but that 85% of consumers would store their FMPs in the refrigerator. Based on the data of Audits International (2000) for processed meats (sold at the delicatessen counter or pre-packaged luncheon meats), temperatures during retail storage were approximated by a Triangular(0, 6, 17) distribution and domestic refrigeration temperatures by a Triangular(-1, 5, 12) distribution. Ambient temperatures at retail or in the home were described by a Triangular(5, 20, 35) distribution.

The overall structure of the model describing inactivation during retail and home storage is shown as an influence diagram in Figure 4.3.

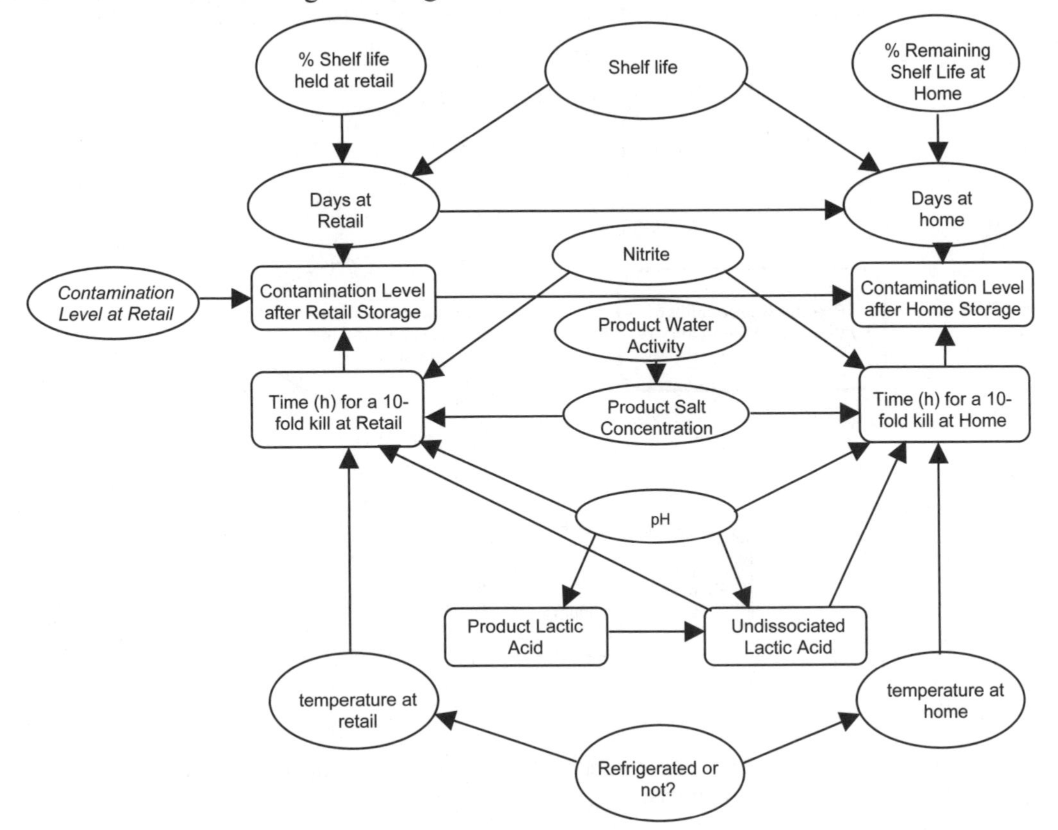

Figure 4.3 Influence diagram describing the modelled inactivation of *Listeria monocytogenes* during retail and home storage prior to consumption.

4.4.4 Risk characterization

The above elements were combined to estimate the risk to public health from *L. monocytogenes* in fermented meat products using stochastic modelling software (Analytica 1.1.1).

The overall model is shown schematically in Figure 4.4 and includes modules for:

- contamination level at retail, the effect of conditions during retail storage, home storage and point of consumption,
- modelled probability of illness per 1 million servings, and

modelled estimates of annual cases of listeriosis per 100 000 population.

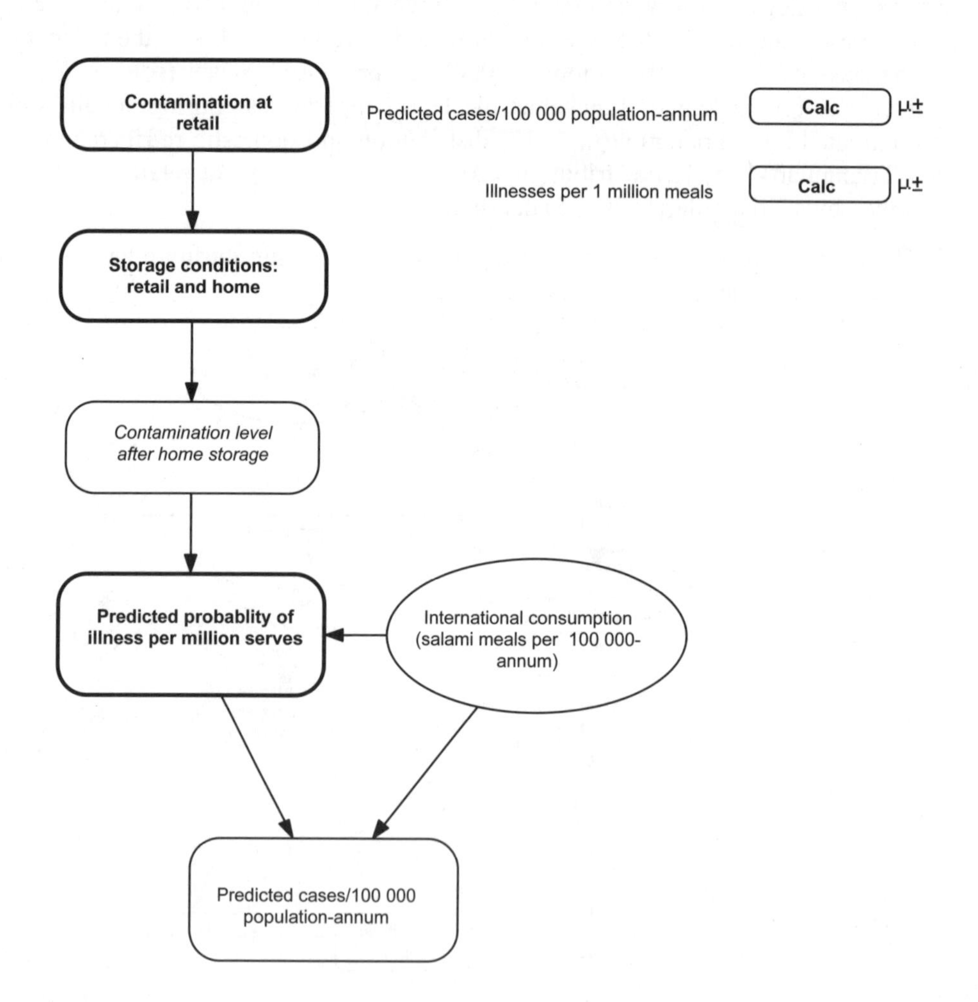

Figure 4.4 Overall structure of the conceptual model used to estimate the public health risk of listeriosis from fermented meat products.

Modelling listeriosis due to consumption of FMPs

The module modelling the relationship between the levels of *L. monocytogenes* in FMPs at the point of consumption and the corresponding prevalence of illness anticipated is shown as an influence diagram in Figure 4.5.

The module contains two sections. The first generates the dose-response relationship, assuming an exponential dose-response model developed and implemented as described in Section 4.1.4, by selecting a value for the parameter R in the exponential dose-response model as outlined in Section 4.1. In the second stage, the dose-response model is combined with serving size data and the modelled contamination level data to predict probability of illness per serving.

Serving size distribution is drawn from the United States of America data presented in FDA/FSIS (2001), and modelled by an empirically derived distribution (Beta(2, 8, 0, 270)). Using this, 80% of meals are predicted to be in the range 20–100 g per serving. A comparison of the original and empirical distribution showing the range of meal sizes is shown in Section A4.6 in Appendix 4.

From the contamination level estimate, serving size and r-value, the probabilty of illness per meal is then estimated.

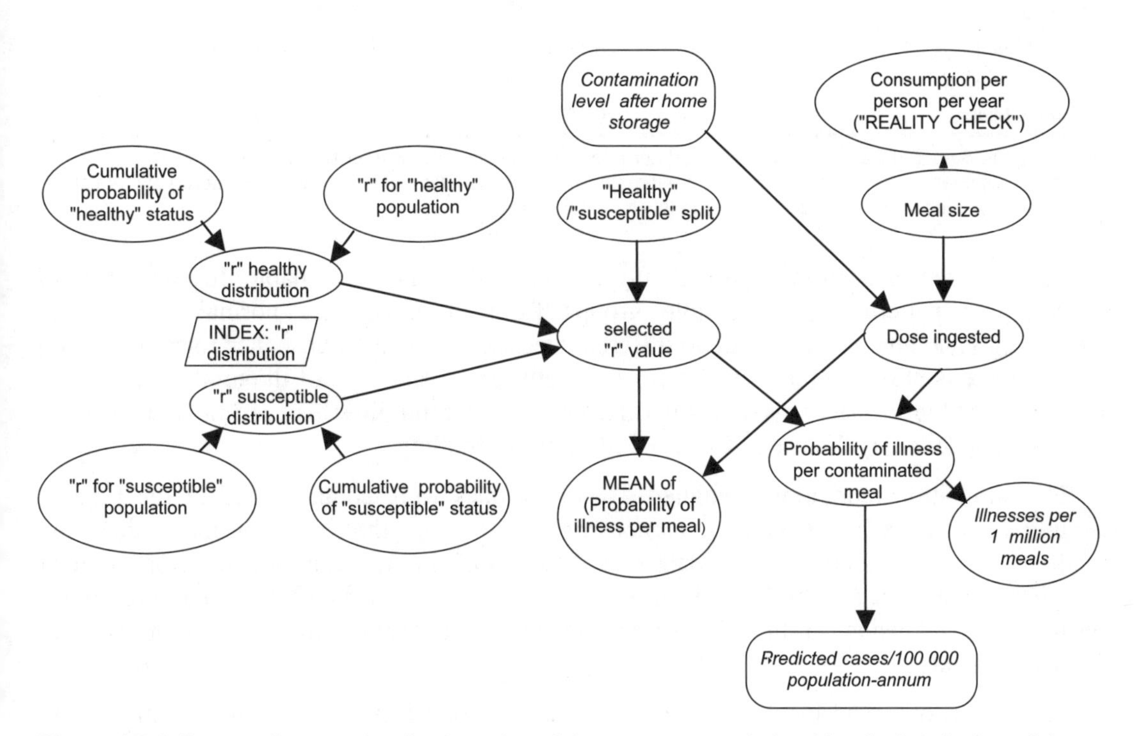

Figure 4.5 Influence diagram showing interplay of dose-response relationships for listeriosis and dose of *Listeria monocytogenes* ingested.

NOTES: Cum = Cumulative. R = r-value.

Calculations

As discussed in Section 4.1.4, the primary measure of risk estimated is the total number of cases per 1 000 000 meals based on the mean per-meal risk estimated from the 32 000 iterations. The final calculation multiplies the risk per serving by the estimate of the number of FMPs servings per annum in a range of nations. This is based on an estimate of the number of servings per person in each of the nations considered, described by a Triangular(6, 25, 62) distribution.

During the simulations it was noted that many calculations resulted in very low predicted concentrations of contamination such that the software was unable to carry through calculations. Accordingly, finite values of 1×10^{-12} CFU were added to intermediate values at some steps. Tests showed that this had no effect on the calculations of means, but did allow additional percentiles to be expressed.

The mean of the mean, minimum and maximum estimates of the results of the 16 trials are shown in Table 4.27, in addition to the standard deviation of the 16 estimates of the mean.

Table 4.27 Predicted risk of listeriosis from fermented meat products based on consumption data from several nations.

Summary statistics for 16 simulation runs	Cases per 100 000 consumers/year	Cases per 1 000 000 meals
Mean of Means	5.47×10^{-6}	2.11×10^{-6}
SD of Means(1)	8.59×10^{-6}	3.66×10^{-6}
Mean of Maxima	0.13	0.055
Mean of Minima	0.00	0.00

NOTE: (1) The Standard Deviation (SD) reported here does not measure the variability in the risk outcome distribution. Rather, it describes how much the estimator of a characteristic (i.e. in this case, the mean) of that risk outcome distribution varies from simulation run to simulation run.

As discussed in Section 4.1.4, no information was available to enable differentiation of consumption patterns of susceptible groups from those of the "normal" population. Accordingly, no attempt was made to differentiate risk for those groups. The total population risk estimate is based on the assumptions that between 15 and 20% of the total population is more susceptible to listeriosis, and the difference in susceptibility between those two broad groups is modelled by the use of two r-values in the modelling.

The simulation modelling predicted 2% of samples were contaminated at the time of serving at levels at >0.04 CFU *L. monocytogenes*, i.e. the threshold for many detection methods. From available survey data (see Table A4.1 in Appendix 4), the proportion of samples contaminated at >0.04 CFU *L. monocytogenes* at retail is 13.2%. This difference results from inclusion in the simulation model of inactivation of *L. monocytogenes* in fermented meats.

The annual observed incidence of listeriosis in the total population of many nations is in the range 0.3–0.5 cases/100 000 population although it has been suggested (Mead et al., 1999) that the true incidence may be twice as high. The above estimate (Table 4.27) suggests that consumption of fermented meat products probably contributes to only a very small proportion of those cases. The results are based on a number of assumptions, as discussed below.

4.4.5 Uncertainty and variability

The model draws together results from disparate studies and national cultures. As such, while the estimates are representative of that total population, they may not simulate very well the risk of listeriosis in any individual nation, because of differences in consumption patterns, for example. Different consumption behaviour between nations is expected and reported indirectly (Holdsworth et al., 2000) but practically no data quantifying national FMPs production were found. From limited data, per capita consumption patterns were estimated (see Section A4.4 in Appendix 4) to range from approximately 300 g to nearly 3 kg per year.

Such differences in estimated levels of consumption suggest that caution is required if using the results of this assessment for risk management actions within a specific nation. Rather, national consumption patterns should be used to estimate the public health risk from fermented meats in specific nations. Similarly, differences in predominant product types and processing methods, handling practices (e.g. storage temperature) would also be expected to lead to systematic differences in individual and population risks of listeriosis from this product type between nations. The model used does not discretely differentiate risk due to different product types.

Similarly, the model does not discretely predict the consequence of process failures (e.g. slow fermentation leading to growth of *L. monocytogenes* during preparation of the FMPs), except as they are represented in the contamination data used. The data used to model contamination levels, however, are sparse and it is unlikely that even low levels of process failure would be represented.

Also, as noted earlier, contamination level estimates were determined at various stages during the product's shelf-life, yet the model assumes that the levels reflect those initially present at the point of production or distribution to retail markets. The net effect of this assumption would be to somewhat underestimate the risk, because some inactivation would be expected to occur between production and the point of sampling at retail, i.e. the assumed starting levels in the modelling are probably lower than the "true" levels

The effect of this assumption was tested by re-running the model, as described above, after reducing the total storage time to 1 day. The effect was to increase the predicted incidence approximately 400-fold (±577 SD), to 6.47×10^{-4} cases per 100 000 per annum. The inference of this recalculation is that, even though the practical effect of sampling location is large, when a "worst-case" calculation is employed, the estimated relative risk of foodborne listeriosis associated with fermented meat products remains very low in comparison with observed international incidence of listeriosis. Another inference is that the model's outputs depend heavily on the validity of the inactivation model used. As indicated above, if no inactivation is modelled, the estimated risk increases several hundred-fold.

Finally, many of the parameter values of the distributions used do not include estimates of uncertainty in those values. As such, the uncertainty in the result is expected to be higher than implied by the spread from minima to maxima in Table 4.27.

Acknowledgements

The assistance of Meat and Livestock Australia for access to the report of Ross and Shadbolt (2001) and the Australian and New Zealand Food Authority for access to product composition data is gratefully acknowledged.

4.5 EXAMPLE 4: COLD-SMOKED FISH

4.5.1 Statement of purpose

Cold-smoked fish products are often found to be contaminated with *L. monocytogenes*. This has caused concern among regulatory agencies, particularly because vacuum-packed cold-smoked fish has a long shelf-life, is known to support the growth of *L. monocytogenes,* and there is indirect epidemiological evidence associating contaminated smoked fish and human cases of listeriosis. Others argue that despite the recognized hazard, the product has never been definitively linked to human systemic listeriosis.

This assessment aims to estimate the risk to a general consumer of vacuum-packed (VP) cold-smoked fish. It does not differentiate risk on the basis of nationality, consumption, age, gender or health status. Instead, it pools data from many nations to estimate a global risk of listeriosis due to consumption of cold-smoked fish. This assessment forms part of an overall assessment of the risk of listeriosis from foods that do, or do not, support the growth of *L. monocytogenes*.

4.5.2 Hazard identification

L. monocytogenes is frequently isolated from VP cold-smoked salmon and other fish products (see Table 4.28). Salmon is the fish type most commonly used for cold-smoked product, and comprises the majority of all cold-smoked fish production globally. The potential for *L. monocytogenes* to grow in RTE seafood products has been demonstrated by many authors (Hudson and Mott, 1993; Bell, Penny and Moorhead, 1995; Dalgaard and Jørgensen, 1998; Jørgensen and Huss, 1998; Thurette et al., 1998; Tienungoon, 1998).

Loncarevic, Tham and Danielsson-Tham (1998) compared *L. monocytogenes* isolates from human cases in Sweden with those isolated from trout and salmon. They found that three strains were isolated from both fish and human cases, suggesting that those fish products may be a source of infections. Similarly, the isolation of identical subclones of *L. monocytogenes* from both human patients and smoked seafoods in Norway (Rørvik et al., 2000) suggested that such products may have been possible sources for listeriosis cases. Conversely, Boerlin et al. (1997) found no such relationship among 47 human isolates and 72 isolates from fish products in Switzerland, nor Norton et al. (2001) between 275 human clinical isolates and 117 isolates from smoked fish and smoked fish processing plants in the United States of America. Kvenberg (1991) reported that there had been no cases of listeriosis in the United States of America that were linked to the consumption of seafood. Similarly, Bean et al. (1996) indicates that, of an average 500 *outbreaks* of foodborne listeriosis in the United States of America during the period 1988–1991, none were attributable to *L. monocytogenes* in seafoods. Recently, *gravad* trout, a lightly preserved (though not smoked) fish product was linked to a small outbreak of listeriosis (Ericsson et al.,

1997). More recently, Miettinen et al. (1999) reported five cases of febrile gastroenteritis (i.e. not invasive listeriosis) linked to cold-smoked trout.

Table 4.28 Incidence of *Listeria monocytogenes* contamination of smoked fish products.

Location	Product and species	No. of samples	% positive for *L. m.*	Levels	Most common serovars	Ref.
Europe	cold-smoked fish at 10 production sites	~340	overall 34–60(1)	*see* Table 4.29		[1]
USA	cold-smoked from plants with known problems	61	78.7			[2]
Norway	smoked salmon	13	33.0			[3]
Cyprus	smoked salmon (at retail)	–	28.6	<20 CFU/g		[4]
Italy	vacuum-packed sliced smoked salmon	100	20.0			[5]
Denmark	preserved fish products (not heated)	335	10.8			[6]
Sweden	gravad fish	58	21.0		1/2, 4	[7]
	cold-smoked fish	26	3.9			
Sweden	hot, cold and gravad					[8]
Canada	hot and cold-smoked fish at retail	258	27.9			[9]
Iceland	smoked salmon	–	29.0	Note (2)		[10]
England and Wales (UK)	smoked mackerel	116	7			[11]
	smoked salmon	86	2			
	other	1	3			
Germany	smoked salmon		7.1			[12]
Japan	smoked salmon	92	5.4		1/2a, b	[13]
USA	smoked finfish	1 210 (3)	12.0–16.3	<10 MPN/g		[14]
Australia	Smoked salmon at final product (one plant only)	285	0.4	Presence in 25 g		[15]
Australia	smoked fish and mussel products, retail, Canberra	49	4.1	4 MPN/g, 460 MPN/g		[16]
Australia	smoked fish	9	10.0	presence in 25 g		[17]
Finland	vacuum-packed cold-smoked salmon	30 (12 producers)	17.0	50%: >100/g	1/2a, 4b	[18]

NOTES: (1) Range of frequency of contamination from individual sites: 1.4–100. (2) 46% of generic *Listeria*-positive samples contained *L. monocytogenes.* (3) Over 6 years (1991–1996). (4) MPN = most probable number.

SOURCES: [1] Jorgensen and Huss, 1998. [2] Eklund et al., 1995. [3] Rørvik et al., 1997. [4] Data sumbmitted to FAO in 2000 by Director, State General Laboratory, Ministry of Health, Cyprus. [5] Cortesi et al., 1997. [6] Andersen and Nørrung, 1995. [7] Loncarevic, Tham and Danielsson-Tham, 1996. [8] Lindqvist and Westöö, 2000. [9] Dillon, Patel and Ratnam, 1994. [10] Hartemink and Georgsson, 1991. [11] McLaughlin and Nichols, 1994. [12] Teufel and Bendzulla, 1993. [13] Inoue et al., 2000. [14] Jinneman, Wekell and Eklund, 1999. [15] Garland, 1995. [16] Rockliff and Millard, 1996. [17] Dunn, Son & Stone, 1998. [18] Johansson et al., 1999.

4.5.3 Exposure assessment

4.5.3.1 Production and consumption of smoked fish products

Data were not readily available for total cold-smoked fish production. Instead, data for cold-smoked salmon production were used as a surrogate for total cold-smoked fish. Those data are presented in Appendix 5. In summary, the data suggest that, in the 15 nations considered, the mean consumption amount is approximately 60 g per serving, with average consumption frequency ranging from <1 to 18 servings per person per year. Mean annual consumption across those nations is estimated at about 144 g per person.

4.5.3.2 Contamination rates and levels

Numerous studies have demonstrated that smoked fish products are frequently contaminated with *L. monocytogenes* at rates varying from 0.4 to 78.7%, but more typically in the range 4 to 30%. Table 4.28 summarizes the results of many of those studies.

From the data in Table 4.28, an overall average contamination rate of 18.6% is estimated based on the mean (weighted according to sample size) of all surveys for which both contamination rate and sample size are available. The unweighted mean is 18.0%.

While several publications provide information on the level of contamination of cold-smoked salmon at retail (Teufel and Bendzulla, 1993; McLaughlin and Nichols, 1994; Jørgensen and Huss, 1998; Nørrung, Andersen and Schlundt, 1999; Inoue et al., 2000), this assessment begins at the point of completion of processing, prior to distribution to retailers. Only one data source (Jørgensen and Huss, 1998) provided information on levels of *L. monocytogenes* at the point of completion of processing. Those data are shown in Table 4.29.

Other reports support the general conclusions of Jørgensen and Huss (1998) that, at production, contamination levels are usually less than 10 CFU/g. Dalgaard and Jørgensen (1998) reported that most of the positive samples in their study had an average contamination of ≤ 8 MPN/g, for samples taken 4 to 12 days after production and held at 5°C. They noted, however, that some samples were in the range 10–100 MPN/g. Similarly, Pelroy et al. (1994) reported median levels at production of 0.5 - 11.7 CFU/g. Due to the paucity of data it was assumed for the purposes of this report that the results of Jørgensen and Huss (1998) are representative of all cold-smoked fish at the final point of processing, and those data were used as the initial contamination level in the modelling.

Table 4.29 Contamination levels at end of production for cold-smoked salmon in Denmark.

Point of Testing	Storage time at 5±1°C between initial and final analyses	No. (%) of 25-g samples positive	Contamination levels. No. of positive samples (% of total samples)				Total number of samples
			<10/g	10 – 100/g	100 – 1000/g	>1000/g	
Initial	0	64 (34)	53 (28)	9 (5)	2 (1)	0	190
Final	14–20 days	46 (40)	11(10)	23 (20)	10 (9)	2 (2)	115
Final	21–50 days	32 (43)	17 (23)	11 (15)	2 (3)	2 (3)	75

SOURCE: Data of Jørgensen and Huss, 1998

4.5.3.3 Time and temperature of storage

The nominal shelf lives for vacuum-packed smoked fish are in the range of 3 to 6 weeks at a storage temperature of 4–5°C. Several studies have assessed the *sensory* acceptability of smoked salmon (Truelstrup Hansen, Drewes Røntved and Huss, 1998; Jørgensen, Dalgaard and Huss, 2000; Leroi et al., 2001) and found that that sensory shelf-life at 5°C for cold-smoked salmon is highly variable (from 3 to 9 weeks) and that there is no single indicator for the onset of spoilage.

Storage temperature data were derived from Audits International (2000) data for refrigerated cabinets at retail used for storage of cold-smoked fish. The data is tabulated in Section A5.2 in Appendix 5. In the modelling, storage temperature was described by Beta(4, 7.5, -5, 22) which generated a mean temperature of 4.4°C, and ranged from ~ -4.5 to +19°C (from 32 000 iterations).

4.5.3.4 Physico-chemical parameters of cold-smoked fish

The physico-chemical composition of cold-smoked fish products are assumed to be similar to those for cold-smoked salmon reported in Ross, Dalgaard and Tienungoon (2000) who noted that the postmortem pH of the fish muscle drops to ~ 6.0–6.4 due to the catabolism of muscle glycogen resulting in lactic acid production. At that pH, the muscle contains from 65 to 130 mM lactic acid. Lower pH is correlated to higher lactate concentration, and the lactate present will enhance the inhibitory affect of the reduced pH.

Cold-smoked fish products typically have low salt levels, in the range of 1.5 to 4% (NaCl in the aqueous phase) and would be expected to have water activity in the range 0.977 to 0.99 (Dalgaard, 1997; Leroi et al., 2001).

Leroi et al. (2001) and Thurette et al. (1998) reported a wide variation in phenol levels in cold-smoked salmon (n = 13) produced in several countries and sampled in France. The levels ranged from 2.7 to 10.8 mg phenol/kg fish (i.e. 2.7 to 10.8 ppm), with an average of 5.5 ±1.5 (SD) mg phenol/kg. These levels are consistent with the levels (5 to 10 ppm) reported by Leblanc et al. (2000) and Eklund et al. (1995; 8 to 13 ppm). However, phenol concentration is not included directly in the modelling (see Section A5.3.1 in Appendix 5), but was included indirectly by manipulation of the "Other growth inhibiting factors" input.

4.5.3.5 Growth potential and microbial ecology of vacuum-packed products

Growth Rate Model

As shown in Table 4.29, *L. monocytogenes* can grow on cold-smoked fish products. FDA/FSIS (2001) collated results of published growth rate studies. In this study, a predictive model for growth rate of *L. monocytogenes* as a function of temperature, pH, water activity and lactic acid concentration is used to calculate growth rate. That model and its basis are discussed in Section A5.3 in Appendix 5. To assess its utility for cold-smoked salmon, the model's predictions for growth

Table 4.30 Comparison of model predictions of *Listeria monocytogenes* growth rate at 5°C on cold-smoked fish and those collated from published literature

	Growth rate estimate (logCFU/day)	
	Predictive model	FDA/FSIS (2001) literature collation
Mean	0.113	0.155
SD	0.055	0.100

rate at 5°C were compared with the growth rates collated by FDA/FSIS (2001). The results in Table 4.30 indicate that the two approaches produce consistent, though not identical, results at 5°C.

The effects of product parameters on the *potential* for growth was also considered, both on the basis of individual factor limits (see Table 3.1) and using the growth/no-growth model of Tienungoon et al. (2000) because in some iterations of the Monte Carlo simulation model the factor combinations sampled may, in fact, preclude growth. The implementation of this is described in Section A5.3 (in Appendix 5).

Effect of lag time and lactic acid bacteria on growth potential

Lag times and the effect of growth of lactic acid bacteria on the shelf-life of the product and the growth potential of *L. monocytogenes* in vacuum-packed cold-smoked fish were also considered explicitly and implemented in the modelling. Maximum population densities were also modelled. The importance of these factors, and their implementation in the modelling, is described in Section A5.3 (in Appendix 5).

4.5.3.6 Exposure assessment model

The overall structure of the exposure assessment model is shown in Figure 4.6. Each box in the diagram represents a module or sub-model. The structure of these sub-models is described below.

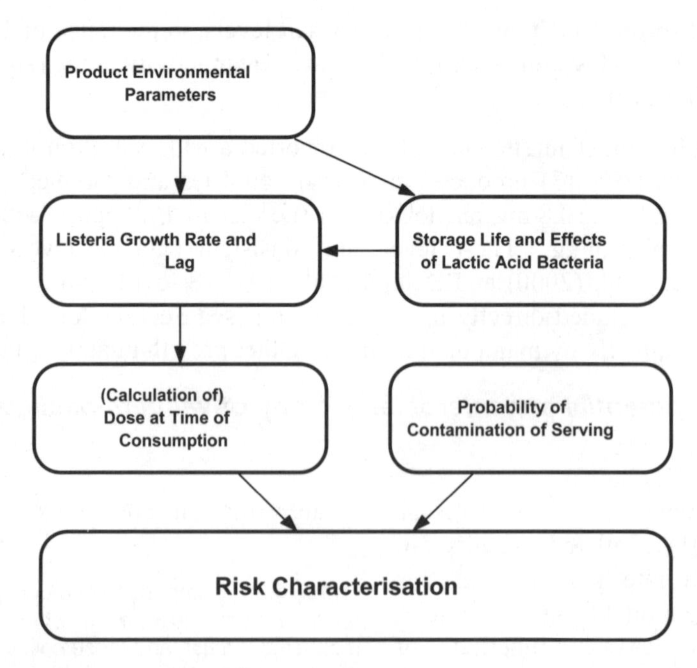

Figure 4.6 Exposure assessment model used shown as an influence diagram. The structure of each of the sub-models is described in the text.

Product environmental parameters

Product environmental parameters (water activity, pH and lactic acid) were modelled as described in Section A5.3 (in Appendix 5) Other components of the product, though not explicitly modelled (e.g. phenol, spices), were considered to reduce growth rate by 10% from that predicted on the basis of product parameters and temperature alone (see below). Storage temperature was modelled as described in Section A5.2 (in Appendix 5)

Listeria Growth Rate and Lag

The growth rate module is fully described in Section A5.3 (in Appendix 5). In summary, growth rate is predicted from storage temperature and product composition values sampled during each iteration. The conditions are first evaluated to determine whether the combinations of pH, temperature and water activity would permit growth, using the model of Tienungoon et al. (2000), or whether any individual parameter value is beyond the range that permits growth of *L. monocytogenes* (see Table 3.1). If growth is predicted to be possible, two growth rates are estimated, i.e. growth before lactic acid bacteria reach their stationary phase and, if appropriate, subsequent growth, but before the product is predicted to spoil at the sampled temperature. The time at which lactic acid bacteria reach stationary phase is predicted from the values in Product Environmental Parameters, as described below.

Storage life and effects of lactic acid bacteria

A nominal storage life of the product at 5°C is specified as described in Section 4.5.3.4. Nominal shelf-life is adjusted for other temperature scenarios using a square root type relative rate function as described in Appendices 2 and 3. Details are given in Section A5.3 in Appendix 5.

In each iteration of the model, the total possible growth of *L. monocytogenes* is calculated from the growth rate and storage time, and, after deducting the contribution of the lag time, is expressed as potential number of generations of growth.

Calculation of dose at the time of consumption

The dose at the time of consumption is calculated from the initial contamination level distribution (described in Section 4.5.3.2), to which is added the predicted growth as described above. The predicted concentration of *L. monocytogenes* in the serving is then compared with the maximum concentration level. If the predicted concentration exceeds the maximum concentration level ($10^{9.5}$ CFU/g) it is changed to the maximum concentration, otherwise the original modelled contamination level is used. That concentration level is combined with the serving size estimate (see Appendix 5 for details of modelling, and Section 4.5.3.1 for summary) to estimate the dose ingested by the consumer from that scenario. These interrelationships are depicted as an influence diagram in Figure 4.7.

Probability of contamination of serving

The probability of consuming a contaminated serving is derived from data presented in Table 4.28 and described empirically in the model as Beta(2, 6.6, 0.004, 0.787), which models a mean contamination rate of 18.6%, a minimum contamination rate of 0.46% and a maximum rate of 67.8%.

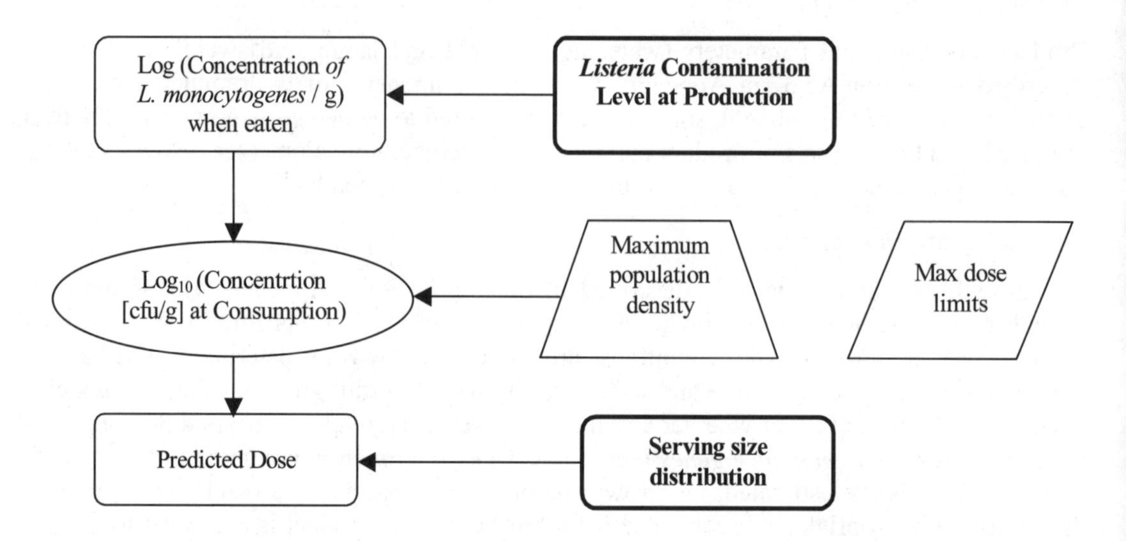

Figure 4.7 Influence diagram showing calculation of *Listeria monocytogenes* dose per contaminated serving of cold-smoked fish.

4.5.4 Risk characterization

4.5.4.1 Introduction

The risk characterization combines the dose-response model described previously (Section 4.1) with the exposure assessment model described above. Outputs of the model are as described in Section 4.1. Figure 4.8 depicts the risk characterization model as an influence diagram.

4.5.4.2 Assumed variables

The model was executed 16 times with 32 000 iterations per execution to generate a set of "baseline" values. Those values are based on a number of assumptions including that:

- the storage life of the product ranges from 1 to 42 days, with a most likely storage time of 28 days;
- the maximum population density of *L. monocytogenes* on the product is 3×10^9 CFU/g;
- the lag time ranges between the equivalent of 0 and 35 generation times, with a most likely lag time equivalent to three generation times at the storage temperature sampled and for the product parameters sampled; and
- when the lactic acid bacteria reach stationary phase, they will reduce the growth rate of *L. monocytogenes* by 85–100% (described in the model as the growth rate predicted as described above, multiplied by Uniform(0.00, 0.15)).

Consumption pattern is described in Section A5.1 in Appendix 5.

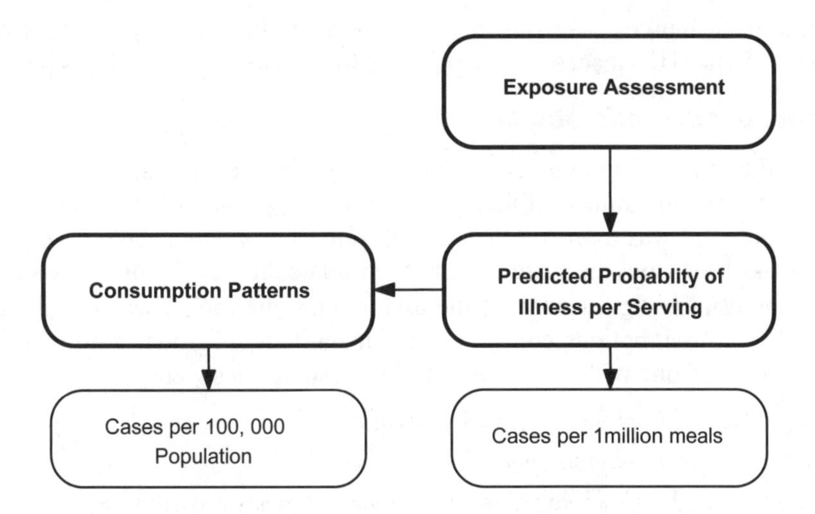

Figure 4.8 Influence diagram depicting the interrelationship of factors governing the risk estimates.

4.5.4.3 Results

The results, based on the assumptions considered above, are presented in Table 4.31. The results are the mean values of the mean and maximum values of the distribution of risk estimates for each of 16 runs of the Monte Carlo model. Maximum values are included to give some indication of the range of risk estimated. In both sets of estimates of the two measures of risk, the minimum value estimates were of the order of 10^{-11} to 10^{-12}.

Table 4.31a Cases per 100 000 consumers per year in cold-smoked fish.

Mean of Means	0.0163
SD of Means	0.0117
Mean of Maxima	186
SD of Maxima	253

Table 4.31b Cases per 1 000 000 meals in cold-smoked fish.

Mean of Means	0.0530
SD of Means	0.0355
Mean of Maxima	555
SD of Maxima	745

4.5.5 Uncertainty and variability

Uncertainty and variability concerning the dose-response component of the assessment are as discussed in the other three case studies.

4.5.5.1 Consumption

There are a number of uncertain variables specific to the exposure assessment component of this model, including the true level of consumption of cold-smoked fish, and factors affecting the level of contamination with *L. monocytogenes*.

From the data presented in Section A5.1 in Appendix 5, consumption was suggested to vary widely between nations, but it was also apparent that different estimates would arise from different data sources. It was also shown that the model possibly overpredicted consumption by 50%. This inaccuracy is expected to contribute a relatively small error.

4.5.5.2 Effect of other microbiota

The effect of other microbiota (e.g. lactic acid bacteria) in vacuum-packed, cold-smoked fish is well known, but the magnitude of the effect could not be quantified with certainty. In the modelling presented, it was *assumed* that growth rate of *L. monocytogenes* was inhibited by between 85% and 100%, but there are no data to evaluate the validity of this assumed level of inhibition. To assess the significance of the assumption, the model was re-run (16 × 32 000 iterations) with four assumptions concerning the magnitude of growth rate inhibition due to the Jameson effect and one with no inhibition. The assumptions were:

- total inhibition of *L. monocytogenes* growth rate;
- 95% inhibition of *L. monocytogenes* growth rate;
- between 80% and 100% inhibition of *L. monocytogenes* growth rate;
- 70% inhibition of *L. monocytogenes* growth rate; and
- no inhibition of *L. monocytogenes* growth rate.

The results (means of mean values of 16 simulated distributions) are compared in Table 4.32, and indicate that the differences can be profound compared with the situation where no inhibition is modelled, indicating the importance of this aspect of the microbial ecology of VP RTE foods for the estimation of the risk of listeriosis. When inhibition was modelled, the differences in the risk estimates were about 2- to 5-fold. Further experimentation with the model, however, suggested that very large increases in risk occurred if inhibition less than 80% were assumed. For example, assuming that growth rate was reduced to 30% (i.e. 70% inhibition) resulted in risk estimates that were thousands of times higher than when 95% inhibition of growth rate was assumed.

Table 4.32 Effect of assumptions concerning the effect of growth rate inhibition of *Listeria monocytogenes* due to high levels of lactic acid bacteria on estimates of the risk of listeriosis in smoked fish.

	Assumed magnitude of *L. monocytogenes* growth rate inhibition				
	No Inhibition	Fixed at 70%	Variable (80–100%)	Fixed at 95%	Complete 100%
Cases per 100 000 popn.	366	5.65	0.010	0.004	0.002
Cases per million servings	1136	–	0.033	0.011	0.005

4.5.5.3 Reality check

There have been a handful of cases of listeriosis reported in the last decade that possibly have been related to cold-smoked fish. Production volumes of cold-smoked fish in the late 1990s were around 80 000 tonne/year. If it is assumed that this production level is representative of the last ten years, then those cases are due to ~800 000 tonnes, or 1.33×10^{10} servings (60 g each). Assuming a factor of 10 for cases of listeriosis due to smoked fish products that are not recognized, or not reported, one case would equate to approximately 0.008 cases per million meals. That estimate is closer to the low end of the estimates derived from the simulation model in this assessment (~90% inhibition) and might suggest that the more stringent inhibition assumptions concerning the inhibition of growth of *L. monocytogenes* by other organisms present in VP cold-smoked fish products are more consistent with actual experience.

Varying the upper population limit (maximum population density – MPD) had no effect on the risk estimate, indicating that in virtually all cases other factors described in the model controlled growth of *L. monocytogenes* to the extent that it never reached MPD.

Not all parameters in the model included estimates of variability, e.g. variability in growth rates. Equally, the conversion of per-meal risk to risk per 100 000 is based on multiplying the mean of the per-serving risk by population estimates. While the mean risk estimate is unaffected, the model does not accurately portray the extent of variability in the estimates. As these risk assessment are all characterized by mean values of population or per-meal risk, the risk estimates are not affected. Note, however, that the standard deviations quoted in Table 4.31 reflect the variability in the simulation modelling procedure, i.e. between-run variation, not the variability in system being modelled.

Finally, because of the combination and pooling of data from many diverse sources, the risk estimates are not nation-specific, and so may not accurately represent the situation in any nation.

Table 4.33 Parameter values for Triangular distributions used for storage time scenarios tested for their effect on risk estimates.

Description	Days before consumption			Mean value of the distribution (days)
	Minimum	Most Likely	Maximum	
"Reduced"	1	14	21	12
"Good"	1	21	28	14
"Realistic"	1	28	42	24
"Product Abuse"	1	28	180	70

4.6 SUMMARY

The risk characterizations for all four examples are summarized in Table 4.34.

Table 4.34 Estimated risks of listeriosis per 100 000 population and per million servings for the four selected foods.

Food	Cases of listeriosis per 100 000 consumers	Cases of listeriosis per 1 million servings
Milk	0.091	0.005
Ice Cream	0.00012	0.000014
Cold-Smoked Fish	0.016	0.053
Fermented Meat Products	0.0000055	0.0000021

Part 5.

Risk Characterization: Response to Codex Questions

5.1 INTRODUCTION

This section addresses the three risk questions posed by CCFH in 2001 in relation to the risk from *L. monocytogenes* in RTE foods. The specific question addressed is given in each case.

5.2 QUESTION 1

Estimate the risk from L. monocytogenes *in food when the number of organisms range from absence in 25 grams to 1000 colony forming units per gram, or millilitre or does not exceed specified levels at the point of consumption.*

The question posed by the CCFH primarily requires a consideration of how the relative risk of acquiring listeriosis is affected by the level of *L. monocytogenes* present in a serving of food at the time of consumption. The ability to answer this question is dependent on the ability to articulate and interpret dose-response relationships for *L. monocytogenes*. However, there are a number of potentially confounding factors that could influence the approach taken and the complexity of the answer provided. In view of the generic nature of the CCFH question and the fact that this is one of the first microbial risk assessments requested by CCFH, it was decided that the response to this question should focus on communicating the key risk assessment concepts. It is also important to note that this question implies a series of comparisons based on relative risks and does not require the much more daunting task of calculating absolute risk. Accordingly, consideration of potential confounding factors was limited and a detailed consideration of uncertainty and variability was not undertaken in addressing this question. An introduction to issues related to the uncertainty and variability associated with dose-response models is provided in the hazard characterization section of this document. In addition to not explicitly addressing uncertainty and variability, a number of simplifying assumptions were made in developing the examples used to answer the question posed by CCFH. For instance, to calculate the ingested dose, knowledge of the size of the serving is needed. A fixed serving size of 31.6 g was assumed for convenience to simplify the calculations because it approximates a typical serving size and because dose levels were estimated in 0.5 $\log_{10}$ increments ($10^{0.5} = 3.16$). To calculate the concentrations for other serving sizes in the tables that follow, the dose levels would have to be divided by the serving size.

As discussed in the hazard characterization, the exponential model was selected to describe the relationship between the dose of *L. monocytogenes* ingested and the probability of developing systemic listeriosis. Dose-response curves were developed for both the healthy population and the susceptible population and include the entire range of ingested doses (i.e. not restricted to 1000 CFU/g food). These curves are population based and describe the average dose-response relationship. A specific outbreak that involves a strain with high virulence or an unusually susceptible population may still result in a significant number of cases from food containing comparatively low numbers of *L. monocytogenes*. For the purposes of this example, only the dose-response curve for the susceptible population was used, and it was assumed that all cases of listeriosis were restricted to that population. The specific dose-response curve selected was the one where the maximum level to which *L. monocytogenes* could grow in a food was assumed to be $10^{7.5}$ CFU/serving. The end result of these assumptions is that the most "conservative" dose-response model was used, i.e. the maximum virulence of *L. monocytogenes* was assumed. The r-value for this relationship was 5.85×10^{-12} (Table 2.18). The dose ingested is a function of the level of the microorganism in the food (CFU/g) multiplied by the size of the serving. Thus, the equation for calculating the probability of listeriosis was:

$$P = 1 - e^{(5.85 \times 10^{-12})\,(31.6\mathrm{g} \times n)}$$

where n is the number of *L. monocytogenes* per gram. By substituting different values for n, the likelihood of listeriosis at levels between 0.04 (1 CFU/25 g) and 1000 CFU/g was calculated.

The overall affect on the number of cases of listeriosis was estimated by multiplying the likelihood of listeriosis per serving by the total number of servings. For this calculation, the total number of RTE servings was assumed to be 6.41×10^{10} servings, i.e. the estimated total number of servings per year consumed in the United States of America for the 20 classes of RTE food considered in FDA/FSIS (2001). The corresponding number of listeriosis cases for the susceptible population was considered to be 2130 (FDA/FSIS, 2001), and will be used to represent the current incidence of listeriosis when comparing the effect of changes to incidence under different theoretical scenarios.

As a simple, worst-case scenario, the predicted risk per serving and predicted number of annual listeriosis cases were estimated by assuming that all 6.41×10^{10} servings had the maximum contamination level being considered. The effects on the incidence of listeriosis of six levels of pathogen were evaluated (0.04, 0.1, 1, 10, 100 and 1000 CFU/g) (Table 5.1).

A more realistic approach would be to use a distribution of *L. monocytogenes* levels in foods when consumed. To explore that more complex approach, the overall distribution of *L. monocytogenes* levels in 20 classes of RTE foods from the FDA/FSIS (2001) risk assessment was used (see Table 5.2) to calculate the probability of listeriosis and the predicted number of cases. At each maximum *L. monocytogenes* level considered, the number of servings from the distribution exceeding the designated contamination level was added to that maximum level. For example, for an upper limit of 1000 CFU/g, the number was 1.18×10^8 servings, i.e. 6.23×10^7 (servings originally predicted to be at 1000 CFU/g) + 2.94×10^7 (servings originally predicted to be at 10 000 CFU/g) + 1.39×10^7 (servings originally predicted to be at 10^5 CFU/g) + 3.88×10^6 (servings originally predicted to be at $10^{5.5}$ CFU/g) + 8.55×10^6 (servings originally predicted to be at $>10^6$ CFU/g). The predicted annual numbers of listeriosis cases were calculated and summed, and the predicted number of cases for each maximum level is given in Table 5.3.

Table 5.1 Probability of illness per serving for the susceptible population estimated for different levels of *Listeria monocytogenes* at the time of consumption and the estimated number of cases per year in the United States of America if all RTE meals were contaminated at that level.

Level (CFU/g)	Dose[(1)] (CFU)	Log_{10} dose (log_{10} CFU/ serving)	Probability of illness per serving	Relative risk[(2)]	Estimated annual number of cases[(3)]
<0.04	1	0	7.39×10^{-12}	1	0.54
0.1	3	0.5	1.85×10^{-11}	2.5	1
1	32	1.5	1.85×10^{-10}	25	12
10	316	2.5	1.85×10^{-9}	250	118
100	3 160	3.5	1.85×10^{-8}	2500	1 185
1000	31 600	4.5	1.85×10^{-7}	25000	11 850

NOTES: (1) Serving size of 31.6 g. (2) Using the risk from a dose of 1 CFU as reference. (3) A total of 6.41×10^{10} servings per year assumed.

Table 5.2 Predicted distribution of levels of *Listeria monocytogenes* occurring in RTE foods.

Level of *L. monocytogenes* in a food at consumption (CFU/g)	Number of servings at the specified dose
<0.04	6.18×10^{10}
0.1	1.22×10^{9}
1	5.84×10^{8}
10	2.78×10^{8}
100	1.32×10^{8}
1000	6.23×10^{7}
10000	2.94×10^{7}
100000	1.39×10^{7}
316000	3.88×10^{6}
>1000000	8.55×10^{6}
Total	6.41×10^{10}

SOURCE: FDA/FSIS, 2001.

Table 5.3 Predicted annual number of listeriosis cases in the susceptible population when the level of *Listeria monocytogenes* was assumed not to exceed a specified maximum value and the levels of *L. monocytogenes* in the food are distributed as indicated in Table 5.2.

Level (CFU/g)	Maximum Dose[(1)] (CFU)	Percentage of servings when maximum level[(2)]	Estimated number of listeriosis cases per year[(3)]
0.04	1	100	0.5
0.1	3	3.6	0.5
1	32	1.7	0.7
10	316	0.8	1.6
100	3160	0.4	5.7
1000	31 600	0.2	25.4

NOTES: (1) Serving size of 31.6 g. (2) Number of servings in the highest *L. monocytogenes* level assumed divided by 6.41×10^{10} times 100. (3) Levels of *L. monocytogenes* per serving used to calculate predicted number of cases based on the overall distribution from the FDA/FSIS risk assessment (2001) (see Table 5.2). A total of 6.41×10^{10} servings per year was assumed.

Tables 5.1 and 5.3 show vast differences in the estimated number of cases for the worst-case answer to the question (Table 5.1) compared with that estimated when an attempt is made to consider the frequency and extent of contamination actually encountered in RTE foods (Table 5.3). While either set of predictions can be challenged on the basis of the assumptions used, such scenarios are useful in framing the extent of the risk likely to be encountered.

These two scenarios (Tables 5.1 and 5.3) demonstrate that when dealing with an infectious agent where a non-threshold model is assumed, where either the frequency of contamination (percentage of contaminated samples) or the extent of contamination (*L. monocytogenes* levels in a contaminated food) increases, then so does the risk and the predicted number of cases. Thus, if all RTE foods went from having 1 CFU/serving to 1000 CFU/serving, the risk of listeriosis would increase 1000-fold (assuming a fixed serving size). Conversely, the effect of introducing into the food supply 10 000 servings contaminated with *L. monocytogenes* at a level of 1000 CFU/g could theoretically be compensated by removing from the food supply a single serving contaminated at a level of 10^7 CFU/g.

In interpreting these results and in attempting to predict the actual effect of a change in the regulatory limits for *L. monocytogenes* in RTE foods, one also has to take into account the extent to which deviations from established limits occur. The current example is based on data from the United States of America, where the current allowable limit for *L. monocytogenes* in RTE foods is effectively 0.04 CFU/g (1 CFU/25 g), a level that if consistently achieved would be expected to result in less than one case of listeriosis per year in the United States of America. However, the baseline level for the United States of America population was 2130 cases (Mead et al., 1999). Both the current risk assessment and the United States of America FDA/FSIS draft risk assessment (2001) indicate that a portion of RTE food contain a substantially greater number of the pathogen than the stated limit and that the public health impact of *L. monocytogenes* is, most probably, almost exclusively a function of the foods that greatly exceed the current limit. Thus, in addressing the question posed by CCFH, the current risk assessment indicates that increasing the level of *L. monocytogenes* in RTE foods from 0.04 to 1000 CFU/g would increase the risk of foodborne listeriosis, provided that the current rate of deviations above the established limit remained proportionally the same. However, it could also be asked whether public health could be improved if a less stringent microbiological limit for RTE foods resulted in a substantial decrease in the number of servings that greatly exceeded the established limit, e.g. if the change encouraged manufacturers to routinely screen for *L. monocytogenes* in the plant environment and to take appropriate remedial actions. Models developed during the current risk assessment could be used estimate the extent of control over deviations from established limits that would be needed to improve public health if regulatory limits were relaxed, provided that sufficient data on the rate and extent of deviations were available for individual RTE foods.

As a means of further examining this concept, a simple hypothetical "what-if" scenario was developed based on the information provided in Tables 5.2 and 5.3. It examines the impact that compliance with a microbiological limit (i.e. defect rates) has on public health. Two potential limits, 0.04 CFU/g and 100 CFU/g, were examined in conjunction with different defect rates, i.e. the percentage of servings that exceed the specified limit. As a means of simplifying the what-if scenario and dramatizing the impact of compliance, a single level of *L. monocytogenes*, 10^6 CFU/g, was assumed for all "defective" servings. Thus, if a serving of food was in compliance, it had a level of *L. monocytogenes* at or below the specified microbiological limit based on the distribution of *L. monocytogenes* levels (Table 5.2) used to calculate the 100% compliance values depicted in Table 5.3. Conversely, if a serving of food was out of compliance, it was assumed to have a set level of *L. monocytogenes* of 10^6 CFU/g, or since the assumed serving size was 31.6 g, a consumed dose of 3.16×10^8 CFU. The predicted number of cases as a function of the percentage of defective servings is provided in Table 5.4.

As noted in Table 5.3, at 100% compliance the number of predicted cases for both limits is low, with an approximate 10-fold differential between the two microbiological limits. As expected, the number of predicted cases increases with an increasing frequency of defective servings. At defect rates >0.0001% a 10-fold increase in the defect rate results in an approximate 10-fold increase in the number of predicted cases, regardless of the microbiological limits (i.e. 0.04 CFU/g versus 100 CFU/g). It is interesting to note that based on the conditions and assumptions of this simple what-if scenario, the defect rate that yielded a value approximately equivalent to the baseline value of 2130 cases used in the FDA/FSIS draft risk assessment (2001) was 0.018%.

A more detailed consideration of compliance could be achieved by incorporation of distributions reflecting the levels of *L. monocytogenes* observed in variety of foods. However, such a detailed consideration of compliance rates was beyond the scope of the current risk assessment. Furthermore, the simple hypothetical what-if scenario presented adequately demonstrates key concepts related to how compliance rates can strongly influence the actual risk associated with a microbiological criterion. In fact, it could be argued that the rate of compliance is a more significant risk factor than the numeric value of the criterion within the range that CCFH asked the risk assessment team to consider. The what-if scenario also demonstrates the concept that a less stringent microbiological limit could lead to an improvement in public health if new criteria lead to a substantive decrease in defect rates. For example, the model (Table 5.4) predicts that if a microbiological limit of 0.04 CFU/g with a 0.018% defect rate (2133 cases) was replaced with a 100 CFU/g limit and a 0.001% defect rate (124 cases), the predicted result based on the scenario is an approximate 95% reduction in foodborne listeriosis.

Table 5.4 Hypothetical "what-if" scenario demonstrating the effect of "defect" rate on the number of predicted cases of foodborne listeriosis.

Assumed percentage of "Defective" servings(1)	**Predicted number of listeriosis cases(2)**	
	Initial standard of 0.04 CFU/g	Initial standard of 100 CFU/g
0	0.5	5.7
0.00001	1.7	6.9
0.0001	12.3	17.4
0.001	119	124
0.01	1 185	1 191
0.018	2 133	2 133
0.1	11 837	11 848
1	117 300	117 363

NOTES: (1) For the purposes of this scenario, all defective servings were assumed to contain 10^6 CFU/g.
(2) For the purposes of this scenario, an r-value of 5.85×10^{-12} was employed and a standard serving size of 31.6 g was assumed. In the case of the 100 CFU/g calculations, the defective servings were assumed to be proportionally distributed according to the number of servings within each cell concentration bin.

5.3 QUESTION 2

Estimate the risk for consumers in different susceptible population groups.

As noted in Section 5.2, listeriosis is primarily a disease of certain subpopulations with impaired or altered immune function (e.g. pregnant women and their fetuses, the elderly, individuals with chronic diseases, AIDS patients, individuals taking immunosuppressive drugs). Susceptibility varies within the broadly defined susceptible group (e.g. the risk of listeriosis appears to be less for pregnant women than transplant recipients). It has been estimated that various subpopulations may have a 20- to 2500-fold increased risk of acquiring listeriosis (FDA/FSIS, 2001; Marchetti, 1996). CCFH requested that the risk assessment team attempt to estimate the differences in the dose-response relations for the different subpopulations with increased susceptibility. While previous risk assessments had considered the relative susceptibility of the entire population at increased risk, versus the general population, these risk assessments did not develop the type of detailed comparisons of subpopulations with increased susceptibility requested by CCFH. Thus, the current risk assessment had to develop *de novo* a means for addressing the request.

The basic approach taken to developing the requested dose-response relations was to take advantage of epidemiological estimates of the relative rates of listeriosis for different subpopulations. These "relative susceptibility" values were generated by taking the total number of listeriosis cases for a subpopulation and dividing it by the estimated number of people in the total population that have that condition. This value is then divided by a similar value for the general population. While there is a substantial uncertainty associated with these values (i.e. a relative susceptibility value is the ratio of two uncertain estimates and the exposures (diets) of the different subpopulations are assumed to be equivalent), it does provide a useful estimate of the differences in the susceptibility among the different subpopulations and the role that immune status has in determining an individual's risk from *L. monocytogenes* (Table 5.5).

Relating the relative susceptibility values to the dose-response relations for the different subpopulations requires a means of converting these point estimates to a dose-response curve. The unique characteristics of the exponential model allowed this to be done. Being a single parameter model, the exponential model allows the entire dose-response curve to be generated once any point on the curve is known. Thus, the r-value for an exponential dose-response curve can be estimated for a subpopulation using a relative susceptibility ratio and a reference r-value for the general population. Using the relative susceptibility value for cancer patients as an example (Table 5.5), the equation for the relative susceptibility is:

$$\text{Relative susceptibility} = RS = P_{cancer}/P_{healthy} = [1 - \exp(-r_{cancer}*N)]/[1 - \exp(-r_{healthy}*N)]$$

where P_{cancer} and $P_{healthy}$ denote the probability of systemic listeriosis for a cancer patient and a healthy adult, respectively, when exposed to a dose N of *L. monocytogenes*, and where r_{cancer} and $r_{healthy}$ are the r-values of exponential dose-response relationships specific for those population sub-groups.

This equation can be rearranged to:

$$r_{cancer} = - \ln [RS * \exp(-r_{healthy}*N) - (RS - 1)]/N$$

As long as the value for N, the number of *L. monocytogenes* consumed, is much smaller than the maximum assumed dose, the above relationship can be used to estimate the $r_{subpopulation}$ value. Using the above equation, the r-values for different classes of patients were estimated based on epidemiological data from France (Tables 5.5) and the United States of America (Table 5.6).

Table 5.5 r-values (exponential dose-response model) for different susceptible populations calculated using relative susceptibility information from France. Relative susceptibilities for the different subpopulations are based on the incidence of listeriosis cases (outbreak and sporadic) in these groups in 1992.

Condition	Relative susceptibility	Calculated r-value(1)	Comparable outbreak r-value
Transplant	2 584	1.41×10^{-10}	Finland butter 3×10^{-7}
Cancer – Blood	1 364	7.37×10^{-11}	
AIDS	865	4.65×10^{-11}	
Dialysis	476	2.55×10^{-11}	
Cancer – Pulmonary	229	1.23×10^{-11}	
Cancer – Gastrointestinal and liver	211	1.13×10^{-11}	
Non-cancer liver disease	143	7.65×10^{-12}	
Cancer – Bladder and prostate	112	5.99×10^{-12}	
Cancer – Gynaecological	66	3.53×10^{-12}	
Diabetes, insulin dependent	30	1.60×10^{-12}	
Diabetes, non-insulin dependent	25	1.34×10^{-12}	
Alcoholism	18	9.60×10^{-13}	
Over 65 years old	7.5	4.01×10^{-13}	
Less than 65 years, no other condition (reference population)	1	5.34×10^{-14}	

NOTES: (1) The r-value assumed for the reference population – "Less than 65 years, no other medical condition" – was 5.34×10^{-14}, which is the median of the r-value calculated assuming a maximum level of 8.5 $\log_{10}$ CFU per serving.

SOURCE: Marchetti, 1996.

Table 5.6 Dose-response curves for different susceptible populations calculated using relative susceptibility information from the United States of America. Relative susceptibilities for the different sub-populations are based on the incidences of listeriosis cases (outbreak and sporadic) in these groups.

Condition	Relative susceptibility	Calculated r-value(1)	Comparable outbreak r-value
Perinatal	14	4.51×10^{-11}	Los Angeles cheese 3×10^{-11}
Elderly (60 years and older)	2.6	8.39×10^{-12}	
Intermediate-age population (reference population)	1	5.34×10^{-14}	

NOTES: (1) The r-value assumed for the reference population – "Intermediate-age population" – was 5.34×10^{-14}, which is the median of the r-values calculated under the assumption of a maximum level of 8.5 $\log_{10}$ CFU per serving.

SOURCE: FDA/FSIS, 2001.

Comparison of the relative susceptibility values and corresponding r-values are consistent with the physiological observation that as an individual's immune system is increasingly compromised, the risk of listeriosis at any given dose increases and this is reflected in a corresponding increase in the r-value of the dose-response curve. The most compromised group in the French data, transplant patients, has an r-value approximately 4 orders of magnitude greater than the reference population (i.e. individuals less than 65 years old with no other medical conditions). The relative susceptibility values for the elderly population showed close agreement, 7.5 and 2.6 for the French and United States of America data, respectively. The differences reflect, in part, the different definitions of the age corresponding to the category "elderly" and the reference population. The United States of America intermediate-age population includes the patients that are separated out from the less-than-65-years-of-age group in the French data and the two reference populations are not expected, therefore, to have the same r-values. Nevertheless, the two tables indicate the magnitude of the impact that the impairment of the immune system by the specific conditions and disease states has on susceptibility to listeriosis.

The two outbreak r-values provide an indication of the validity of the models. The r-value for the Los Angeles outbreak in pregnant women from consumption of Hispanic cheese was very close to that estimated (Table 5.6). The r-value for the Finland outbreak from butter in hospitalized transplant patients differed from the values based on transplant patients by 1000-fold (Table 5.5). This may have resulted from the smaller number of individuals exposed, the extremely compromised and highly variable immunological status of the population, or the involvement of a highly virulent strain of *L. monocytogenes*. There is a clear need in future outbreaks for exposure levels, immune status of the patients and strain characteristics to all be investigated so that these dose-response models can be further refined and validated.

5.4 QUESTION 3

Estimate the risk from L. monocytogenes *in foods that support growth and foods that do not support growth at specific storage and shelf-life conditions.*

L. monocytogenes growth on foods is not the only determinant of risk of listeriosis. Additional factors that affect the risk associated with any food, regardless of whether it does or does not support *L. monocytogenes* growth, include:

- frequency of contamination;
- level of contamination;
- frequency of consumption; and
- susceptibility of consuming population.

This question suggests a number of alternative approaches to a simple growth/no-growth evaluation, such as a consideration of the effect on consumer risk of limiting the storage temperature and shelf-life of a product that supports the growth of *L. monocytogenes*. The risk assessment team has attempted to also consider these approaches while formulating its answer to the question.

As was discussed in the response to Question 1 (Section 5.2), it is possible that a food that does not permit the growth of *L. monocytogenes* but that is frequently contaminated at moderate levels could pose a greater risk than a food infrequently contaminated, or contaminated at low levels, but that does support growth of *L. monocytogenes*. Also, as noted previously, it is clear that an increase in the *total* numbers of *L. monocytogenes* in a food (whether through growth or increased frequency of contamination) will lead to increased consumer risk because, for *L. monocytogenes*, the dose-response model used indicates that public health risk is proportional to total number of *L. monocytogenes* in the food when consumed. Furthermore, as bacterial growth is exponential, the risk might be expected to increase exponentially with storage time.

Three approaches for answering this question are provided:

(i) general consideration of the impact of the ingested dose on the risk of listeriosis;

(ii) comparison of four foods that were selected, in part, to evaluate the effect of growth on risk; and

(iii) comparison of what-if scenarios for the foods evaluated that do support *L. monocytogenes* growth if they did *not* support *L. monocytogenes* growth. Each of the approaches is discussed below.

5.4.1 Growth rates in foods

L. monocytogenes is able to grow in many RTE foods, even if stored under appropriate refrigeration conditions. Factors affecting the growth of *L. monocytogenes* in foods are discussed in detail in Section 3.5. These include product formulation, storage time and temperature, and interactions with other microorganisms present in the product. In vacuum-packed foods, lactic acid bacteria can reach stationary phase without causing product spoilage. This can slow, or even prevent, the subsequent growth of *L. monocytogenes*. Table 5.7 presents representative generation times for different products as a function of product type and storage temperature. For every three generations of growth, there is approximately a 10-fold increase in the bacterial population. As discussed in Section 5.2, a 10-fold increase in the levels of *L. monocytogenes* ingested produces a corresponding 10-fold increase in risk to human health. Thus, the risk from a food that supports the growth of *L. monocytogenes* increases with increasing storage time. However, the degree that the risk increases is dependent on the extent of growth in the food, which, in turn, is largely a function of *L monocytogenes'* growth rate in the food and the storage duration and conditions.

L. monocytogenes has been reported to grow in foods at temperatures as low as 0°C, water activities as low as 0.91–0.93 and pH as low as 4.2 (see Table 3.1). Combinations of suboptimal levels reduce the growth rate and can prevent growth at less extreme conditions than any of these factors acting alone. This principle, often referred to as hurdle technology or combination treatment, is exploited in food processing to prevent or limit the growth of bacteria in RTE foods.

The potential extent of growth varies among different foods, depending on the pathogen's growth rate in a specific food, which is a function of the product's composition and storage conditions, and on shelf-life of the product. From Table 5.7 it is evident that the growth of *L. monocytogenes* within the normal shelf-life of products could be substantial. For example, fresh cut vegetables have a relatively short shelf-life and do not support as rapid growth of *L. monocytogenes* as some other foods, such as milk or deli-meats. Thus, it would be expected that extent of growth in fresh cut vegetables would not be as great as those in other foods, resulting in a lower risk for given initial contamination rates and levels.

The example of the effect of storage time and temperature on the growth of *L. monocytogenes* and the subsequent risk of listeriosis can be considered a worst-case scenario in that it only considers the effect of temperature on generation times. Additional factors that act to delay the initiation of growth of *L. monocytogenes* (e.g. consideration of the lag phase), reduce the rate of growth (e.g. modified-atmosphere packaging), or suppress the maximum level reached by *L. monocytogenes* (e.g. growth of lactic acid bacteria) would decrease the extent of growth within a specified period of a product's shelf-life, with a corresponding decrease in risk. The actual calculation of risk would also have to consider that different servings would be consumed at various times within the total product shelf-life, as typically only a small fraction of a product is consumed near the end of its declared shelf-life.

5.4.2 Comparison of four foods

As discussed above, the four foods evaluated in the risk assessment (milk, ice cream, cold-smoked fish, and fermented meat products) were selected, in part, to compare the effect of various product characteristics on growth. This included specific consideration of the ability of foods to support growth. Thus, milk and ice cream were compared because they have similar compositions, servings sizes, frequencies of consumption, and rates and extents of initial contamination. However, milk supports *L. monocytogenes* growth while ice cream does not. Similarly, cold-smoked fish and fermented meat products have similar rates of initial contamination, serving sizes and frequencies of consumption, but the former supports the growth of *L. monocytogenes* while the latter does not.

Table 5.7 Representative generation times (hours) and growth potential of *Listeria monocytogenes* at different temperatures and shelf lives at 5°C in various RTE foods.

Temperature (°C)	Milk	Vacuum-packed cold-smoked fish	Vacuum-packed processed meats	Sliced vegetables
	Generation time (hours)			
5[1]	27.6	46.6	29.6	111
(95% confidence interval)	(14–226)	(20–infinite)	(14–infinite)	(28–infinite)
5[2]	25–30	40–49	16–48	–
10[2]	5–7	8–11	7–10	–
25[2]	0.7–1.0	1.2–1.7	1- 1.6	–
	Growth potential[3]			
5	~2–3	~4–5	~8–9	~0.3
	Advisory shelf-life (weeks)			
5	1–2	4–6	6–8	1

NOTES: (1) Values based on data collated in FDA/FSIS, 2001.
(2) Representative predictions and ranges from several published predictive models developed for growth rate of *L. monocytogenes.* No predictions were possible for vegetables because none of the published models were developed, or validated, for use with sliced vegetables.
(3) Log increase ignoring lag phase or suppression of growth by lactic acid bacteria.

Comparisons of the predicted risk per million servings (Table 4.34) between milk and ice cream, and cold-smoked fish and fermented meat products, indicate that the ability of a product to support growth within its shelf-life can increase substantially the risk of that product being a vehicle for foodborne listeriosis. Thus, the predicted risk per million servings of milk was approximately 100-fold greater than that for ice cream, and the risk for cold-smoked fish was approximately 10 000-fold greater than the corresponding risk for fermented meat products.

5.4.3 What-if scenarios

One of the useful features of a quantitative risk assessment is that the underlying mathematical models can be modified to allow various what-if scenarios to be run to evaluate the likely impact of different risk management options. Accordingly, a limited number of what-if scenarios were evaluated for milk and cold-smoked seafood, the two foods that supported the growth of *L. monocytogenes* and considered in the risk assessment. The results of these analyses were then compared to the predicted baseline risks to determine the impact of the intervention.

5.4.3.1 Milk

The initial assessment of risk associated with recontaminated pasteurized milk considered the likely growth of *L. monocytogenes* during the shelf-life of the product (see Section 4.2), using Canadian consumption characteristics as an example. To help answer CCFH Question 3, the model was re-executed after being modified so that the effect of growth was ignored, i.e. no growth during storage was modelled. The results of the two calculations were then compared to estimate the effect of growth on risk (Table 5.8).

The results suggest that an approximately 1000-fold increase in risk can be attributed to the predicted growth of *L. monocytogenes* in pasteurized milk by either measure of risk, i.e. risk per 1 million meals or risk per 100 000 population. The uncertainty measures associated with the comparison suggested that the predicted increase in risk attributable to growth could be as little as 100-fold, or as much as >10 000-fold.

Several what-if scenarios were calculated for milk to illustrate the interactions of the various factors in determining the risks (Table 5.9). In one scenario, if all milk was consumed immediately after purchase at retail, the risks per serving and cases per population in both susceptible and healthy populations would decrease approximately 1000-fold. In contrast, if the contamination levels of milk were truncated at 100 CFU/g at retail but with growth still allowed, the incidence of listeriosis is predicted to be reduced by only about 70%. Two scenarios examined the impact of storage temperatures and times. When the temperature distribution was shifted so the median increased from 3.4 to 6.2°C, the mean number of illnesses increased over 10-fold for both populations. When the storage time distribution was shifted from a median of 5.3 days to 6.7 days, the mean rate of illnesses increased 4.5-fold and 1.2-fold for the healthy and susceptible populations, respectively.

5.4.3.2 Smoked Fish

The assumptions used with the cold-smoked fish model differ slightly from those used with the pasteurized milk example. The cold-smoked fish model also considers the effect of the growth of indigenous lactic acid bacteria in the product, which, when they grow to high numbers, suppress the growth of *L. monocytogenes* (see Section 4.5). The extent of that growth suppression is not known with certainty. In the baseline model, two assumptions concerning the growth rate suppression by lactic acid bacteria were tested. In the what-if scenario the growth rate inhibition of *L. monocytogenes* by the lactic acid bacteria was set to zero. Table 5.10 compares the risk estimates when growth was modelled to occur or not, including the effect of different assumptions about the magnitude of the inhibition of *L. monocytogenes* growth rate due to the growth of lactic acid bacteria.

Table 5.8 Estimates of the increase in risk of listeriosis from growth during storage of pasteurized milk between purchase and consumption.

	Normal-risk population		High-risk population		Mixed population	
	Mean	(s.e.)[a]	Mean	(s.e.)	Mean	(s.e.)
With growth (baseline model)						
Cases per 100 000 population	1.6×10^{-2}	(5.0×10^{-4})	5.2×10^{-1}	(3.1×10^{-2})	9.1×10^{-2}	(4.7×10^{-3})
Cases per 1 000 000 servings	1.0×10^{-3}	(1.0×10^{-4})	2.2×10^{-2}	(9.0×10^{-4})	5.0×10^{-3}	(2.0×10^{-4})
Without growth						
Cases per 100 000 population	1.3×10^{-5}	(6.7×10^{-8})	3.8×10^{-4}	(1.6×10^{-6})	6.7×10^{-5}	(2.4×10^{-7})
Cases per 1 000 000 servings	5.9×10^{-7}	(3.1×10^{-9})	1.7×10^{-5}	(7.5×10^{-8})	3.6×10^{-5}	(1.4×10^{-8})
Increased risk with growth relative to that without growth (n-fold increase)						
Cases per 100 000 population	1 231		1 366		1 358	
Cases per 1 000 000 servings	1 695		1 294		139	

KEY: s.e. = Standard error of the mean.

Table 5.9 Three what-if scenarios that illustrate the impact of contamination and storage on the estimated risks of listeriosis per 100 000 population and per 1 000 000 servings for milk under typical conditions of storage and use.

Food	Estimated mean cases of listeriosis per 100 000 people	Estimated mean cases of listeriosis per 1 000 000 servings
Milk baseline	9.1×10^{-2}	4.6×10^{-3}
No growth	6.7×10^{-5}	
With contamination truncated at 100 CFU/g	2.8×10^{-2}	
Increase storage temperature (from 3.4 to 6.2°C)	1.2×10^{0}	
Increase storage time (from 5.3 to 6.7 days)	2.0×10^{-1}	

With either assumption concerning the effect of lactic acid bacteria on *L. monocytogenes* growth potential, growth greatly increased the risk of listeriosis. Assuming that 80 to 100% suppression occurred, it allowed more growth than the assumption of 95% growth rate suppression, a result of the faster overall growth rate after lactic acid bacteria have achieved maximum population growth. The risk per serving and cases per 100 000 population increased 700- to 1000-fold in the first assumption (80–100% growth rate suppression) and 67- to 85-fold under the latter assumption (95%) from the "no *L. monocytogenes* growth" to the baseline (growth) scenarios.

For the cold-smoked fish model, between 15 and 20% of the population were assumed to be in the high-risk category, but the cases attributable to the normal and high-risk categories were not estimated discretely. Rather, as in the previous example, the predicted number of cases is a weighted mean of the normal and high-risk populations. It is known that the population with increased susceptibility to listeriosis experiences between 80 and 98% of total reported cases of listeriosis. Also, in this example, no attempt to differentiate consumption between these two susceptibility classes was made, unlike that undertaken in the assessment of milk (Section 4.2). These differences do not affect the interpretation of the results with a food but some caution must be exercised in comparing the impact of growth on the risk *between* the foods. However, the differences in the modelling are relatively minor and the predicted increase in risk due to growth in the two examples is roughly comparable. For example, in the case of pasteurized milk (Table 5.9), the modelling also suggests that the increase in risk due to the growth of *L. monocytogenes* within the normal shelf-life of the product is between approximately 100- and 1000-fold, similar to the risk increase predicted for cold-smoked fish due to *L. monocytogenes* growth during storage.

A further what-if scenario was performed to estimate the effect on risk of reducing the shelf-life of smoked fish by 50%. This was tested by replacing the original shelf-life distribution of 1–28 days, with a most likely value of 14 days, by a shelf-life distribution of 1–14 days, with a most likely value of 7 days. The effect of this change resulted in an 80% reduction in the predicted increase in risk due to growth. The fact that the change was not greater is probably due to the effect of lactic acid bacteria, which is modelled to begin to suppress *L. monocytogenes* growth after approximately 3 weeks of storage at 5°C (see Section 4.5.3.7).

Table 5.10 Impact of the growth of *Listeria monocytogenes* during storage of cold-smoked fish between purchase and consumption on the risk of listeriosis under typical conditions of storage and use.

Growth rate inhibition due to growth of lactic acid bacteria	Cases per 1 000 000 meals		Cases per 100 000 population	
	No Growth	Growth Modelled	No Growth	Growth Modelled
80–100%	4.51×10^{-4}	4.59×10^{-1}	9.60×10^{-5}	6.57×10^{-2}
	(3.09×10^{-5})[1]	(3.29×10^{-1})	(1.07×10^{-5})	(3.78×10^{-2})
Difference[2]		1020-fold		684-fold
95%		3.82×10^{-2}		6.48×10^{-3}
		(1.96×10^{-2})		(2.26×10^{-3})
Difference[2]		85-fold		67-fold

NOTE: (1) Values in parentheses are standard deviations. (2) Increase in risk of listeriosis in the growth versus the no-growth scenarios

5.4.4 Summary

Three different approaches were taken to demonstrate the effect of growth of *L. monocytogenes* on the risk of listeriosis associated with RTE foods. It is apparent that the potential for growth strongly influences risk, though the extent of that increase is dependent on the characteristics of the food and the conditions and duration of refrigerated storage. However, using the examples provided in the risk assessment, the ability of these RTE foods to support the growth of *L. monocytogenes* appears to increase the risk of listeriosis on a per-serving basis by 100- to 1000-fold over what the risk would have been if the foods did not support growth. While it is not possible to present a single value for the increased risk for all RTE foods because of the different properties of the various foods, the range of values here provide some insight into the magnitude of the increase in risk that may be associated with the ability of a food to support the growth of *L. monocytogenes.*

Part 6.

Key Findings and Conclusions

This risk assessment reflects the state of knowledge on listeriosis and on contamination of foods with *L. monocytogenes* when the work was undertaken, in 2002. It provides an insight into some of the issues to be addressed in order to control the problems posed by *L. monocytogenes,* and approaches for modelling a system to evaluate potential risk management options. It addresses the specific questions posed by the CCFH and provides a valuable resource for risk managers in terms of the issues to be considered when managing the problems associated with *L. monocytogenes,* and alternative or additional factors or means to consider when addressing a problem.

A number of important findings have come out of this work. Firstly, the probability of illness as a result of consuming a specified number of *L. monocytogenes* is appropriately conceptualized by the disease triangle, where the food matrix, the virulence of the strain and the susceptibility of the consumer are all important factors. However, little information was found on food matrix effects for *L. monocytogenes*. In animal studies the impact of strain variation on virulence has been shown to be large, but it is not currently possible to determine the human virulence for any individual strain and explicitly include that in the model. However, the epidemiologically-based models used in the risk assessment implicitly consider the variation in virulence among strains. Population-based models were developed that estimate the likelihood of illness for various immunocompromised human populations after consuming specified numbers of *L. monocytogenes*. Although the maximum levels of contamination at consumption are uncertain, different models based on different values all lead to the same general findings.

An important finding of the risk assessment was that, based on the predictions of the models developed, nearly all cases of listeriosis result from the consumption of high numbers of the pathogen. Conversely, the models predict that the consumption of low numbers of *L. monocytogenes* has a low probability of causing illness. Old age and pregnancy increase susceptibility and thus the risk of acquiring listeriosis when exposed to *L. monocytogenes*. Likewise, diseases and medical interventions that severely compromise the immune system greatly increase the risks. The risk of acquiring listeriosis from the consumption of contaminated food appears to be adequately described by the type of "probabilistic statement" that underlies the exponential dose-response relationship used in the risk assessment, namely, that there is a finite, albeit exceedingly small, possibility that a case of listeriosis could occur if an unusually susceptible consumer ingested low numbers of an unusually virulent strain

The data used in this risk assessment came from a number of different countries, although these were predominantly industrialized countries. Based on this available data there is no evidence that the risk from consuming a specific number of *L. monocytogenes* varies from one country to another for the equivalent population. Differences in manufacturing and handling practices in various countries may affect the contamination pattern and therefore the risk per serving for a food. The public health impact of a food can be evaluated by both the risk per serving (considers the frequency of contamination and the distribution of contamination levels within that particular food), and the annual number of cases per population (considers the number of servings of the food consumed by the population and the size of that population). A food may have a relatively high risk per serving but, if a minor component of the national diet, it may have a relatively small impact on public health as defined by the number of cases per year attributable to that food. Conversely, a food that has a relatively small risk per serving but that is consumed frequently and in large quantities may account for a greater portion of the cases within a population.

With regard to the outcome of the modelling work undertaken, this risk assessment indicates that control measures that reduce the frequencies of contamination with *L. monocytogenes* bring about proportional reductions in the rates of illness, provided the proportions of high contaminations are reduced similarly. Control measures that prevent the occurrence of high levels of contamination at consumption would be expected to have the greatest impact on reducing the rates of listeriosis. Contamination with high numbers of *L. monocytogenes* at manufacturing and retail is rare, and foods such as ice cream and fermented meat products that do not permit growth during storage have relatively low risks per serving and low annual risks per population. In foods that permit growth during storage, particularly if stored at higher temperatures or for longer duration, the low numbers of *L. monocytogenes* at manufacture and retail may increase during storage to levels that represent substantially elevated relative risks of causing listeriosis.

Although high levels of contamination at retail are relatively rare, improved public health could be achieved by reducing these occurrences at manufacture and retail in foods that do not permit growth. In foods that permit growth, control measures, such as better temperature control or limiting the length of storage periods, will reduce the increase in risk that occurs due to growth of *L. monocytogenes*. Re-formulating foods so they do not support growth would be expected to also reduce the occurrence of high doses and thus reduce the risk of listeriosis.

Finally, based on the risk assessment it is concluded that the vast majority of cases of listeriosis are associated with the consumption of foods that do not meet current standards for *L. monocytogenes* in foods, whether the standard is zero tolerance or 100 CFU/g. Raising a zero tolerance standard to a higher value (e.g. changing the standard from 1 CFU/25 g to 100/g) would be expected to result in increased incidence of listeriosis. However, if by relaxing the standard, there was a greater level of compliance with that standard through the improved adoption of control measures that significantly decreased the incidence of RTE food servings that exceeded the standard, particularly the number of servings with elevated levels of *L. monocytogenes,* then increasing the standard would actually have a positive impact on public health.

While this risk assessment has documented a number of important findings and addressed specific risk management questions from Codex it is not without its weaknesses. It is important that these are recognized, acknowledged and documented. This facilitates better understanding of the risk assessment as well as its correct interpretation and use. Transparency in this area can actually help minimize the weaknesses. There are a number of limitations and caveats to this current risk assessment that the end user should be aware of so that he/she can make optimal use of the work in the appropriate manner. These are outlined below.

- The risk assessment focuses on four RTE foods and only examines them from retail to consumption. This limits the application of the risk assessment particularly with regard to the consideration of risk management options at the primary production and processing stages.
- The risk characterization results are subject to uncertainty associated with a modelled representation of reality involving simplification of the relationships among prevalence, cell number, growth, consumption characteristics and the adverse response to consumption of some number of *L. monocytogenes* cells. However, the modelling is appropriate to quantitatively describe uncertainty and variability related to all kinds of factors and attempts to provide estimates of the uncertainty and variability associated with each of the predicted levels of risk.
- The amount of quantitative data available on *L. monocytogenes* contamination was limited and restricted primarily to European foods.
- Data on the prevalence and number of *L. monocytogenes* in foods came from many different sources, which adds to uncertainty and variability. Also, assumptions had to be made with regard to distribution of the pathogen in foods.
- The data used for prevalence and cell numbers may not reflect changes in certain commodities that have occurred in the food supply chain during the past ten years.
- The consumption characteristics used in the risk assessment were primarily those for Canada or the United States of America.
- The r-values and their distributions were developed using epidemiological data on the current frequency of *L. monocytogenes* strain diversity observed, with their associated virulence. If that distribution of virulence were to change (as reflected by new epidemiological data), the r-values would have to be re-calculated.
- There is uncertainty associated with the form of the dose-response function used, and with the parameterization. Also, the dose-response section of the hazard characterization is entirely a product of the shape of the distribution of predicted consumed doses in the exposure assessment component of the *Listeria* risk assessment undertaken in the United States of America (FDA/FSIS, 2001). Therefore its validity is dependant on the validity of the FDA/FSIS exposure assessment, and changes to that exposure assessment should lead directly to changes in the parameter, r.
- Predictive modelling was used to model the growth of *L. monocytogenes* in RTE foods, between the point of retail and the point of consumption, and the exposure assessment was based on information derived from those models. It is known that models may overestimate growth in food, and so reliance on such a model can result in an overestimation of the risk.

While the available data were considered adequate for the current purposes, the risk assessment could be improved with additional data of better quality for every factor in the assessment. The uncertainty ranges about the risks per serving and number of cases in a population indicate the effect of data gaps on the estimates.

The consumption data available are usually collected for nutritional purposes and lack critical information relevant to microbial quality. Contamination data were often neither recent, systematic, quantitative nor representative for different countries. In particular, the frequencies of high levels of contamination need to be better known. Additional knowledge on modelling growth would improve the estimates of the levels of *L. monocytogenes* consumed. Specific areas where further investigations would be beneficial include the maximum levels of growth, interactions with the indigenous spoilage flora (including the lactic acid bacteria), distributions of storage times, and interactions of storage times and temperatures with spoilage.

The dose-response models are all based upon pairing population consumption patterns with epidemiological statistics. Improved investigation of outbreaks of foodborne disease to determine the food involved, the amount of food consumed, number of *L. monocytogenes* consumed, the number of people exposed, number of people ill, the immunological status of all exposed people, and the virulence properties of the causative strain together would eventually lead to more accurate and specific dose-response models.

New data is constantly becoming available, but in order to complete this work it was not possible to incorporate the very latest data in the risk assessment. A future iteration of the work would incorporate such new data

This risk assessment reflects the current state of knowledge about the contamination of foods with *L. monocytogenes* and rates of listeriosis. Implementation of systematic surveys to determine the handling, consumption and contamination of foods would improve future risk assessments. Research to further the understanding of microbial growth dynamics would increase the ability to estimate final levels of contamination. More complete investigation of outbreaks of foodborne disease and determination of the virulence characteristics of *L. monocytogenes* will make the dose-response relationships more accurate and precise. Nevertheless, the dose-response models used in the current risk assessment should be applicable to all countries. Conversely, the exposure assessments are unique to each country and depend upon specific data on the factors that affect that population's exposure.

This risk assessment did not attempt to evaluate the factors that lead to the contamination of a food at retail. Additional product pathway exposure assessments for selected foods would provide additional understanding of how these foods become contaminated and the factors that have the greatest impact on preventing or eliminating that contamination. Creating valid product pathway assessments would then permit testing the impact on the incidences of listeriosis of various mitigations or postulated effects of regulatory changes. The critical factor in evaluating the risk from a food is the frequency distribution of the levels of contamination when that food is consumed. Estimating the actual effect of a proposed regulatory programme or risk mitigation strategy on this distribution is highly uncertain, yet determining the resulting change in the distribution is fundamental to reducing the occurrence of listeriosis.

This risk assessment should improve our overall understanding of the issue of *L. monocytogenes* in foods and associated listeriosis and it is anticipated that it can therefore pave the way for risk management action to address this problem at the international level.

Part 7

References cited

ABS [Australian Bureau of Statistics]. 1995. Australian National Nutrition Survey 1995. Australian Government Publishing Service, Canberra, Australia.

Andersen, J. K. & Nørrung, B. 1995. Occurrence of *Listeria monocytogenes* in Danish retail foods. pp. 241–244, *in: Proceedings of the XII International Symposium on Problems of Listeriosis*, Perth, Australia, 2–6 October 1995. Promaco Conventions Pty Ltd.

Andre, P., Roose, H., van Noyen, R., Dejaegher, L., Uyttendaele, I. & De Schrijver, K. 1990. Neuro-meningeal listeriosis associated with consumption of an ice cream. *Médecine et maladies infectieuses,* **20**: 570–572.

Arumugaswamy, R.K., Ali, G.R.R. & Hamid, S.N.B.A. 1994. Prevalence of *Listeria monocytogenes* in food in Malaysia. *International Journal of Food Microbiology*, **23**: 117–121.

Audits International. 2000. 1999 U.S. Food Temperature Evaluation. Design and Summary Pages. Audits International and U.S. Food and Drug Administration. 13p.

Audurier, A., Pardon, P., Marly, J. & Lantier, F. 1980. Experimental infection of mice with *Listeria monocytogenes* and *L. innocua. Annales Microbiologie (Inst. Pasteur)*, **131**: 47–57.

Augustin, J.C. & Carlier, V. 2000a. Mathematical modelling of the growth rate and lag time for *Listeria monocytogenes. International Journal of Food Microbiology*, **56**: 29–51.

Augustin, J.C. & Carlier, V. 2000b. Modelling the growth rate of *Listeria monocytogenes* with a multiplicative type model including interactions between environmental factors. *International Journal of Food Microbiology*, **56**: 53–70.

Augustin, J.C., Carlier, V. & Rozier J. 1998. Mathematical modelling of the heat resistance of *Listeria monocytogenes. Journal of Applied Microbiology*, **84**: 185–191.

Aureli, P., Fiorucci, G.C., Caroli, D., Marchiaro, B., Novara, O., Leone, L. & Salmoso, S. 2000. An outbreak of febrile gastroenteritis associated with corn contaminated by *Listeria monocytogenes*. *New England Journal of Medicine*, **342**: 1236–1241.

Avery, S.M. & Buncic, S. 1997. Differences in pathogenicity for chick embryos and growth kinetics at 37°C between clinical and meat isolates of *Listeria monocytogenes* previously stored at 4°C. *International Journal of Food Microbiology*, **34**: 319–327.

Babic, I., Watada, A.E. & Buta, J.G. 1997. Growth of *Listeria monocytogenes* restricted by native microorganisms and other properties of fresh-cut spinach. *Journal of Food Protection*, **60**: 912–917.

Baek, S.Y., Lim, S.Y., Lee, D.H., Min, K.H. & Kim, C.M. 2000. Incidence and characterization of *Listeria monocytogenes* from domestic and imported foods in Korea. *Journal of Food Protection*, **63**: 186–189.

Baker, M., Brett, M., Calder, L., & Thornton, R. 1993. Listeriosis and mussels. *Communicable Diseases New Zealand*, **93**: 13–14.

Ballen, P.H., Loffredo, F.R. & Painter, B. 1979. Listeria endophthalmitis. *Archives Ophthalmology*, **1**: 101–102.

Barker, J. & Brown, M.R.W. 1994. Trojan-horses of the microbial world – protozoa and the survival of bacterial pathogens in the environment. *Microbiology-UK*, **140**: 1253–1259.

Bean, N.H., Goulding, J.S., Lao, C. & Angulo, F.J. 1996. Surveillance for foodborne disease outbreaks – United States, 1988-1991. CDC Surveillance Summaries, October 25, 1996. *Morbidity and Mortality Weekly Report,* **45** (SS-5): 1–66.

Bell, C. & Kyriakides, A. 1998. *Listeria – a practical approach to the organism and its control in foods*. London: Blackie Academic and Professional.

Bell, R.G., Penny, N. & Moorhead, S.M. 1995. Growth of the psychrotrophic pathogens *Aeromonas hydrophila, Listeria monocytogenes* and *Yersinia enterolitica* on smoked blue cod (*Parapercis colias*) packed under vacuum or carbon dioxide. *International Journal of Food Science and Technology,* **30**: 515–521.

Bemrah, N., Sana, M., Cassin, M.H., Griffiths, M.W. & Cerf, O. 1998. Quantitative risk assessment of human listeriosis from consumption of soft cheese made from raw milk. *Preventative Veterinary Medicine*, **37**: 129–145.

Bille, J. 1990. Epidemiology of listeriosis in Europe, with special reference to the Swiss outbreak. pp. 71-74, *in:* Miller, Smith & Somkuti, 1990, q.v.

Black, R.E., Levine, M.M., Clements, M.L., Hughes, T.P. & Blaser, M.J. 1988. Experimental *Campylobacter jejuni* infection in humans. *Journal of Infectious Diseases*, **157**: 472–479.

Blaser, M.J. & Newman, L.S. 1982. A review of human Salmonellosis: I. Infective dose. *Review of Infectious Diseases*, **4**: 1096–1105.

Boerlin, P. & Piffaretti, J.C. 1991. Typing of human, animal, food and environmental isolates of *Listeria monocytogenes* by multilocus enzyme electrophoresis. *Applied and Environmental Microbiology*, **57**: 1624–1629.

Boerlin, P., Boerlin-Petzold, F., Bannerman, E., Bille, J. & Jemmi, T. 1997. Typing *Listeria monocytogenes* isolates from fish products and human listeriosis cases. *Applied and Environmental Microbiology*, **63**: 1338–1343.

Bourgeois, N., Jacobs, F., Tavares, M.L., Rickaert, F., Deprez, C., Liesnard, C., Moonens, F., Van de Stadt, J., Gelin, M. & Adler, M. 1993. *Listeria monocytogenes* hepatitis in a liver transplant recipient: A case report and review of the literature. *Journal of Hepatology*, **18**: 284289.

Bracegirdle, P., West, A.A., Lever, M.S., Fitzgeorge, R.B. & Baskerville, A. 1994. A comparison of aerosol and intragastric routes of infection with *Listeria* spp. *Epidemiology and Infection,* **112**: 69–76.

Brackett, R.E. 1999. Incidence and behaviour of *Listeria monocytogenes* in products of plant origin. pp. 631–654, *in:* Ryser & Marth, 1999, q.v.

Brackett, R.E. & Beuchat, L.R. 1990. Pathogenicity of *Listeria monocytogenes* grown on crabmeat. *Applied and Environmental Microbiology*, **56**: 1216–1220.

Braun, T.I., Travis, D., Dee, R.R & Nieman, R.E. 1993. Liver abscess due to *Listeria monocytogenes*: Case report and review. *Clinical Infectious Diseases*, **17**: 267–269.

Bréand, S., Fardel, G., Flandrois, J.P., Rosso, L. & Tomassone R. 1997. A model describing the relationship between lag time and mild temperature increase duration. *International Journal of Food Microbiology*, **38**: 157–167.

Bréand, S., Fardel, G., Flandrois, J.P., Rosso, L. & Tomassone, R. 1999. A model describing the relationship between regrowth lag time and mild temperature increase for *Listeria monocytogenes*. *International Journal of Food Microbiology*, **46**: 251–261.

Brehm, K., Kreft, J., Ripio, M.T. & Vazquez-Boland, J.A. 1996. Regulation of virulence gene expression in pathogenic *Listeria*. *Microbiología SEM*, **12**: 219–236.

Brett, M.S., Short, P. & McLauchlin, J. 1998. A small outbreak of listeriosis associated with smoked mussels. *International Journal of Food Microbiology*, **43**: 223–229.

Broome, C.V., Gellin, B. & Schwartz, B. 1990. Epidemiology of listeriosis in the United States. pp. 61–65, *in:* Miller, Smith & Somkuti, 1990, q.v.

Buchanan, R.L. & Bagi, L.K. 1997. Microbial competition: Effect of culture conditions on the suppression of *Listeria monocytogenes* Scott A by *Carnobacterium piscicola*. *Journal of Food Protection*, **60**: 254–261.

Buchanan, R.L. & Golden, M.H. 1994. Interaction of citric acid concentration and pH on the kinetics of *Listeria monocytogenes* inactivation. *Journal of Food Protection*, **57**: 567–570.

Buchanan, R.L. & Golden, M.H. 1995. Model for the non-thermal inactivation of *Listeria monocytogenes* in a reduced oxygen environment. *Food Microbiology*, **12**: 203–212.

Buchanan, R.L. & Golden, M.H. 1998. Interactions between pH and malic acid concentration on the inactivation of *Listeria monocytogenes*. *Journal of Food Safety*, **18**: 37–48.

Buchanan, R.L., Golden, M.H. & Phillips, J.G. 1997. Expanded models for the non-thermal inactivation of *Listeria monocytogenes*. *Journal of Applied Microbiology,* **82**: 567–577.

Buchanan, R.L., Golden, M.H., & Whiting, R.C. 1993. Differentiation of the effects of pH and lactic and acetic acid concentration on the kinetics of *Listeria monocytogenes* inactivation. *Journal of Food Protection*, **56**: 474–478.

Buchanan, R.L. & Phillips, J.G. 1990. Response surface model for predicting the effects of temperature, pH, sodium chloride content, sodium nitrite concentration and atmosphere on the growth of *Listeria monocytogenes*. *Journal of Food Protection*. **53**: 370–376.

Buchanan, R.L. & Phillips, J.G. 2000. Updated models for the effects of temperature, initial pH, NaCl, and $NaNO_2$ on the aerobic and anaerobic growth of *Listeria monocytogenes*. *Quantitative Microbiology*, **2**(2): 103–128.

Buchanan, R.L. & Whiting, R.C. 1997. Risk assessment – A means for linking HACCP plans and public health. *Journal of Food Protection*, **61**: 1531–1534.

Buchanan, R.L., Damert, W.G., Whiting, R.C. & van Schothorst, M. 1997. Use of epidemiologic and food survey data to estimate a purposefully conservative dose-response relationship for *Listeria monocytogenes* levels and incidence of listeriosis. *Journal of Food Protection*, **60**: 918–922.

Buchanan, R.L., Golden, M.H., Whiting, R.C., Phillips, J.G., & Smith, J.L. 1994. Model for the non-thermal inactivation of *Listeria monocytogenes*. *Journal of Food Science,* **59**: 179–188.

Buchanan, R.L., Klawitter, L.A., Bhaduri, S. & Stahl, H.G. 1991. Arsenite resistance in *Listeria monocytogenes*. *Food Microbiology*, **8**: 161–166.

Buchanan, R.L., Smith, J.L. & Long, W. 2000. Microbial risk assessment: Dose-response relations and risk characterization. *International Journal of Food Microbiology*, **58**: 159–172

Buchanan, R.L., Stahl, H.G., Bencivengo, M.M., & del Corral, R. 1989. Comparison of lithium chloride-phenylethanol-moxalactain and modified Vogel Johnson agars for detection of *Listeria* species in retail-level meats, poultry and seafood. *Applied and Environmental Microbiology*, **55**: 599–603.

Bula, C.J., Bille, J. & Glausser, M.P. 1995. An epidemic of foodborne listeriosis in western Switzerland: Description of 57 cases involving adults. *Clinical Infectious Disease,* **20**: 66–72.

Burmaster, D.E. & Anderson, P.D. 1994. Principles of good practice for the use of Monte Carlo techniques in human health and ecological risk assessments. *Risk Analysis*, **14**: 477–481.

Campanini, M., Pedrazonnoni, I., Barbuti, S. & Baldini, P. 1993. Behaviour of *Listeria monocytogenes* during the maturation of naturally and artificially contaminated salami: Effect of lactic acid bacteria starter cultures. *International Journal of Food Microbiology*, **20**: 169–175.

Cantoni, C., D'Aubert, S. & Valenti, M. 1989. *Listeria* spp. in formaggi ed insaccati crudi [*Listeria* spp. in cheese and uncooked dry sausages]. *Industrie Alimentari*, **28**: 1068–1070.

Carlos, V.S., Oscar, R.S. & Irma, Q.R.E. 2001. Occurrence of *Listeria* species in raw milk in farms on the outskirts of Mexico city. *Food Microbiology*, **18**: 177–181.

Casarotti, V.T., Gallo, C.R. & Camargo, R. 1994. [Occurrence of *Listeria monocytogenes* in raw milk, pasteurized type C milk and minas frescal cheese commercialized in Piracicaba-Sao Paulo] (abstract, Portuguese). *Archivos Latinoamericanos de Nutricion*, **44**: 158–163.

Cassin, M.H., Lammerding, A.M., Todd, E.C.D., Ross, W. & McColl, R.S. 1998. Quantitative risk assessment for *Escherichia coli* O157:H7 in ground beef hamburgers. *International Journal of Food Microbiology*, **41**: 21–44.

Cassin, M.H., Paoli, G.M., McColl, R.S. & Lammerding, A.M. 1996. A comment on Hazard assessment of *Listeria monocytogenes* in the processing of bovine milk. (*Journal of Food Protection,* **57**: 689–697 (1994)). *Journal of Food Protection*, **59**: 341–343.

CCFH [Codex Committee on Food Hygiene]. 1999. Discussion paper on management of *Listeria monocytogenes* in foods. Paper presented at Thirty-second Session, Washington DC, 29 November – 4 December 1999. Document no, CX/FH 99/10. FAO/WHO Rome. 15p.

CDC [U.S. Centers for Disease Control and Prevention]. 1998. 1997 Annual Report. CDC/USDA/FDA Foodborne Diseases Active Surveillance Network. CDC's Emerging Infections Program (Revised 14 March 2000).

CDC. 2000. FoodNet 2000. Foodborne Diseases Active Surveillance Network. CDC's Emerging Infections Program. 1999 surveillance results. Preliminary report.

CDC. 2001. Preliminary FoodNet data on the incidence of foodborne illnesses – selected sites, United States, 2000. *Morbidity and Mortality Weekly Report,* [April 06, 2001] **50**(13): 241-246. Complete data also published in *2000 Annual Report.* See: www.cdc.gov/foodnet/annual/2000/2000_annual_report_foodnet.htm

Cerf, O. 1977. Tailing of survival curves of bacterial spores. *Journal of Applied Bacteriology*, **42**: 1–19.

CFPNS [Canadian Federal-Provincial Nutrition Surveys]. 1992–1995. Bureau of Biostatistics and Computer Applications, Food Directorate, Health Canada. See also: Karpinski & Nargundkar, 1992; Junkins & Karpinski, 1994; Junkins, 1994; Junkins & Laffey, 2000; Junkins, Laffey & Weston, 2001.

Chakraborty, T., Ebel, F., Wehkland, J., Dufrenne, J. & Notermans, S. 1994. Naturally occurring virulence-attenuated isolates of *Listeria monocytogenes* capable of inducing long-term protection against infection by virulent strains of homologous and heterologous serotypes. *FEMS Immunology and Medical Microbiology*, **10**: 1–10.

Chasseignaux, E., Gerault, P., Toquin, M.T., Salvat, G., Colin, P. & Ermel, G. 2002. Ecology of *Listeria monocytogenes* in the environment of raw poultry meat and raw pork meat processing plants. *FEMS Microbiology Letters*, **210**: 271–275.

Cheers, C. & McKenzie, I.F.C. 1978. Resistance and susceptibility of mice to bacterial infection: Genetics of listeriosis. *Infection and Immunity*, **19**: 755–762.

Coleman, M., & Marks, H. 1998. Topics in dose-response modeling. *Journal of Food Protection*, **61**: 1550–1559.

Conlan, J.W. & North, R.J. 1992. Roles of *Listeria monocytogenes* virulence factors in survival: Virulence factors distinct from listeriolysin are needed for the organism to survive an early neutrophil-mediated host defense mechanism. *Infection and Immunity*, **60**: 951–957.

Cortesi, M.L., Sarli, T., Santoro, A., Murru, N. & Pepe, T. 1997. Distribution and behaviour of *Listeria monocytogenes* in three lots of naturally-contaminated vacuum-packed smoked salmon stored at 2 and 10°C. *International Journal of Food Microbiology*, **37**: 209–214.

Cox, N.E., Bailey, J.S. & Ryser, E.T. 1999.Incidence and behaviour of *Listeria monocytogenes* in poultry and egg products. pp. 565–600, *in:* Ryser & Marth, 1999, q.v.

Czuprynski, C.J., Brown, J.F. & Roll, J.T. 1989. Growth at reduced temperatures increases the virulence of *Listeria monocytogenes* for intravenously but not intragastrically inoculated mice. *Microbial Pathology*, **7**: 213–223.

Czuprynski, C.J., Theisen, C. & Brown, J.F. 1996. Treatment with the antigranulocyte monoclonal antibody RB6-8C5 impairs resistance of mice to gastrointestinal infection with *Listeria monocytogenes*. *Infection and Immunity*, **64**: 3946–3949.

Dalgaard, P. 1997. Predictive microbiological modelling and seafood quality. pp. 431–443, *in:* J. Luten, T. Børresen and J. Oehlenschläger (eds). *Seafood from Producer to Consumer, Integrated Approach to Quality.* Amsterdam, The Netherlands: Elsevier.

Dalgaard, P. & Jørgensen, L.V. 1998. Predicted and observed growth of *Listeria monocytogenes* in seafood challenge tests and in naturally contaminated cold-smoked salmon. *International Journal of Food Microbiology*, **40**: 105–115.

Dalgaard, P., Ross, T., Kamperman, L., Neumeyer, K. & McMeekin, T.A. 1994. Estimation of bacterial growth rates from turbidimetric and viable count data. *International Journal of Food Microbiology*, **23**: 391–404.

Dalton, C.B., Austin, C.C., Sobel, J., Hayes, P.S., Bibb, W.F., Graves, L.M. & Swaminathan, B. 1997. An outbreak of gastroenteritis and fever due to *Listeria monocytogenes* in milk. *New England Journal of Medicine,* **336**: 100–105.

Davies, A.R. 1997. Modified-atmosphere packaging of fish and fish products. pp. 200–223, *in:* G.M. Hall (ed). *Fish Processing Technology*. 2nd edition. London: Blackie Academic & Professional.

del Corral, F., Buchanan, R.L., Bencivengo, M.M. & Cooke, P. 1990. Quantitative comparison of selected virulence-associated characteristics in food and clinical isolates of *Listeria*. *Journal of Food Protection*, **53**: 1003–1009.

de Valk, H., Vaillant, V., Jacquet, C., Rocourt, J., Le Querrec, F., Stainer, F., Quelquejeu, N., Pierre, O., Pierre, V., Desenclos, J.C., & Goulet, V. 2001. Two consecutive nationwide outbreaks of Listeriosis in France, October 1999-February 2000. *American Journal of Epidemiology*, **154**(10): 944–950.

de Valk, H., Jacquet, C., Goulet, V., Vaillant, V., Perra, A., Desenclos, J-C., Martin, P. & the *Listeria* Working Group. 2003. Feasibility study for a collaborative surveillance of *Listeria* infections in Europe. Report to the European Commission, DG SANCO.

Delignette-Muller, M.L. 1998. Relation between the generation time and the lag time of bacterial growth kinetics. *International Journal of Food Microbiology*, **43**: 97–104.

Destro, M.T., Serrano, A.D. & Kabuki, D.Y. 1991. Isolation of *Listeria* species from some Brazilian meat and dairy products. *Food Control,* **2:** 110–112.

Dhanashree, B., Otta, S.K., Karunasagar, I., Goebel, W. & Karunasagar, I. 2003. Incidence of *Listeria* spp. in clinical and food samples in Mangalore, India. *Food Microbiology*, **20**: 447–453.

Di Lorenzo, G., Balistreri, C.R., Candore, G., Cigna, D., Colombo, A., Romano, G.C., Listi, F., Potestio, M. & Caruso C. 1999. Granulocyte and natural killer activity in the elderly. *Mechanisms Aging Development*, **108**: 25–38.

Dillon, R., Patel, T. & Ratnam, S. 1994. Occurrence of *Listeria* in hot and cold smoked seafood products. *International Journal of Food Microbiology*, **22**: 73–77.

Dowe, M.J., Jackson, E.D., Mori, J.G. & Bell, J.G. 1997. *Listeria monocytogenes* survival in soil and incidence in agricultural soils. *Journal of Food Protection*, **60**: 1201–1207.

Duffes, F., Leroi, F., Boyaval, P. & Dousset, X. 1999. Inhibition of *Listeria monocytogenes* by *Carnobacterium* spp. strains in a simulated cold-smoked fish system stored at 4°C. *International Journal of Food Microbiology*, **47**: 33–42.

Dunn, Son & Stone. 1998. Victorian Food Surveillance Annual Report. North Melbourne, Victoria: Dunn, Son & Stone.

Dupont, H.L., Hornick, R.B., Snyder, M.J., Libonati, J.P., Formal, S.B. & Gangarosa, E.J. 1972. Immunity in Shigellosis II. Protection induced by oral live vaccine or primary infection. *Journal of Infectious Diseases,* **125**(1): 12–16.

EC [European Commission]. 1999. Opinion of the Scientific Committee on Veterinary Measures Relating to Public Health on *Listeria monocytogenes*, 23 September 1999. European Commission, Health & Consumer Protection Directorate-General.

Eklund, M.W., Poysky, F.T., Pardnjpye, R.N., Lashbrook, L.C., Peterson, M.E. & Pelroy, G.A. 1995. Incidence and sources of *Listeria monocytogenes* in cold-smoked fishery products and processing plants. *Journal of Food Protection,* **58**: 502–508.

Eleftheriadou, M., Varnava-Tello, A., Metta-Loizidou, M., Nikolaou, A.S. & Akkelidou, D. 2002. The microbiological profile of foods in the Republic of Cyprus: 1991–2000. *Food Microbiology*, **19**: 463–471.

Ellis, L.C., Segreti, J., Gitelis, S. & Huber, J.F. 1995. Joint infections due to *Listeria monocytogenes*: Case report and review. *Clinical Infectious Diseases*, **20**: 1548–1550.

Encinas, J.P., Sanz, J.J., Garcia-Lopez, M.-L. & Otero, A. 1999. Behaviour of *Listeria* spp. in naturally contaminated *chorizo* (Spanish fermented sausage). *International Journal of Food Microbiology*, **46**: 167–171.

EPA [United States Environmental Protection Agency]. 1997. Guiding principles for Monte Carlo analysis. Doc. no. EPA/630/R-91/001. Washington DC, USA.

Erdenlia, S., Ainsworth, A.J. & Austin, F. W. 2000. Pathogenicity and production of virulence factors by *Listeria monocytogenes* isolates from channel catfish. *Journal of Food Protection*, **63**: 613–619.

Ericsson, H., Eklöw, A., Danielson-Tham, M.L, Loncarevic, S., Mentzing, L.O., Persson, I., Unnerstad, H. & Tham, W. 1997. An outbreak of listeriosis suspected to have been caused by rainbow trout. *Journal of Clinical Microbiology*, **35**: 2904–2907.

FAO. 1999. Report of the FAO Expert Consultation on the Trade Impact of *Listeria* in Fish Products. Amherst, MA, USA. 17–20 May 1999. *FAO Fisheries Report*, No. 604. 34p.

FAO/WHO. 2003. Hazard characterization for pathogens in food and water: Guidelines. *FAO/WHO Microbiological Risk Assessment Series*, No. 3. 61p.

FAO/WHO. 2004. Exposure assessment of microbiological hazards in food and water: Guidelines. *FAO/WHO Microbiological Risk Assessment Series*, No. 7 (*in press*).

Farber, J.M. 1991. *Listeria monocytogenes*. *Journal of the Association of Official Analytical Chemists*, **74**: 701–704.

Farber, J.M. & Pagotto, F. 1992. The effect of acid shock on the heat resistance of *Listeria monocytogenes*. *Letters in Applied Microbiology*, **15**: 197–201.

Farber, J.M. & Peterkin, P.I. 1991. *Listeria monocytogenes*: A foodborne pathogen. *Microbiology Reviews*, **55**: 476–511.

Farber, J.M. & Peterkin, P.I. 1999. Incidence and behaviour of *Listeria monocytogenes* in meat products. pp. 505–564, *in:* Ryser & Marth, 1999, q.v.

Farber, J. M., & Peterkin, P.I. 2000. *Listeria monocytogenes*. pp. 1178–1232, *in:* B.M. Lund, T.C. Baird-Parker and G.W. Gould (eds). *The Microbiological Safety of Food.* Vol. 2. Gaithersburg, Maryland: Aspen Publishers.

Farber, J.M., Daley, E., Coates, F., Beausoleil, N. & Fournier, J. 1991. Feeding trials of *Listeria monocytogenes* with a nonhuman primate model. *Journal of Clinical Microbiology*, **29**: 2606–2608.

Farber, J.M., Daley, E., Holley, R., & Usborne, W.R. 1993. Survival of *Listeria monocytogenes* during the production of uncooked German, American and Italian-style fermented sausages. *Food Microbiology*, **10**: 123–132.

Farber, J.M., Ross, W.H. & Harwig, J. 1996. Health risk assessment of *Listeria monocytogenes* in Canada. *International Journal of Food Microbiology*, **30**: 145–156.

Farber, J.M., Sanders, G.W. & Johnston, M.A. 1989. A survey of various foods for the presence of *Listeria* species. *Journal of Food Protection*, **52**: 456–458.

FDA/FSIS [U.S. Food and Drug Administration/Food Safety and Inspection Agency (USDA)]. 2001. Draft Assessment of the relative risk to public health from foodborne *Listeria monocytogenes* among selected categories of ready-to-eat foods. Center for Food Safety and Applied Nutrition (FDA) and Food Safety Inspection Service (USDA) (Available at: www.foodsafety.gov/~dms/lmrisk.html). [Report published September 2003 as: Quantitative assessment of the relative risk to public health from foodborne *Listeria monocytogenes* among selected categories of ready-to-eat foods. Available at: www.foodsafety.gov/~dms/lmr2-toc.html].

Feldman, M., Cryer, B., Sammer, D., Lee, E. & Spechler, S. J. 1999. Influence of *H. pylori* infection on meal-stimulated gastric acid secretion and gastroesophageal acid reflux. *AJP-Gastrointestional and Liver Physiology*, **277**: G1159–G1164.

Fernandez-Garayzabal, J.F., Dominguez-Rodriguez, L., Vazquez-Boland, J.A., Blanco-Cancelo, J.L. & Suarez-Fernandez, G. 1986. *Listeria monocytogenes* dans le lait pasteurizé. *Canadian Journal of Microbiology*, **32:** 149–150.

Fleming, D.W., Cochi, S.L., MacDonald, K.L., Brondum, J., Hayes, P.S., Plikaytis, B.D., Holmes, M.B., Auduruer, A., Broome, C.V. & Reingold, A.L. 1985. Pasteurized milk as a vehicle of infection in an outbreak of listeriosis. *New England Journal Medicine*, **312**: 404–407.

FSIS [Food Safety and Inspection Service]. 1995. Focus on: Sausages. Consumer Education and Information, Food Safety and Inspection Service, United States Department of Agriculture, Washington, DC. See: www.fsis.usda.gov/OA/pubs/sausages.htm

Fyfe, W.M., Campbell, D.M., Galea, P., Gellin, B.G. & Broome, C.V. 1991. Neonatal listeriosis in Scotland. *Annals Academy Medicine*, **20**: 236–240.

Gahan, C.G., O'Driscoll, B. & Hill, C. 1996. Acid adaptation of *Listeria monocytogenes* can enhance survival in acidic foods and during milk fermentation. *Applied and Environmental Microbiology*, **62**: 3128–3132.

Gaillard, J.L., Berche, P. & Sansonetti, P. 1986. Transposon mutagenesis as a tool to study the role of haemolysin in the virulence of *Listeria monocytogenes*. *Infection and Immunity*, **52**: 50–55.

Gallagher, P.G. & Watanakunakorn, C. 1988. *Listeria monocytogenes* endocarditis: A review of the literature 1950–1986. *Scandinavian Journal of Infectious Diseases*, **20**: 359–368.

Garland, C.D. 1995. Microbiological quality of agriculture products with special reference to *Listeria monocytogenes* in Atlantic salmon. pp. 261–275, *in: Proceedings of the XII International Symposium on Problems of Listeriosis*, Perth, Australia, 2–6 October 1995. Promaco Conventions Pty Ltd.

Gauto, A.R., Cone, L.A., Woodard, D.R., Mahler, R., Lynch, R.D. & Stoltzman, D.H. 1992. Arterial infections due to *Listeria monocytogenes*: Report of four cases and review of world literature. *Clinical Infectious Diseases*, **14**: 23–28.

Gellin, B.G. & Broome, C.V. 1989. Listeriosis. *Journal of the American Medical Association*, **261**: 1313–1320.

Genovese, F., Mancuso, G., Cuzzola, M., Biondo, C. Beninati, C., Delfino, D. & Teti, G. 1999. Role of IL-10 in a neonatal mouse listeriosis model. *Journal of Immunology*, **163**: 2777–2782.

George, S.M., Richardson, L.C.C. & Peck, M.W. 1996. Predictive models of the effect of temperature, pH and acetic and lactic acids on the growth of *Listeria monocytogenes*. *International Journal of Food Microbiology,* **32**: 73–90.

Gianella, R.A., Broitman, S.A. & Zamcheck, N. 1973. Influence of gastric acidity on bacterial and parasitic enteric infections. A perspective. *Annals of Internal Medicine*, **78**: 271–276.

Glaser, P., Frangeul, L., Buchrieser, C., Rusniok, C. and 51 others. 2001. Comparative genomics of *Listeria* species. *Science,* **294**(5543): 849–852.

Glynn, J.R. & Bradley, D.J. 1992. The relationship between infecting dose and severity of disease in reported outbreaks of salmonella infections. *Epidemiology and Infection,* **109**: 371–388.

Glynn, J.R., & Palmer, S.R. 1992. Incubation period, severity of disease, and infecting dose: Evidence from a salmonella outbreak. *American Journal of Epidemiology*, **136**: 1369–1377.

Gohil, V.S., Ahmed, M.A., Davies, R. & Robinson, R.K. 1995. Incidence of *Listeria* spp. in retail foods in the United Arab Emirates. *Journal of Food Protection*, **58**: 102–104.

Golden, M.H., Buchanan, R.L. & Whiting, R.C. 1995. Effect of sodium acetate or sodium propionate with EDTA and ascorbic acid on the inactivation of *Listeria monocytogenes*. *Journal of Food Safety,* **15**: 53–65.

Golnazarian, C.A., Donnelly, C.W., Pintauro, S.J. & Howard, D.B. 1989. Comparison of infectious dose of *Listeria monocytogenes* F5817 as determined for normal versus compromised C57B1/6J mice. *Journal of Food Protection*, **52**: 696–701.

Gould, G.W. 1989. Drying, raised osmotic pressure and low water activity. pp. 97–117, *in*: G.W. Gould (ed). *Mechanisms of Action of Food Preservation Procedures*. London: Elsevier Applied Science.

Goulet, V. 1995. Investigation en cas d'épidémie de listériose. *Médecine Maladies Infectieuses*, **25**: 184–190.

Goulet, V. & Marchetti, P. 1996. Listeriosis in 225 non-pregnant patients in 1992: clinical aspects and outcome in relation to predisposing conditions. *Scandinavian Journal of Infectious Diseases*, **28**: 367–374.

Goulet, V., Jacquet, C., Vaillant V., Rebiere, I., Mouret, E., Lorente, C., Maillot, E., Stanïer, F. & Rocourt, J. 1995. Listeriosis from consumption of raw-milk cheese. *Lancet*, **345**(8964): 1581–1582.

Goulet, V., Rocourt, J., Rebiere, I., Jacquet, C. Moyse, C., Dehaumont, P., Salvat, G. & Veit, P. 1998. Listeriosis outbreak associated with the consumption of rillettes in France in 1993. *Journal of Infectious Disease*, **177**: 155–160.

Goulet, V., de Valk, H., Pierre, O., Stanier, F., Rocourt, J., Vaillant, V., Jacquet, C., & Desenclos, J-C. 2001a. Effect of prevention measures on incidence of human listeriosis, France 1987–1997. *Emerging Infectious Diseases*, **7**: 983–989.

Goulet, V., Jacquet, C., Laurent, E., Rocourt, J., Vaillant, V., & de Valk, H. 2001b. La surveillance de la listériose humaine en France en 1999. *Bulletin Epidemiologique Hebdomadaire*, **34**: 161–165.

Grau, F.H. 1993. Processed meats and *Listeria monocytogenes*. pp. 13–24, *in*: Prevention of Listeria in Processed Meats. Proceedings of a series of workshops. CSIRO Division of Food Science and Technology, Meat Research Laboratory, Queensland, Australia.

Grau, F.H. & Vanderlinde, P.B. 1992. Occurrence, numbers, and growth of *Listeria monocytogenes* on some vacuum-packaged processed meats. *Journal of Food Protection*, **55**: 4–7.

Gray, M.L. & Killinger, A.H. 1966. *Listeria monocytogenes* and Listeric infections. *Bacteriological Reviews*, **30**: 309–382.

Greenwood, M.H., Roberts, D. & Burden, P. 1991. The occurrence of *Listeria* species in milk and dairy products: a national survey in England and Wales. *International Journal of Food Microbiology*, **12**: 197–206.

Haas, C.N. 1983. Estimation of the risk due to low doses of microorganisms: A comparison of alternative methodologies. *American Journal of Epidemiology*, **118**: 573–582.

Haas, C.N., Madabusi, A.T., Rose, J.B. & Gerba, C.P. 1999. Development and validation of dose-response relationship from *Listeria monocytogenes*. *Quantitative Microbiology*, **1**: 89–102.

Haas, C.N., Rose, J.B. & Gerba, C.P. 1999. *Quantitative Microbial Risk Assessment*. New York NY: Wiley.

Hall, R., Shaw, D., Lim, I., Murphy, F., Davos, D., Lanser, J., Delroy, B., Tribe, I., Holland, R. & Carman, J. 1996. A cluster of listeriosis cases in South Australia. *Communicable Diseases Intelligence*, **20**: 465.

Hartemink, R. & Georgsson, F. 1991. Incidence of *Listeria* species in seafood and seafood salads. *International Journal of Food Microbiology*, **12**: 189–196.

Hartung, M. 2000. Ergebnisse der Zoonosenerhebungen 1999 bei Lebensmitteln. pp. 85–94, *in:* Ergebnisprotokoll der 53. Arbeitstagung des Arbeitskreises Lebensmittelhygienischer Tierärztlicher Sachverständiger (ALTS). Berlin, Germany, 14–16 June 2000.

Harvey, J. & Gilmour, A. 1992. Occurrence of *Listeria* species in raw milk and dairy products produced in Northern Ireland. *Journal of Applied Bacteriology* **72:** 119–125.

Heitmann, M., Gerner-Smidt, P. & Heltberg, O. 1997. Gastroenteritis caused by *Listeria monocytogenes* in a private day-care facility. *Pediatric Infectious Disease Journal*, **16**: 827–828.

Hird, D.W. & Genigeorgis, C. 1990. Listeriosis in food animals: Clinical signs and livestock as a potential source of direct (nonfoodborne) infection for humans. pp. 31–39, *in:* Miller, Smith & Somkuti, 1990, q.v.

Hitchins, A.D. 1996. Assessment of alimentary exposure to *Listeria monocytogenes*. *International Journal of Food Microbiology,* **30**: 71-85.

Ho, J.L., Shands, K.N., Friedland, G., Eckind, P. & Fraser, D.W. 1986. An outbreak of type 4b *Listeria monocytogenes* infection involving patients from eight Boston hospitals. *Archives of Internal Medicine*, **146**: 520–524.

Hof, H. & Rocourt, J. 1992. Is any strain of *Listeria monocytogenes* detected in food a health risk? *International Journal of Food Microbiology,* **16**: 173–182.

Holcomb, D.L., Smith, M.A., Ware, G.O., Hung, Y.C., Brackett, R.E. & Doyle, M.P. 1999. Comparison of six dose-response models for use with foodborne pathogens. *Risk Analysis*, **19**: 1091–1100.

Holdsworth, M., Gerber, M., Haslam, C., Scali, J., Bearsdworth, A., Avallone, M.H. & Sherratt, E. 2000. A comparison of dietary behaviour in Central England and a French Mediterranean region. *European Journal of Clinical Nutrition*, **54**: 530–539.

Holland, S., Alfonso, E., Gelender, H., Heidemann, D., Mendelsohn, A., Ullman, S. & Miller, D. 1987. Corneal ulcer due to *Listeria monocytogenes*. *Cornea*, **6**: 144–146.

Houtsma, P.C., De Wit, J.C. & Rombouts, F.M. 1993. Minimum inhibitory concentration (MIC) of sodium lactate for pathogens and spoilage organisms occurring in meat products. *International Journal of Food Microbiology*, **20**: 247–257.

Huang, S., Hendriks, W., Althage, A., Hemmi, S., Bluethmann, H., Kamijo, R., Vilcek, J., Zinkernagel, R.M. & Aguet, M. 1993. Immune response in mice that lack the interferon-gamma receptor. *Science*, **259**: 1742–1745.

Hudson, J.A. 1993. Effect of pre-incubation temperature on the lag time of *Aeromonas hydrophila. Letters in Applied Microbiology*, **16**: 274–276.

Hudson, J.A. & Mott, S.J. 1993. Growth of *Listeria monocytogenes*, *Aeromonas hydrophila* and *Yersinia enterocolitica* on cold smoked salmon under refrigeration and mild temperature abuse. *Food Microbiology*, **10**(1): 61–68.

Huismans, H. 1986. Acute chorioretinitis caused by *Listeria monocytogenes*. *Klinische Monatsblatter fur Augenheilkunde*, **189**: 48–50.

Humpheson, L., Adams, M.R., Anderson, W.A. & Cole, M.B. 1998. Biphasic thermal inactivation kinetics in *Salmonella enteritidis* PT4. *Applied and Environmental Microbiology*, **64**: 459–464.

ICD [Industry Council for Development of the Food and Allied Industries]. 2000. Industry data on contamination of final product and manufacturing environment by *Listeria monocytogenes*. Kent, United Kingdom.

ICMSF [International Commission on the Microbiological Specification of Foods]. 1994. Choice of sampling plan and criteria for *Listeria monocytogenes*. *International Journal of Food Microbiology*, **22**: 89–96.

ICMSF. 1996. *Microorganisms in Foods, Microbiological Specifications of Food Pathogens.* Vol. 5. London: Blackie Academic and Professional. 513p.

Ingham, S.C., Escude, J.M. & McCown, P. 1990. Comparative growth rates of *Listeria monocytogenes* and *Pseudomonas fragi* on cooked chicken load stored under air and two modified atmospheres. *Journal of Food Protection*, **53**: 289–291.

Inoue, S., Nakama, A., Arai, Y., Kokubo, Y., Maruyama, T., Saito, Y., Yoshida, T., Terao, M. & Yamamoto, S. 2000. Prevalence and contamination levels of *Listeria monocytogenes* in 12 retail foods in Japan. *International Journal of Food Microbiology*, **59**: 73–77.

Jacquet, C., Brouillé, F., Saint-Cloment, C., Catimel, B. & Rocourt, J. 1999. La listériose humaine en France en 1998. *Bulletin Epidémiologique Hebdomadaire*, **37**: 153–154.

Jacquet, C., Catimel, B., Brosch, R., Buchrieser, C., Dehaumont, P., Goulet, V., Lepoutre, A., Veit, P. & Rocourt, J. 1995a. Investigations related to the epidemic strain involved in the French listeriosis outbreak in 1992. *Applied and Environmental Microbiology*, **61**: 2242–2246.

Jacquet, C., Catimel, B., Brosch, R., Goulet, V., Lepoutre, A., Veit, P., Dehaumont, P. & Rocourt, J. 1995b. Typing of *Listeria monocytogenes* during epidemiological investigations of the French listeriosis outbreaks in 1992, 1993 and 1995. pp. 161–176, *in: Proceedings of the XII International Symposium on Problems of Listeriosis*, Perth, Australia, 2–6 October 1995. Promaco Conventions Pty Ltd.

Jameson, J.F. 1962. A discussion of the dynamics of *Salmonella* enrichment. *Journal of Hygiene [Cambridge]*, **60**: 193–207.

Jensen, A., Frederiksen, W. & Gerner-Smidt, P. 1994. Risk factors for listeriosis in Denmark, 1989-1990. *Scandinavian Journal of Infectious Diseases*, **26**: 171–178.

Jinneman, K.C., Wekell, M.M. & Eklund, M.W. 1999. Incidence and behaviour of *Listeria monocytogenes* in fish and seafood products. pp. 601–630, *in:* Ryser & Marth, 1999, q.v.

Johansson, T., Rantala, L., Palmu, L. & Honkanen-Buzalski, T. 1999. Occurrence and typing of *Listeria monocytogenes* strains in retail vacuum-packed fish products and in a production plant. *International Journal of Food Microbiology,* **47**: 111–119.

Johnson, A.E., Donkin, A.J., Morgan, K., Lilley, J.M., Neale, R.J., Page, R.M. & Silburn, R. 1998. Food safety knowledge and practice among elderly people living at home. *Journal of Epidemiology and Community Health*, **52**: 745–748.

Jørgensen, L.V. & Huss, H.H. 1998. Prevalence and growth of *Listeria monocytogenes* in naturally contaminated seafood. *International Journal of Food Microbiology,* **42**: 127–131.

Jørgensen, L.V., Dalgaard, P. & Huss, H.H. 2000. Multiple compound quality index for cold-smoked salmon (*Salmo salar*) developed by multivariate regression of biogenic amines and pH. *Journal of Agriculture and Food Chemistry*, **48**: 2448–2453.

Junkins, E. 1994. Saskatchewan Nutrition Survey 1993/94. Methodology for estimating usual intake. BBCA Technical Report E451311-005. Bureau of Biostatistics and Computer Applications, Food Directorate, Health Canada.

Junkins, E. & Karpinski, K. 1994. Enquéte québécoise sur la nutrition. Méthodologie pour estimer l'apport habituel, les statistiques sommaires et les erreurs-types. Bureau of Biostatistics and Computer Applications, Food Directorate, Health Canada.

Junkins, E. & Laffey, P. 2000. Alberta Nutrition Survey 1994. Methodology for estimating usual intake. BBCA Technical Report E451311-006. Bureau of Biostatistics and Computer Applications, Food Directorate, Health Canada.

Junkins, E., Laffey, P. & Weston, T. 2001. Prince Edward Island Nutrition Survey 1995. Methodology for estimating usual intake. BBCA Technical Report E451311-007. Bureau of Biostatistics and Computer Applications, Food Directorate, Health Canada.

Kapperud, G., Gustavsen, S., Hellesnes, I., Hansen, A.H., Lassen, J. Hirn, J., Jahkola, M., Montenegro, M.A. & Helmuth, R. 1990. Outbreak of *Salmonella typhimurium* infection traced to contaminated chocolate and caused by a strain lacking the 60-megadalton virulence plasmid. *Journal of Clinical Microbiology*, **28**: 2597–2601.

Karpinski, K. & Nargundkar, M. 1992. Nova Scotia Nutrition Survey. Methodology Report. BBCA Technical Report E451311-001. Bureau of Biostatistics and Computer Applications, Food Directorate, Health Canada.

Kiss, R., Papp, N.E., Vamos, G.Y. & Rodler, M. 1996. *Listeria monocytogenes* isolation from food in Hungary. *Acta Alimentaria*, **25**: 83–91.

Kittson, E. 1992. A case cluster of listeriosis in Western Australia with links to paté consumption. pp. 39–40, *in:* Proceedings of the 11th *International Symposium on the problem of listeriosis* (ISOPOL XI). Statens Seruminstitut, Copenhagen, Denmark, 11–14 May 1992.

Klein, G. 1999. FAO/WHO: Exposure assessment of microbiological hazards in foods. Prevalence of *Listeria monocytogenes* in foods (Germany). BgVV, Berlin, Germany.

Kocks, C., Goulin, E., Tabouret, M., Berche, P., Ohayon, H. & Cossart, P. 1992. *L. monocytogenes*-induced actin assembly requires the actA gene product, a surface protein. *Cell*, **68**: 521–531.

Kovacs-Domjan, D.H. 1991. Occurrence of Listeria infection in meat industry raw materials and end products. *Magyar Allatorvosok Lapja*, **46**: 229–233.

Kozak, J., Balmer, T., Byrne, R. & Fisher, K. 1996. Prevalence of *Listeria monocytogenes* in foods: Incidence in dairy products. *Food Control*, **7:** 215–221.

Kroll, R.G. & Pratchett, R.A. 1992. Induced acid tolerance *inListeria monocytogenes. Letters in Applied Microbiology*, **14**: 224–227.

Kvenberg, J.E. 1991. Non-indigenous bacterial pathogens. pp. 267–284, *in:* D.R. Ward and C. Hackney (eds). *Microbiology of Marine Products*. New York, NY: Van Nostrand Reinhold.

Lammerding, A.M., Glass, K.A., Gendron-Frizpatrick, A. & Doyle, M.P. 1992. Determination of virulence of different strains of *Listeria monocytogenes* and *Listeria innocua* by oral inoculation of pregnant mice. *Applied and Environmental Microbiology*, **58**: 3991–4000.

Lamont, R.J. & Postlethwaite, R. 1986. Carriage of *Listeria monocytogenes* and related species in pregnant and non-pregnant women in Aberdeen, Scotland. *Journal of Infection*, **19**: 263–266.

Laukova, A., Czikkova, S., Laczkova, S. & Turek, P. 1999. Use of enterocin CCM 4231 to control *Listeria monocytogenes* in experimentally contaminated dry fermented Hornad salami. *International Journal of Food Microbiology*, **52**: 115–119.

Leblanc, I., Leroi, F., Hartke, A. & Auffray, Y. 2000. Do stresses encountered during the smoked salmon process influence the survival of the spoiling bacterium *Shewanella putrefaciens*? *Letters in Applied Microbiology*, **30**: 437–442.

Lebrun, M., Mengaud, J., Ohayon, H., Nato, F. & Cossart, P. 1996. Internalin must be on the bacterial surface to mediate entry of *Listeria monocytogenes* into epithelial cells. *Molecular Microbiology*, **21**: 579–592.

Lecuit, M., Dramsi, S., Gottardi, C., Fedor-Chaiken, M., Gumbiner, B. & Cossart, P. 1999. A single amino acid in E-cadherin responsible for host specificity towards the human pathogen *Listeria monocytogenes*. *EMBO Journal*, **18**: 3956–3963.

Lecuit, M., Vandormael-Pournin, S., Lefort, J., Huerre, M., Gounon, P., Dupuy, C., Babinet, C. & Cossart, P. 2001. A transgenic model for listeriosis: role of internalin in crossing the intestinal barrier. *Science,* **292**(5522): 1722–1725.

Lennon, D., Lewis, B., Mantell, C., Becroft, D., Dove, B., Farmer, K., Tonkin, S., Yates, N., Stamp, R. & Mickleson, K. 1984. Epidemic perinatal listeriosis. *Pediatric Infectious Disease*, **3**: 30–34.

Leroi, F., Joffraud, J.J., Chevalier F. & Cardinal, M. 2001. Research of quality indices for cold-smoked salmon using a stepwise multiple regression of microbiological counts and physico-chemical parameters. *Journal of Applied Microbiology*, **90**: 578–587.

Levine, M.M., Dupont, H.L., Formal, S.B., Hornick, R.B., Takeuchi, A., Gangarosa, E.J., Snyder, M.J. & Libonati, J.P. 1973. Pathogenesis of *Shigella dysenteriae* 1 (Shiga) Dysentery. *Journal of Infectious Diseases*, **127**: 261–269.

Lewis, D.B., Larsen, A. & Wilson, C.B. 1986. Reduced interferon-gamma mRNA levels in human neonates. *Journal of Experimental Medicine*, 163: 1018–1023.

Lindqvist, R. & Westöö, A. 2000. Quantitative risk assessment for *Listeria monocytogenes* in smoked or gravad salmon/rainbow trout in Sweden. *International Journal of Food Microbiology,* **58**: 181–196.

Linnan, M.J., Mascola, L., Lou, X.D., Goulet, V., May, S., Salminen, C., Hird, D.W., Yonekura, M.L. Hayes, P., Weaver R., et al. 1988. Epidemic listeriosis associated with Hispanic-style cheese. *New England Journal of Medicine*, **319**: 823–828.

Loncarevic, S., Tham, W. & Danielsson-Tham, M.-L. 1996. The clones of *Listeria monocytogenes* detected in food depend on the method used. *Letters in Applied Microbiology*, **22**: 381–384.

Loncarevic, S., Tham, W. & Danielsson-Tham, M.-L. 1998. Changes in serogroup distribution among *Listeria monocytogenes* human isolates in Sweden, p. 20, *in:* Program and Abstracts of the XIIIth International Symposium on Problems of Listeriosis. Halifax, Nova Scotia, Canada, 1998.

Lorber, B. 1990. Clinical listeriosis – implications and pathogenesis. pp. 41–49, *in:* Miller, Smith & Somkuti, 1990, q.v.

Lou, Y. & Yousef, A.E. 1999. Characteristics of *Listeria monocytogenes* important to food processors. pp. 131–225, *in:* Ryser & Marth, 1999, q.v.

Lou, Y. & Yousef, A.E. 1997. Adaptation to sublethal environmental stresses protects *Listeria monocytogenes* against lethal preservation factors. *Applied and Environmental Microbiology*, **63**: 1252–1255.

Louthrenoo, W. & Schumacher, H.R. Jr. 1990. *Listeria monocytogenes* osteomyelitis complicating leukemia: Report and literature review of Listeria osteoarticular infections. *Journal of Rheumatology*, **17**: 107–110.

Lücke, F-K. 1995. Fermented meats. pp. F-1–F-23, *in: LFRA Microbiology Handbook: Meat Products*. Leatherhead, Surrey, UK: Food Research Association.

Luisjuanmorales, A., Alanizde, R., Vazquezsandoval, M.E. & Rosasbarbosa, B.T. 1995. Prevalence of *Listeria monocytogenes* in raw milk in Guadalajara, Mexico. *Journal of Food Protection*, **58**: 1139–1141.

Lyytikäinen, O., Autio, T., Maijala, R., Ruutu, P., Honkanen-Buzalski, T., Miettinen, M., Hatakka, M., Mikkola, J., Anttila, V.J., Johansson, T., Rantala, L., Aalto, T., Korkeala, H. & Siitonen, A.J. 2000. An outbreak of *Listeria monocytogenes* serotype 3a infections from butter in Finland. *Journal of Infectious Diseases,* **181**(5): 1838–1841.

MacGowan, A.P., Bowker, K., McLauchlin, J., Bennett, P.M. & Reeves, D.S. 1994. The occurrence and seasonal changes in the isolation of *Listeria* spp. in shop bought food stuffs, human faeces, sewage and soil from urban sources. *International Journal of Food Microbiology*, **21**: 325–334.

MacGowan, A.P., Marshall, R.J., MacKay, I.M. & Reeves, D.S. 1991. *Listeria* faecal carriage by renal transplant recipients, haemodialysis patients and patients in general practice: its relation to season, drug therapy, foreign travel, animal exposure and diet. *Epidemiology and Infection*, **106**: 157–166.

Mackey, B.M. 1999. Injured Bacteria. Chapter 15, *in:* B.M. Lund, A.C. Baird-Parker and G.W. Gould (eds). *The Microbiological Quality and Safety of Food.* Maryland, USA: Aspen Publishing.

Mackey, B.M., Boogard, E., Hayes, C.M. & Baranyi, J. 1994. Recovery of heat-injured *Listeria monocytogenes. International Journal of Food Microbiology*, **22**: 227–237.

Maijala, R., Lyytikainen, O., Autio, T., Aalto, T., Haavisto, L. & Honkanen-Buzalski, T. 2001. Exposure of *Listeria monocytogenes* within an epidemic caused by butter in Finland. *International Journal of Food Microbiology,* **70** (1-2): 97–109.

Mainou-Fowler, T., MacGowan, A.P. & Postlethwaite, R. 1988. Virulence of *Listeria* spp: course of infection in resistant and susceptible mice. *Journal of Medical Microbiology,* **27**: 131–140.

Marchetti, P. 1996. Étude de 225 cas de listériose non materno-fœtal survenus en France en 1992: Influennce des conditions prédisposantes sur les manifestations cliniques et la pronostic de l'infection. Thèse pour le diplôme d'État de doctorat en médicine. Faculte de Medecine, Universite de Nantes, Nantes, France.

Marks, H.M., Coleman, M.E., Jordan Lin, C.T. & Roberts, T. 1998. Topics in microbial risk assessment: Dynamic flow tree process. *Risk Analysis*, **18**: 309–328.

Marron, L., Emerson, N., Gahan, C.G. & Hill, C. 1997. A mutant of *Listeria monocytogenes* LO28 unable to induce an acid tolerance response displays diminished virulence in a murine model. *Applied and Environmental Microbiology*, **63**: 4945–4947.

Mascola, L., Sorvillo, F., Goulet, V., Hall, B., Weaver, R. & Linnan, M. 1992. Fecal carriage of *Listeria monocytogenes*: Observations during a community-wide, common-source outbreak. *Clinical Infectious Diseases*, **15**: 557–558.

Mazzulli, T. & Salit, I.E. 1991. Pleural fluid infection caused by *Listeria monocytogenes*: Case report and review. *Reviews Infectious Diseases*, **13**: 564–570.

Mbawuike, I.N., Acuna, C.L., Walz, K.C., Atmr, R.L., Greenberg, S.B. & Couch, R.B. 1997. Cytokines and impaired CD8$^+$ CTL activity among elderly persons and the enhancing effect of IL-12. *Mechanisms Aging and Development*, **94**: 25–39.

McCarthy, S.A. 1990. Listeria in the environment. pp. 25-29, *in:* Miller, Smith & Somkuti, 1990, q.v.

McLauchlin, J. 1990a. Human listeriosis in Britian, 1967-85: A summary of 722 cases. 1. Listeriosis during pregnancy and in the newborn. *Epidemiology and Infection*, **104**: 181–189.

McLauchlin, J. 1990b. Distribution of serovars of *Listeria monocytogenes* isolated from different categories of patients with listeriosis. *European Journal of Clinical Microbiology and Infectious Diseases*, **9**: 210–213.

McLauchlin, J. 1993. Listeriosis and *Listeria monocytogenes*. *Environmental Policy and Practice*, **3**: 210–214.

McLauchlin, J. 1996. The relationship between *Listeria* and listeriosis. *Food Control*, **7**: 187–193.

McLauchlin, J. 1997. The pathogenicity of *Listeria monocytogenes*: A public health perspective. *Reviews in Medical Microbiology*, **8**: 1–14.

McLauchlin, J., Hall, S.M., Velani, S.K. & Gilbert, R.J. 1991. Human listeriosis and pâté: A possible association. *British Medical Journal*, **303**: 773–775.

McLaughlin, J. & Nichols, G.N. 1994. Listeria and seafood. *PHLS Microbiology Digest*, **11**: 151–154.

McMeekin, T.A., Olley, J., Ross, T. & Ratkowsky, D.A. 1993. *Predictive Microbiology. Theory and Application*. Taunton, UK: Research Studies Press. 340p.

Mead, P.S. 1999. Multistate outbreak of listeriosis traced to processed meats, August 1998-March 1999. 27 May 1999, CDC unpublished report. 23p.

Mead, P.S., Slutsker, L., Dietz, V., McCraig, L.F., Bresee, S., Shapiro, C., Griffin, P.M. & Tauxe, R.V. 1999. Food-related illness and death in the United States. *Emerging Infectious Diseases,* **5**: 607–625.

Mengaud, J., Dramsi, S., Gouin, E., Vazquez-Boland, J.A., Milon, G. & Cossart, P. 1991. Pleiotropic control of *Listeria monocytogenes* virulence factors by a gene that is autoregulated. *Molecular Microbiology*, **5**: 2273–2283

Mellefont, L.A., McMeekin, T.A. & Ross, T. 2003. The effect of abrupt osmotic shifts on the lag phase duration of foodborne bacteria. *International Journal of Food Microbiology,* **83**: 281–293.

Mellefont, L.A. & Ross, T. 2003. The effect of abrupt shifts in temperature on the lag phase duration of *Escherichia coli* and *Klebsiella oxytoca. International Journal of Food Microbiology,* **83**: 295–305.

Miettinen, M.K., Bjorkroth, K.J. & Korkeala, H.J. 1999. Characterization of *Listeria monocytogenes* from an ice cream plant by serotyping and pulsed-field gel electrophoresis. *International Journal of Food Microbiology,* **46**: 187–192.

Miettinen, M.K., Siitonen, A., Heiskanen, P., Haajanen, H., Björkroth, K.J. & Korkeala, H.J. 1999. Molecular epidemiology of an outbreak of febrile gastroenteritis caused by *Listeria monocytogenes* in cold-smoked rainbow trout. *Journal of Clinical Microbiology,* **37**: 2358–2360.

Miller, A.J. 1992. Combined water activity and solute effects on growth and survival of *Listeria monocytogenes* Scott A. *Journal of Food Protection,* 55: 414–418.

Miller, A.J., Smith, J.L. & Somkuti, G.A. (eds). 1990. *Topics in Industrial Microbiology: Foodborne Listeriosis*. New York NY: Elsevier Science Pub.

Miller, A.J., Whiting, R.C. & Smith, J.L. 1997. Use of risk assessment to reduce listeriosis incidence. *Food Technology*, **51**: 100–103.

Misrachi, A., Watson, A.J. & Coleman, D. 1991. *Listeria* in smoked mussels in Tasmania. *Communicable Diseases Intelligence*, **15**: 427.

Mitchell, D.L. 1991. A case cluster of listeriosis in Tasmania. *Communicable Diseases Intelligence*, **15**: 427.

MLA [Meat and Livestock Australia]. 1999. *Safe Meat Retailing.* Australia: Meat and Livestock Australia.

Morgan, M.G. 1993. Risk analysis and management. *Scientific American*, July 1993: 32–41.

Mossel, D.A., & Oei, H.Y. 1975. Letter: Person to-person transmission of enteric bacterial infection. *Lancet*, 29 March 1975 1: 751.

Moura, S.M., Destro, T.M. & Franco, B.D. 1993. Incidence of *Listeria* species in raw and pasteurized milk produced in Sao Paulo, Brazil. *International Journal of Food Microbiology,* **19**: 229–237.

Murray, E.G.D., Webb, R.A. & Swann, M.B.R. 1926. A disease of rabbits characterised by a large mononuclear leucocytosis, caused by a hitherto undescribed bacillus *Bacterium monocytogenes* (n. sp.). *Journal of Pathology and Bacteriology,* **29**: 407–439.

NACMCF [National Advisory Committee on Microbiological Criteria for Foods]. 1991. *Listeria monocytogenes*: Recommendations by the National Advisory Committee on Microbiological Criteria for Foods. *International Journal of Food Microbiology,* **14**: 185–246.

Nakane, A., Nishikawa, S., Sasaki, S., Miura, T., Asano, M., Kohanawa, M., Ishiwata, K. & Minagawa, T. 1996. Endogenous interleukin-4, but not interleukin-10, is involved in suppression of host resistance against *Listeria monocytogenes* infection in gamma interferon-depleted mice. *Infection and Immunity*, **64**: 1252–1258.

Neumeyer, K., Ross, T. & McMeekin, T.A. 1997. Development of a predictive model to describe the effects of temperature and water activity on the growth of spoilage pseudomonads. *International Journal of Food Microbiology,* **38**: 45–54.

Neumeyer, K., Ross, T., Thompson, G. & McMeekin, T.A. 1997. Validation of a model describing the effects of temperature and water activity on the growth of psychotrophic pseudomonads. *International Journal of Food Microbiology,* **38**: 55–63.

NGS [National Geographic Society]. 1999. The National Geographic desk reference. Edited by R.M. Downs, F.A Day, P.L. Knoxw, P.H. Meserve and B. Warf. Washington DC: National Geographic Society. 700p.

Nguyen, M.H. & Yu, V.L. 1994. *Listeria monocytogenes* peritonitis in cirrhotic patients. Value of ascitic fluid gram stain and a review of literature. *Digestive Diseases and Sciences,* **39**: 215–218.

Nieman, R.E. & Lorber, B. 1980. Listeriosis in adults: A changing pattern. *Reviews in Infectious Diseases*, **2**: 207–227.

Nilsson, L., Gram, L. & Huss, H.H. 1999. Growth control of *Listeria monocytogenes* on cold-smoked salmon using a competitive lactic acid bacterial flora. *Journal of Food Protection*, **62**: 336–342.

Nilsson, L., Huss, H.H. & Gram, L. 1997. Inhibition of *Listeria monocytogenes* on cold-smoked salmon by nisin and carbon dioxide atmosphere. *International Journal of Food Microbiology,* **38**: 217–228.

Nørrung, B., Andersen, J.K. & Schlundt, J. 1999. Incidence and control of *Listeria monocytogenes* in foods in Denmark. *International Journal of Food Microbiology,* **53**: 195–203.

Norton, D.M., McCamey, M.A., Gall, K.L., Scarlett, J.M., Boor, K.J. & Wiedmann, M. 2001. Molecular studies on the ecology of *Listeria monocytogenes* in the smoked fish processing industry. *Applied and Environmental Microbiology*, **67**: 198–205.

Notermans, S., Dufrenne, J., Teunis, P., Beumer, R., te Giffel, M. & Weem, P. 1997. A risk assessment study of *Bacillus cereus* present in pasteurized milk. *Food Microbiology,* **14**: 143–151.

Notermans, S., Dufrenne, J., Teunis, P. & Chackraborty, T. 1998. Studies on the risk assessment of *Listeria monocytogenes*. *Journal of Food Protection*, **61**: 244–248.

O'Brien, G.D. 1997. Domestic refrigerator air temperatures and the public's awareness of refrigerator use. *International Journal of Environmental Health Research*, **7**: 141–148.

O'Driscoll, B., Gahan, C.G. & Hill, C. 1996. Adaptive acid tolerance response in *Listeria monocytogenes*: isolation of an acid-tolerant mutant which demonstrates increased virulence. *Applied and Environmental Microbiology*, **62**: 1693–1698.

Patchett, R.A., Watson, N., Fernandez, P.S. & Kroll, R.G. 1996. The effect of temperature and growth rate on the susceptibility of *Listeria monocytogenes* to environmental stress conditions. *Letters in Applied Microbiology*, **22**: 121–124.

Paul, M.L., Dwyer, D.E., Chow, C., Robson, J., Chambers, I., Eagles, G. & Ackerman, V. 1994. Listeriosis – a review of eighty-four cases. *Medical Journal of Australia*, **160**: 489–493.

Peeler, J.T. & Bunning, V.K. 1994. Hazard assessment of *Listeria monocytogenes* in the processing of bovine milk. *Journal of Food Protection*, **57**: 689–697.

Peleg, M. & Cole, M.B. 1998. Reinterpretation of microbial survival curves. *Critical Reviews in Food Science and Nutrition,* **38**: 353–380.

Pelroy, G.A., Peterson, M.E., Paranjpye, R.N., Almond, J. & Eklund, M.W. 1994. Inhibition of *Listeria monocytogenes* in cold-process (smoked) salmon by sodium nitrite and packaging method. *Journal of Food Protection*, **57**: 114–119.

Peterson, W.L., MacKowiak, P.A., Barnett, C.C., Marling-Cason, M., & Haley, M.L. 1989. The human gastric bactericidal barrier: Mechanisms of action, relative antibacterial activity, and dietary influences. *Journal of Infectious Diseases*, **159**: 979–983.

Phan-Thanh, L. & Montagne, A. 1998. Physiological and biochemical aspects of the acid survival of *Listeria monocytogenes*. *Journal of General and Applied Microbiology*, **44**: 183–191.

Pine, L., Kathariou, S., Quinn, F., George, V., Wenger, J.D. & Weaver, R.E. 1991. Cytopathogenic effects in enterocytelike Caco-2 cells differentiate virulent from avirulent *Listeria* strains. *Journal of Clinical Microbiology*, **29**: 990–996.

Pine, L., Malcolm, G.B. & Plikaytis, B.D. 1990. *Listeria monocytogenes* intragastric and intraperitoneal approximate 50% lethal doses for mice are comparable, but death occurs earlier by intragastric feeding. *Infection and Immunity*, **58**: 2940–2945.

Pinner, R.W., Schuchat, A., Swaminathan, B., Hayes, P.S., Deaver, K.A., Weaver, R.E. & Plikaytis, B.D. 1992. Role of foods in sporadic listeriosis. *Journal of the American Medical Association*, **267**: 2046–2050.

Pirie, J.H.H. 1927. A new disease of veldt rodents "Tiger Rover Disease". *Publications of the South African Institute of Medical Research,* **3**: 163–186.

Pitt, W.M, Harden, T.J. & Hull, R.R. 1999. *L. monocytogenes* in milk and dairy products. *Australian Journal of Dairy Technology*, **54**: 49–65.

Polanco, A., Giner, C., Canton, R., Leon, A., Garcia Gonzalez, M., Baquero, F. & Meseguer, M. 1992. Spontaneous bacterial peritonitis caused by *Listeria monocytogenes*: Two case reports and literature review. *European Journal of Clinical Microbiological Infectious Diseases*, **11**: 346–349.

Potel, J. 1953. Aetiologie der Granulomatosis Infantisepticum. *Wissenshaftliche Zeitschrift der Martin Luther Universitat Halle-Wittenberg*, **2**: 341–349.

Presser, K.A., Ross, T. & Ratkowsky, D.A. 1998. Modelling the growth limits (growth/no growth interface) of *Escherichia coli* as a function of temperature, pH, lactic acid concentration, and water activity. *Applied and Environmental Microbiology*, **64**: 1773–1779.

Proctor, M.E., Brosch, R., Mellen, J.W., Garrett, L.A., Kaspar, C.W. & Luchansky, J.B. 1995. Use of pulsed-field gel electrophoresis to link sporadic cases of invasive listeriosis with recalled chocolate milk. *Applied and Environmental Microbiology*, **61**: 3177–3179.

Pron, B., Boumaila, C., Jaubert, F., Berche, P., Milon, G., Geissmann, F. & Gaillard, J.L. 2001. Dendritic cells are early cellular targets of *Listeria monocytogenes* after intestinal delivery and are involved in bacterial spread in the host. *Cellular Microbiology,* **3**: 331–340.

Ratkowsky, D.A., Olley, J., McMeekin, T.A. & Ball, A. 1982. Relationship between temperature and growth rate of bacterial cultures. *Journal of Bacteriology*, **149**: 1–5.

Ratkowsky, D.A., Lowry, R.K., McMeekin, T.A., Stokes, A.N. & Chandler, R.E. 1983. Model for bacterial culture growth rate throughout the entire biokinetic temperature range. *Journal of Bacteriology*, **154**: 1222–1226.

Ravishankar, S., Harrison, M.A. & Wicker, L. 2000. Protein profile changes in acid adapted *Listeria monocytogenes* exhibiting cross-protection against an activated lactoperoxidase system in tryptic soy broth. *Journal of Food Safety*, **20**: 27–42.

Reed, R.W. 1958. Listeria and Erysipelothrix. pp. 453–462, *in:* R.J. Dubos (ed). *Bacterial and Mycotic Infections in Man*. 3rd ed. Philadelphia PA: J. B. Lippincott.

Renzoni, A., Cossart, P. & Dramsi, S. 1999. PrfA, the transcriptional activator of virulence genes is upregulated during interaction of *Listeria monocytogenes* with mammalian cells and in eukaryotic cell extracts. *Molecular Microbiology*, **34**: 552–561.

Riedo, F.X., Pinner, R.W., De Lourdes Tosca, M., Cartter, M.L., Graves, L.M., Reeves, M.W., Weaver, R.E., Plikaytis, B.D. & Broome, C.V. 1994. A point-source foodborne outbreak: Documented incubation period and possible mild illness. *Journal of Infectious Diseases*, **170**: 693–696.

Rink, L., Cakman, I. & Kirchner, H. 1998. Altered cytokine production in the elderly. *Mechanisms Ageing Development,* **102**: 199–209.

Roberts, T., Ahl, A. & McDowell, R. 1995. Risk assessment for foodborne microbial hazards. *In: Tracking Foodborne Pathogens from Farm to Table*. USDA Economic Research Service, Miscellaneous Publication Number 1532.

Robinson, T.P., Ocio, M.J., Kaloti, A. & Mackey, B.M. 1998. The effect of the growth environment on the lag phase of *Listeria monocytogenes*. *International Journal of Food Microbiology*, **44**: 83–92.

Robinson, T.P., Aboaba, O.O., Kaloti, A., Ocio, M.J., Baranyi, J. & Mackey, B.M. 2001. The effect of inoculum size on the lag phase of *Listeria monocytogenes*. *International Journal of Food Microbiology*, **70**: 163–173.

Rockliff, S., & Millard, G. 1996. Microbial quality of smoked products. ACT Health services. Food survey reports, 1996–97. See: http://www.health.act.gov.au/c/health?a=da&did=10017393&pid=1053853752&sid= (Downloaded 22 March 2004).

Rocourt, J. 1991. Human Listeriosis: 1989. WHO/HPP/FOS 91.3.

Rocourt, J. 1996. Risk factors for listeriosis. *Food Control*, **7**: 192–202.

Rocourt, J. & Cossart, P. 1997. *Listeria monocytogenes*. pp. 337–352, *in:* M.P. Doyle, L.R. Beuchat and T.J. Montville (eds). *Food Microbiology. Fundamentals and Frontiers*. American Society of Microbiology. Washington, DC.

Rocourt, J., Goulet, V., Lepoutre-Toulemon, A., Jacquet, Ch., Catimel, B., Rebière, I., Miegeville, A.F., Courtieu, A.L., Pierre, O., Dehaumont, P. & Veit, P. 1993. Epidémie de listériose en France en 1992. *Médecine et maladies infectieuses,* **23**S: 481–484.

Rödel, W., Stiebing, A. & Kröckel, L. 1993. Ripening parameters for traditional dry sausages with a mould covering. *Fleischwirtschaft*, **73**: 848–853.

Roering, A.M., Luchansky, J.B., Ihnot, A.M., Ansay, S.E., Kaspar, C.W. & Ingham, S.C. 1999. Comparative survival of *Salmonella typhimurium* DT 104, *Listeria monocytogenes*, and *Escherichia coli* O157:H7 in preservative-free apple cider and simulated gastric fluid. *International Journal of Food Microbiology*, **46**: 263–269.

Rola, J., Kwiatek, K., Wojton, B. & Michalski, M.M. 1994. Wystepowanie *Listeria monocytogenes* w mleku surowym i produktach mlecznch [Incidence of *Listeria monocytogenes* in raw milk and dairy products]. *Medycyna Weterynaryjna*, **50**(7): 323–325.

Roll, J.T., & Czuprynski, C.J. 1990. Hemolysin is required for extraintestinal dissemination of *Listeria monocytogenes* in intragastrically inoculated mice. *Infection and Immunity*, **58**: 3147–3150.

Rørvik, L.M., Aase, B., Alvestad, T. & Caugant, D.A. 2000. Molecular epidemiological survey of *Listeria monocytogenes* in seafoods and seafood-processing plants. *Applied and Environmental Microbiology*, **66**: 4779–4784.

Rørvik, L.M., Skjerve, E., Knudsen, B.R. & Yndestad, M. 1997. Risk factors for contamination of smoked salmon with *Listeria monocytogenes* during processing. *International Journal of Food Microbiology,* **37**: 215–219.

Rose, J.B., Haas, C.N. & Regli, S. 1991. Risk assessment and control of waterborne giardiasis. *American Journal of Public Health*, **81**: 709–713.

Ross, T. 1999. Predictive food microbiology models in the meat industry. Meat and Livestock Australia, Sydney, Australia. 196p.

Ross, T. & McMeekin, T.A. 1991. Predictive Microbiology. Applications of a square root model. *Food Australia,* **43**: 202–207.

Ross, T. & McMeekin, T.A. 2003. Modeling microbial growth within food safety risk assessments. *Risk Analysis,* **23**: 179–197.

Ross, T. & Sanderson, K. 2000. A risk assessment of selected seafoods in NSW. SafeFood NSW, Sydney, Australia. 275p.

Ross, T. & Shadbolt, C.T. 2001. Predicting *Esherichia coli* inactivation in uncooked comminuted fermented meat products. Meat and Livestock Australia, Sydney, Australia. 66p.

Ross, T., Baranyi, J. & McMeekin, T.A. 1999. Predictive Microbiology and Food Safety. pp. 1699–1710, *in:* R. Robinson, C.A. Batt and P. Patel (eds). *Encyclopaedia of Food Microbiology*. London: Academic Press.

Ross, T., Dalgaard, P. & Tienungoon, S. 2000. Predictive modelling of the growth and survival of *Listeria* in fishery products. *International Journal of Food Microbiology*, **62**: 231–246.

Ross, T., Fazil, A., Paoli, G. & Sumner, J. In press. *Listeria monocytogenes* in Australian processed meat products: risks and their management. Meat and Livestock Australia, Sydney, NSW, Australia.

Rosso, L., Lobry, J.R., Bajard, S. & Flandois, J.P. 1995. Convenient model to describe the combined effects of temperature and pH on microbial growth. *Applied and Environmental Microbiology*, **61**: 610–616.

Ryser, E.T. 1999a. Foodborne listeriosis. pp. 299–358, *in:* Ryser & Marth, 1999, q.v.

Ryser, E.T. 1999b. Incidence and behaviour of *Listeria monocytogenes* in unfermented dairy products. pp. 359–410, *in:* Ryser & Marth, 1999, q.v.

Ryser, E.T. & Marth, E.H. (eds). 1991. *Listeria, Listeriosis, and Food Safety*. New York NY: Marcel Dekker. 632p.

Ryser, E.T. & Marth, E.H. (eds). 1999. *Listeria, Listeriosis, and Food Safety.* 2nd edition, revised and expanded. New York NY: Marcel Dekker. 738p.

Salamah, A.A. 1993. Isolation of *Yersinia enterocolitica* and *Listeria monocytogenes* from fresh vegetables in Saudi Arabia and their growth behavior in some vegetable juices. *Journal of the University of Kuwait-Science*, **20**: 283–291.

Salamina, G., Dalle Donne, E., Niccolini, A., Poda, G., Cesaroni, D., Bucci, M., Fini, R., Maldin, M., Schuchat, A., Swaminathan, B., Bibb, W., Rocourt, J., Binkin, N. & Salmasol, S. 1996. A foodborne outbreak of gastroenteritis involving *Listeria monocytogenes*. *Epidemiology and Infection*, **117**: 429–436.

Salvat, G., Toquin, M.T., Michel, Y. & Colin, P. 1995. Control of *Listeria monocytogenes* in the delicatessen industries: The lessons of a listeriosis outbreak in France. *International Journal of Food Microbiology*, **25**: 75–81.

Samelis, J., Metaxopoulos, J., Vlassi, M. & Pappa, A. 1998. Stability and safety of traditional Greek salami – a microbiological ecology study. *International Journal of Food Microbiology*, **44**: 69–82.

Schillinger, U., Kaya, M. & Lücke, F.-K. 1991. Behaviour of *Listeria monocytogenes* in meat and its control by a bacteriocin-producing strain of *Lactobacillus sake*. *Journal of Applied Bacteriology*, **70**: 473–478.

Schlech, W.F., 3rd. 1993. An animal model of foodborne *Listeria monocytogenes* virulence: Effect of alterations in local and systemic immunity on invasive infection. *Clinical Investigative Medicine*, **16**: 219–225.

Schlech, W.F., 3rd. 1991. Listeriosis: Epidemiology, virulence and the significance of contaminated foodstuffs. *Journal of Hospital Infection*, **19**: 211–224.

Schlech, W.F., 3rd., Chase, D.P. & Badley, A. 1993. A model of foodborne *Listeria monocytogenes* infection in the Sprague-Dawley rat using gastric inoculation: Development and effect of gastric acidity on infective dose. *International Journal of Food Microbiology,* **18**: 15–24.

Schlech, W.F., 3rd., Lavigne, P.M., Bortolussi, R.A., Allen, A.C., Haldane, E.V., Wort, A.J., Hightower, A.W., Johnson, S.E., King, S.H., Nicholls, E.S. & Broome, C.V. 1983. Epidemic listeriosis: Evidence for transmission by food. *Medical Intelligence,* **308**: 203–206.

Schuchat, A., Deaver, K., Wenger, J.D., Plikaytis, B.D., Mascola, L., Pinner, R.W., Reingold, A.L., Broome, C.V. & the Listeria Study Group. 1992. Role of foods in sporadic listeriosis. *Journal of the American Medical Association*, **267**: 2041–2045.

Schuchat, A., Swaminathan, B. & Broome, C.V. 1991. Epidemiology of human listeriosis. *Clinical Microbiological Review*, **4**: 169–183.

Schwartz, B., Broome, C.V., Brown, G.R., Hightower, A.W., Ciesielski, C.A., Gaventa, S., Gellin, B.G. & Mascola, L. 1988. Association of sporadic listeriosis with consumption of uncooked hot dog and undercooked chicken. *Lancet*, **2**: 779–782.

Schwartz, B., Hexter, D., Broome, C.V., Hightower, A.W., Hirschhorn, R.B., Porter, J.D., Hayes, P.S., Bibb, W.F., Lorber, B. & Faris, D.G. 1989. Investigation of an outbreak of listeriosis: New hypotheses for the etiology of epidemic *Listeria monocytogenes* infections. *Journal of Infectious Diseases*, **159**: 680–685.

Schwartz, B., Pinner, R.W. & Broome, C.V. 1990. Dietary risk factors for sporadic listeriosis: association with consumption of uncooked hot dogs and undercooked chicken. pp. 67–69, *in:* Miller, Smith & Somkuti, 1990, q.v.

Schwartz, R.H. 1999. Immunological tolerance. pp. 701–739, *in:* W.E. Paul (ed). *Fundamental Immunology*. 4th ed. Philadelphia PA: Lippincott-Raven Publishers.

Schwarzkopf, A. 1996. *Listeria monocytogenes* – aspects of pathogenicity. *Pathological Biology (Paris)*, **44**: 769–774.

Seeliger, H.P.R. 1961. *Listeriosis*. New York NY: Hafner Publishing.

Sergelidis, D., Abrahim, A., Sarimvei, A., Panoulis, C., Karaioannoglou, P. & Genigeorgis, C. 1997. Temperature distribution and prevalence of *Listeria* spp. in domestic, retail and industrial refrigerators in Greece. *International Journal of Food Microbiology*, **34**: 171–177.

Shadbolt, C.T., Ross, T. & McMeekin, T.A. 1999. Non-thermal death of *Escherichia coli. International Journal of Food Microbiology,* **9**: 129–138.

Shelef, L.A. 1989. Listeriosis and its transmission by food. *Progress in Food and Nutrition Science*, **13**: 363–382.

Singhal, R.S. & Kulkarni, P.R. 2000. Freezing of foods: damage to microbial cells. pp. 840–845, *in;* R.K. Robinson, C. Batt and P.D. Patel (eds). *Encyclopaedia of Food Microbiology*. New York NY: Academic Press.

Skidmore, A.G. 1981. Listeriosis at Vancouver General Hospital, 1965-79. *Canadian Medical Association*, **125**: 1217–1221.

Slutsker, L. & Schuchat, A. 1999. Listeriosis in humans. pp. 75–95, *in:* Ryser & Marth, 1999, q.v.

Smerdon, W.J., Jones, R., McLauchlin, J. & Reacher, M. 2001. Surveillance of listeriosis in England and Wales, 1995-1999. *Communicable Disease and Public Health*, **4**(3): 188–193.

Sprong, R.C., Hulsterin, M.F. & Van der Meer, R. 1999. High intake of milk fat inhibits colonization of *Listeria* but not of *Salmonella* in rats. *Journal of Nutrition*, **129**: 1382–1389.

Stainer, F. & Maillot, E. 1996. Epidemiologie de la listeriose en France. *Epidemiol. sante anim.*, **29**: 37–42.

Stamm, A.M., Smith, S.H., Kirklin, J.K. & McGiffin, D.C. 1990. Listerial myocarditis in cardiac transplantation. *Reviews in Infectious Diseases*, **12**: 820–823.

Starfield, A.M., Smith, K.A. & Bleloch, A.L. 1990. *How to Model It: Problem Solving for the Computer Age*. New York NY: McGraw-Hill. 206p.

Steinmeyer, & Terplan, G. 1990. Listerien in Lebensmitteln - eine aktuelle Ubersicht zu Vorkommen, Bedeutung als Krankheitserreger, Nachweis und Bewertung. *Dtsch. Molkerei Ztg.*, **5**: 150–156; **6**: 179–183. Cited in Klein, 1999, q.v.

Stelma, G.N., Reyes, A.L., Peeler, J.T., Francis, D.W., Hun, J.M., Spaulding, P.L., Hohnson, C.H. & Lovett, J. 1987. Pathogenicity test for *Listeria monocytogenes* using immunocompromised mice. *Journal of Clinical Microbiology*, **25**: 2085–2089.

Stephens, J.C., Roberts, I.S., Jones, D. & Andrews, P.W. 1991. Effect of growth temperature on virulence of strains of *Listeria monoyctogenes* in the mouse: Evidence for a dose dependence. *Journal of Applied Bacteriology*, **70**: 239–244.

Stephens, P.J., Cole, M.B. & Jones, M.V. 1994. Effect of heating rate on the thermal inactivation of *Listeria monocytogenes*. *Journal of Applied Microbiology,* **77**: 702–708.

Stephens, P.J., Joynson, J.A., Davies, K.W., Holbrook, R., Lappin-Scott, H.M. & Humphrey, T.J. 1997. The use of an automated growth analyser to measure recovery times of single heat injured *Salmonella* cells. *Journal of Applied Microbiology,* **83**: 445–455.

Stevenson, M.M., Rees, J.C. & Meltzer, M.S. 1980. Macrophage function in tumor-bearing mice: evidence for lactic dehydrogenase-elevating virus-associated changes. *Journal of Immunology*, **124**: 2892–2899.

Szabo, E.A. & Cahill, M.E. 1998. The combined affects of modified atmosphere, temperature, nisin and ALTA 2341 on the growth of *Listeria monocytogenes*. *International Journal of Food Microbiology*, **18**: 21–31.

Tabouret, M., DeReycke, J., Audurier, A. & Poutrel, B. 1991. Pathogenicity of *Listeria monocytogenes* isolates in immunocompromised mice in relation to listeriolysin production. *Journal of Medical Microbiology*, **34**: 13–18.

Tappero, J.W., Schuchat, A., Deaver, K.A., Mascola, L. & Wenger, J. D. 1995. Reduction in the incidence of human listeriosis in the United States. Effectiveness of prevention efforts? The Listeria Study Group. *Journal of the American Medical Association*, **273**: 1118–1122.

Teufel, P. & Bendzulla, C. 1993. Bundesweite Erhebung zum Vorkommen von *L. monocytogenes* on Lebensmitteln. Bundesinstitut fur gesundheitlichen Verbraucherschutz and Veterinarmedizin, Berlin.

Teunis, P.F.M. 1997. Infectious gastro-enteritis B opportunities for dose-response modelling. National Institute of Public Health and the Environment. Report 284550003. Bilthoven, The Netherlands.

Teunis, P.F.M. & Havelaar, A.H. 2000. The Beta Poisson Dose-Response Model is not a single-hit model. *Risk Analysis*, **20**: 513–520.

Teunis, P.F.M., Nagelkerke, N.J.D. & Haas, C.N. 1999. Dose response models for infectious gastroenteritis. *Risk Analysis*, **19**: 1251–1260.

Teunis, P.F.M., van der Heijden, O.G., van der Giessen, J.W.B. & Havelaar, A.H. 1996. The dose-response relation in human volunteers for gastrointestinal pathogens. National Institute of Public Health and the Environment. Report 284550002. Bilthoven, The Netherlands.

Thurette, J., Membra, J.M., Han Ching, L., Tailliez, R. & Catteau, M. 1998. Behavior of *Listeria* spp. in smoked fish products affected by liquid smoke, NaCl concentration, and temperature. *Journal of Food Protection*, **61**: 1475–1479.

Tienungoon, S. 1998. Some aspects of the ecology of *Listeria monocytogenes* in salmonid aquaculture. Ph.D. Thesis, University of Tasmania, Hobart, Tasmania, Australia.

Tienungoon, S., Ratkowsky, D.A., McMeekin, T.A. & Ross, T. 2000. Growth limits of *Listeria monocytogenes* as a function of temperature, pH, NaCl and lactic acid. *Applied and Environmental Microbiology*, **66**: 4979–4987.

Todd, E.C.D. & Harwig, J. 1996. Microbial risk assessment of food in Canada. *Journal of Food Protection*, S:10–18.

Tripp, C.S., Gately, M.K., Hakimi, J., Ling, P. & Unanue, E.R. 1994. Neutralization of IL-12 decreases resistance to *Listeria* in SCID and C.B-17 mice: Reversal by IFN-gamma. *Journal of Immunology*, **152**: 1883–1887.

Truelstrup Hansen, L., Drewes Røntved, S. & Huss, H.H. 1998. Microbiological quality and shelf life of cold-smoked salmon from three different processing plants. *Food Microbiology*, **15**: 137–150.

Unanue, E.R. 1997. Studies in listeriosis show the strong symbiosis between the innate cellular system and the T-cell response. *Immunological Reviews*, **158**: 11–25.

US FDA 1987. FY86 dairy products program. U.S. Food and Drug Administration, Washington, D.C., United States of America.

van Gerwen, S.J.C., de Wit, J.C., Notermans, S. & Zwietering, M.H. 1997. An identification procedure for foodborne microbial hazards. *International Journal of Food Microbiology*, **38**: 1–15.

van Schaik, W., Gahan, C.G. & Hill, C. 1999. Acid-adapted *Listeria monocytogenes* displays enhanced tolerance against the antibiotics nisin and lacticin 3147. *Journal of Food Protection*, **62**: 536–539.

Vazquez-Boland, J.A., Kuhn, M., Berche, P., Chakraborty, T., Dominguez-Bernal, G., Goebel, W., Gonzalez-Zorn, B., Wehland, J. & Kreft, J. 2001. Listeria pathogenesis and molecular virulence determinants. *Clinical Microbiology Reviews*, **14**: 584–640.

Venables, J. 1989. *Listeria monocytogenes* in dairy products: the Victorian experience. *Food Australia*, **41**: 942–943.

Vorster, S.M., Greebe, R.P. & Nortje, G.L. 1993. The incidence of *Listeria* in processed meats in South Africa. *Journal of Food Protection*, **56**: 169–172.

Vose, D. 1998. The application of quantitative risk assessment to microbial food safety. *Journal of Food Protection*, **61**: 640–648.

Vose, D. 1996. *Quantitative Risk Analysis: a Guide to Monte Carlo Simulation Modelling*. New York NY: John Wiley and Sons.

Walker, R.L., Jensen, L.H., Kinde, H., Alexander, A.V. & Owens, L.S. 1991. Environmental survey for *Listeria* species in frozen milk product plants in California. *Journal of Food Protection*, **54**(3): 178–182.

Walls, I. & Scott, V.N. 1997. Validation of predictive mathematical models describing the growth of *Listeria monocytogenes*. *Journal of Food Protection*, **60**: 1142–1145.

Wang, C., Zhang, L.F., Austin, F.W. & Boyle, C.R. 1998. Characterization of *Listeria monocytogenes* isolated from channel catfish. *American Journal of Veterinary Research*, **59**: 1125–1128.

Warke, R., Kamat, A., Kamat, M. & Thomas, P. 2000. Incidence of pathogenic psychrotrophs in ice creams sold in some retail outlets in Mumbai, India. *Food Control*, **11**: 77–83.

Wessels, S. & Huss, H.H. 1996. Suitability of *Lactococcus lactis* subsp. lactis ATCC 11454 as a protective culture for lightly preserved fish products. *Food Microbiology*, **13**: 323–332.

Whiting, R.C. & Bagi, L.K. 2002. Modeling the lag phase of *Listeria monocytogenes*. *International Journal of Food Microbiology,* 73: 291–295.

WHO. 1988. Foodborne listeriosis. Report of the WHO Working Group. *Bulletin of the World Health Organization*, **66**: 421–428.

Wiedman, M., Bruce, J.L., Keating, C., Johnson, A.E., McDonough, P.L. & Batt, C.A. 1997. Ribotypes and virulence gene polymorphisms suggest three distinct *Listeria monocytogenes* lineages with differences in pathogenic potential. *Infection and Immunity*, **65**: 2707–2716.

Willocx, F., Hendrickx, M. & Tobback, P. 1993. Temperatures in the distribution chain. pp. 80–99 [oral presentation], *in:* Proceedings of the First European "Sous vide" cooking symposium. Leuven, Belgium, 25–26 March 1993.

Woods, L.F.J., Wood, J.M. & Gibbs, P.A. 1989. Nitrite. pp. 225–246, *in:* G.W. Gould (ed). *Mechanisms of Action of Food Preservation Procedures*. London: Elsevier Applied Science

Zhao, L., Montville, T.J. & Schaffner, D.W. 2000. Inoculum size of *Clostridium botulinum* 56A spores influences time-to-detection and percent growth-positive samples. *Journal of Food Science*, **65**: 1369–1375.

Zwietering, M.H., De Wit, J.C. & Notermans, S. 1996. Application of predictive microbiology to estimate the number of *Bacillus cereus* in pasteurized milk at the time of consumption. *International Journal of Food Microbiology,* **30**: 55–70.

Appendices

Appendix 1 – Glossary of terms used

Appendix 2 – Simulation modelling for the four risk assessment examples

Appendix 3 – Predictive microbiology: concepts, applications and sources

Appendix 4 – Prevalence and incidence of *Listeria monocytogenes* in fermented meat products

Appendix 5 – Background for the cold-smoked fish risk assessment

Appendix 1. Glossary of Terms

Beta distribution

The Beta distribution is defined as

$$f(x) = \frac{\Gamma(\alpha+\beta)}{\Gamma(\alpha)\Gamma(\beta)} x^{\alpha-1}(1-x)^{\beta-1},$$

where $0 \leq x \leq 1$, $\alpha > 0$ and $\beta > 0$. There are generalizations to a random variable defined on any interval [a, b].

(Source: http://www.statsoft.com/textbook/glosfra.html)

Binomial distribution

The binomial distribution is used when each trial has exactly two mutually exclusive possible outcomes, often labelled success and failure. The binomial distribution is the probability of obtaining x successes in N trials where the probability of success on a single trial is π. The binomial distribution assumes that π is fixed for all trials. The formula for the binomial probability mass function is

$$p(x;n,\pi) = \binom{n}{x} \pi^{x}(1-\pi)^{n-x}, \text{ for } x = 0, 1, 2, \ldots, n>0.$$

(Source: www.itl.nist.gov/div898/handbook/eda/section3/eda366h.htm)

Confidence interval

A range of values believed to include an unknown population parameter. Associated with the interval is a measure of the confidence we have that the interval contains the parameter of interest, the *confidence level*, (depending on interpretation) the probability that the parameter of interest will fall within the specified confidence interval. Where a point estimate is a specific numerical value that estimates a parameter, an interval estimate such as a confidence interval is a numeric range that estimates a parameter, generally with an associated probability. A confidence interval for a parameter generalizes to a *confidence set* for more than one parameter at a time. (Source: www2.spsu.edu/tmgt/richardson/Statistics/)

It should be noted that, under the frequentist definition, the confidence level is the probability of the interval covering the true unknown value. The true value is fixed and it is the interval that is random in repeated experimentation.

Continuous random variable

A continuous random variable is one that takes an infinite number of possible values. Continuous random variables are usually measurements. Examples include height, weight, and the concentration of *L. monocytogenes* in a sample. Examples of probability distributions for continuous random variables are the Normal distribution and the Gamma distribution.

(Source: www.stats.gla.ac.uk/steps/glossary/)

Correlation

Correlation is a measure of the relation between two or more variables. Correlation coefficients can range from -1 to +1. The value of -1 represents a perfect negative correlation while a value of +1 represents a perfect positive correlation. A value of 0 represents a lack of correlation.

(Source: www.statsoft.com/textbook/stathome.html)

Convolution

Consider X and Y are non-negative, independent, integer-valued random variables with probability distributions $\Pr\{X=j\}=a_j$ and $\Pr\{Y=j\}=b_j$ and $\Pr\{X=y, Y=k\}=a_jb_k$. The sum S=X+Y is a random variable also, and we recognize that the event S=r is the union of events (X=0, Y=r), (X=1, Y=r-1), …, (X=r, Y=0). So, the probability distribution for S is $\Pr\{S=r\}=c_r$ where $c_r=a_0b_r + a_1b_{r-1} + \ldots + a_rb_0 = \sum^r_{k=0}a_kb_{r-k}$. Feller (1968: 266 et ff.) names this operation convolution (German *Faltung*, French *composition*) and extends the definition to any 2 sequences $\{a_k\}$ and $\{b_k\}$, not necessarily probability distributions. Combinations like this appear in much of the simulation.

Deterministic

Commonly, *deterministic* is an antonym for *stochastic.*

Discrete random variable

A discrete random variable is one which may take on only a countable number of distinct values such as 0, 1, 2, 3, 4, ... Discrete random variables are usually, but not necessarily, counts. If a random variable can take only a finite number of distinct values, then it must be discrete. Examples of discrete random variables include the number of children in a family, the Friday night attendance at a cinema, the number of *L. monocytogenes* organisms in a serving of food. Examples of probability distributions for discrete random variables are the Binomial distribution and the Poisson distribution.

(Source: www.stats.gla.ac.uk/steps/glossary/)

Distribution function

The distribution of a variable is a description of the relative numbers of times each possible outcome in the domain of the variable will occur in a number of trials. The function describing the distribution is called the probability function (probability mass function if the random variable takes only discrete values; probability density function if the random variable is continuous). The *cumulative distribution function* describes the probability that a trial takes on a value less than or equal to a number, commonly $F(x) = \Pr\{X \leq x\}$. The cumulative distribution function is monotone increasing whereas the probability density is not. (Source: mathworld.wolfram.com)

Empirical distribution function

Given data $\{x_k, k = 1, \ldots, n\}$ sorted from smallest to largest, $\{x_{(k)}, k = 1, \ldots, n\}$, $x_{(1)} \leq x_{(2)} \leq \ldots \leq x_{(n)}$, the empirical (cumulative) distribution function (e.c.d.f. or e.d.f.) is the function defined by $\hat{F}(x) = \frac{\text{number of } x_k \leq x}{n}$, a step function with steps of size $1/n$. The values of the e.c.d.f. are the discrete set of cumulative probabilities $\{0, 1/n, \ldots, n/n\}$. When used in a simulation, values between any two consecutive samples, $x_{(k)}$ and $x_{(k+1)}$ cannot be simulated, nor can a value smaller than the minimum, nor can a value larger than the maximum. The e.c.d.f. has mean equal to the sample mean, and variance equal to $(n\text{-}1)/n$ times the sample variance. The e.c.d.f. tends to underestimate the true mean and variance when the underlying distribution is skewed to the right. Expected values of simulated e.c.d.f. quantiles are equal to the sample quantiles. Some variations on the e.c.d.f. appear in simulations: linearly extrapolating between observations; or adding lower and upper tails to the data to reflect a range of the variable outside the observed range, either through expert judgement or by postulating some shape to the tails beyond the sample extremes.

Gamma distribution

The probability density of the Gamma distribution is defined as

$$f(x) = x^{\alpha-1} e^{-x/\beta} [\beta^{\alpha}\Gamma(\alpha)]^{-1},$$

where $x \geq 0$, $\alpha > 0$, $\beta > 0$. α is referred to as the shape parameter. β is referred to as the scale parameter. For integral α, one can recognize the Gamma distribution as the distribution of the waiting time for α Poisson events. As a special case, when $\alpha = 1$, the Gamma distribution is the Exponential distribution.

(Source: www.statsoft.com/textbook/glosfra.html)

Latin Hypercube Sampling

Latin Hypercube Sampling (LHS) is a stratified sampling technique where the random variable distributions are divided into equal probability intervals. A probability is randomly selected from within each interval for each basic event. Generally, LHS will require fewer samples than simple Monte Carlo sampling for similar accuracy. LHS ensures that the entire range of each variable is sampled.

(Source: http://saphire.inel.gov/guest_area/SAF00758.htm)

Lognormal distribution

The lognormal distribution has the probability density function

$$f(x) = \frac{1}{[x\sigma\sqrt{2\pi}]} \exp(-\tfrac{1}{2}[\ln x - \mu]^2 / \sigma^2),$$

where $0 \leq x < \infty$, $\mu > 0$, $\sigma > 0$. If the distribution of a random variable X is lognormal, then the distribution of ln(X) is Normal.

(Source: www.statsoft.com/textbook/glosfra.html)

Maximum likelihood

The method of maximum likelihood is a general method of estimating parameters of a population by values that maximize the *likelihood* (L) of a sample. The likelihood L of a sample of n observations $x_1, x_2, ..., x_n$, is the joint probability function $p(x_1, x_2, ..., x_n)$ when $x_1, x_2, ..., x_n$ are discrete random variables. If $x_1, x_2, ..., x_n$ are continuous random variables, then the likelihood L of a sample of n observations, $x_1, x_2, ..., x_n$, is the joint density function $f(x_1, x_2, ..., x_n)$. When L is a function of parameters, then the maximum likelihood estimates (m.l.e.) of the parameters are the values that maximize L.

(Source: www.statsoft.com/textbook/stathome.html)

Method of moments

This method can be employed to determine parameter estimates for a distribution. The method of matching moments sets the distribution moments equal to the data moments and solves to obtain estimates for the distribution parameters. For example, for a distribution with two parameters, the first two moments of the distribution (the mean μ and variance σ^2 of the distribution) would be set equal to the first two moments of the data (the sample mean and variance, e.g. the unbiased estimators $\bar{x}$ and s^2) and solved for the parameter estimates.

(Source: www.statsoft.com/textbook/glosfra.html)

Monte Carlo

In Monte Carlo methods, the computer uses random number simulation techniques to mimic a statistical population. For each Monte Carlo replication, the computer: simulates a random sample from the population; analyses the sample; and stores the result. After many replications, the stored results will mimic the sampling distribution of the statistic.

(Source: www.statsoft.com/textbook/stathome.html).

Normal distribution

A continuous random variable X has a Normal distribution if its probability density function is $f(x) = \frac{1}{\sqrt{2\pi}\sigma} \exp(-\frac{(x-\mu)^2}{\sigma^2})$, $-\infty<x<\infty$, $\sigma>0$, $-\infty<\mu<\infty$. The normal probability density function has two parameters: μ (mean) and σ (standard deviation). The Normal distribution is sometimes called the *Gaussian* distribution.

(Source: http://ce597n.www.ecn.purdue.edu/CE597N/1997F/students/ michael.a.kropinski.1/project/tutorial#Normal Distribution)

Quantile

The pth quantile of a distribution of values is a number x_p such that a proportion p of the population values are less than or equal to x_p. In a simple random sample of n values, where the sample values ordered in ascending order are $x_{(1)}, \ldots, x_{(n)}$, it is common to use the $x_{(k)}$ as an estimate of the $k/(n+1)^{th}$ quantile, although different software packages use variations of this, $(k-\alpha)(n-\alpha-\beta)^{-1}$ for $\alpha,\beta>0$ (Hyndman and Fan, 1996).

(Source: www.statsoft.com/textbook/glosfra.html)

Quantitative risk assessment

If "risk assessment is generally regarded as a process to scientifically evaluate the probability and severity of known or potential adverse effects attributable to a hazardous agent, process or circumstance" (Cassin et al., 1998), then quantitative risk assessment "implies an estimation of the probability and impact of adverse health outcomes…" (Cassin et al., 1998).

Poisson distribution

The Poisson distribution is defined as $\Pr\{X=k\} = \mu^k e^{-\mu}/k!$, $x = 0, 1, \ldots$, where $\mu>0$ is the average number of occurrences (count) per interval. A Poisson random variable X is a count, interpreted in the context of either distance, area, volume, time or other measure of size (interval) as follows:

- Each non-overlapping interval increment of interest is so small that only one event can occur within it (or at least, the probability of 2 or more events in the interval is negligible), but the sum of the individual increments comprises the entire interval or time period; and
- the probability of an event occurring in the given increment is constant. The number of events observed depends only on the length of the interval considered and not on its end points. If length of interval is 0 and time is 0, the number of events observed is 0. The numbers of changes in non-overlapping intervals are independent for all intervals.

Examples occur in many fields: the number of imperfections (gas trap or cracks) per square metre in rolls of metals; the number of telephone calls per hour received by an office; the number of cashews per can in one can of mixed nuts; the number of bacteria in a given culture; or the number of typing errors per page. The specified region can be an area, a volume, a segment of a line or even a piece of material.

(Sources: http://engineering.uow.edu.au/Courses/Stats/File40.html
http://mathworld.wolfram.com/PoissonDistribution.html)

Rank Correlation

A rank correlation coefficient is a correlation coefficient that is based on the ranks of the sample values and not the actual values. A rank is a consecutive number assigned to a specific observation in a sample of observations sorted by their values. So, ranks reflect the ordered relation of one observation to the others in the sample. The lowest value is assigned a rank of 1; the higher ranks represent the higher values.

(Source: www.statsoft.com/textbook/stathome.html)

Simulation

Etymology: Middle English *simulation*, from Middle French, from Latin simulation-, *simulatio*, from *simulare*

1. the act or process of simulating. 2. a sham object. 3a. the imitative representation of the functioning of one system or process by means of the functioning of another <*a computer simulation of an industrial process*>. 3b. examination of a problem often not subject to direct experimentation by means of a simulating device.

See also: Monte Carlo. (Source: Merriam-Webster Collegiate Dictionary On-line. www.m-w.com/cgi-bin/mweb)

Stochastic

Etymology: Greek *stochastikos* skilful in aiming, from *stochazesthai* to aim at, guess at, from *stochos* target, aim, guess.

- RANDOM; specifically: involving a random variable <a stochastic process>
- involving chance or probability: PROBABILISTIC <a stochastic model of radiation-induced mutation>.

(Source: Merriam-Webster Collegiate Dictionary On-line. http://www.m-w.com/cgi-bin/mweb)

A stochastic process is a family of random variables X(t) indexed by a parameter t, which usually takes values in the discrete set $T = \{0, 1, 2, \ldots\}$ or the continuous set $T = [0, +\infty)$. In many cases t represents time, and X(t) is a random variable observed at time t. Examples are the Poisson process, the Brownian motion process, and the Ornstein-Uhlenbeck process. Considered as a totality, the family of random variables $\{X(t), t \in T\}$ constitutes a "random function".

(Source: www.britannica.com/bcom/eb/article/3/0,5716,117323+26+109439,00.html)

Commonly, *deterministic* is an antonym for *stochastic*.

NOTE: A more extensive glossary of terms related to microbiological risk assessment can be found in MRA 3, an earlier volume in this series (FAO/WHO, 2003).

REFERENCES CITED IN THE GLOSSSARY OF TERMS

Feller, W. 1968. *An Introduction to Probability Theory and its Applications*. 2nd edition. New York NY: Wiley.

Hyndman, R.J. & Fan, Y. 1996. Sample quantiles in statistical packages. *American Statistician,* **50**: 361–365.

Cassin, M.H., Lammerding, A.M., Todd, E.C.D., Ross, W. & McColl, R.S. 1998. Quantitative risk assessment for *Escherichia coli* O157:H7 in ground beef hamburgers. *International Journal of Food Microbiology*, **41**: 21–44.

FAO/WHO. 2003. Hazard characterization for pathogens in food and water: Guidelines. *FAO/WHO Microbiological Risk Assessment Series*, No. 3. 61p.

Appendix 2. Simulation modelling for the four risk assessment examples

A2.1 INTRODUCTION

This appendix serves as documentation for the simulation modelling carried out for the pasteurized milk and ice cream examples, making the work transparent. The model documentation reflects methodological issues, but is not intended to explain the issues in detail. Specific issues related to the development of the two risk assessment models are addressed here. These include a description of the consumption characteristics used and the modelling of home storage conditions, and how non-susceptible and susceptible populations might be defined. How to combine independent data sets to describe prevalence of *L. monocytogenes* in foods and to describe concentration of *L. monocytogenes* in foods are also addressed. The appendix provides a list of references to support the documentation here, and to provide that vast amount of supplementary material that is the background for much of the work. Still other methodological material appears in the main body of the report (Part 3 – Exposure assessment, and Part 4 – Example risk assessments). Implementation of the Monte Carlo simulation for this exposure assessment was performed using Analytica™1.11, 2.0.1 or 2.0.5 (Lumina Decisions) software. Additional computations and preparation of graphs were done using Microsoft® Excel 97 and Microsoft® Excel 2000, and with S-Plus 4.5 Professional, S-Plus 2000 Professional and S-Plus 6 Professional (MathSoft, Inc.).

A2.2 MODELLING THE EXPOSURE ASSESSMENT

A2.2.1 Overview

Objectives for the exposure assessment are to simulate the number of *L. monocytogenes* organisms in a serving of a particular RTE) food, *Lm ingested*, and to determine the annual frequency of servings for individuals in the consuming population, *Annual meals*. Every shape in the influence diagram (Figure A2.1) is termed a node. Different shaped nodes perform different functions. The hexagonal figures, *Lm ingested* and *Annual meals*, represent the stochastic results that answer the questions deriving from the objectives. Elliptical shapes, such as *Food amount eaten*, are chance (stochastic) nodes that hold intermediate calculations that form part of the modelling for the objective nodes. The round-cornered rectangular nodes, such as *Prevalence and concentration*, are organizing modules that contain other nodes. The hexagonal pennant boxes, *Discrete distributions* and *Study indices* are libraries of functions that support some calculations or that contain index nodes that structure the results. Arrows indicate influences and indicate the direction of the influence. For example, the number of *L. monocytogenes* organisms ingested in a serving when that number exceeds zero, *Lm ingested given >0*, depends on the *Concentration in ingested food* and the *Food amount eaten*. The values in *Food frequency* determine what values reside in *Annual meals*.

To model the exposure assessment, the following information is needed:

1. *Prevalence and concentration characteristics*, measured at the same consistent point in the farm-to-fork chain. Prevalence relates how often the food is contaminated with *L. monocytogenes*. The notion is generalized to consider it equivalent to the probability that a serving from a package or unit of product contains any contamination. Concentration defines how many *L. monocytogenes* organisms are in a contaminated portion.
2. *Storage characteristics* and *Growth characteristics* that determine the amount of *Growth* of *L. monocytogenes* in the product from that point in the process to the point of consumption.
3. *Consumption characteristics* that relate how much food consumers eat and how often they eat it. How large a serving the consumer eats determines how many *L. monocytogenes* organisms the consumer ingests.
4. *Non-susceptible and susceptible populations*. Hazard Identification generally indicates that some portions of the consuming population are more susceptible to infection or illness from *L. monocytogenes*.

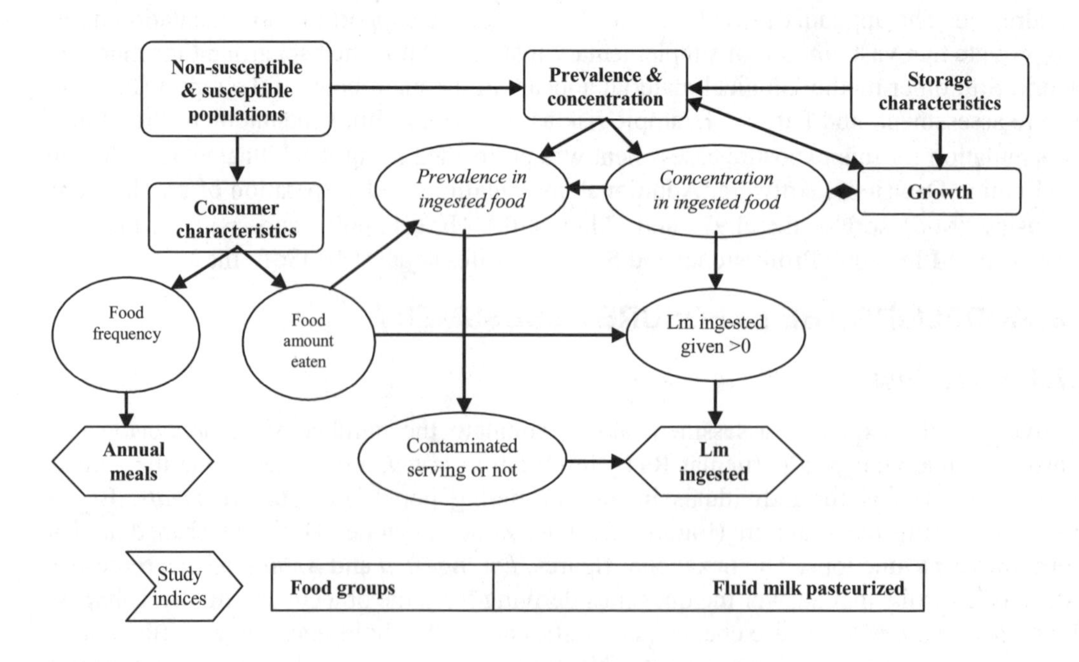

Figure A2.1 Influence diagram for *Listeria monocytogenes* exposure assessment.

Figure A2.1 clarifies the model structure and what information is needed from the exposure assessment. However, it does not make explicit the node's specific parameterization. Also, it does not show all of the interrelationships and dependencies. Accompanying documentation in this Appendix does that (such as Table A2.1), and specific methodological issues are addressed in the various sections.

Table A2.1 Nodes for top level of *Listeria monocytogenes* in RTE eat foods exposure assessment model.

Title Identifier Structure	**Description and Definition**
Non-susceptible and susceptible populations Nonandsusceptible Module	Module holds characteristics that define the allocation of individuals from Gender × Age groups to non-susceptible and susceptible groups. Among adults, for whom we have some information about consumption characteristics, susceptible groups are defined to include all adults 65 and older, pregnant women (1.3% of the population) and individuals with suppressed immune systems and certain medical conditions such as cancer and recent organ transplantation (3.3% of the population) (Miller, Whiting and Smith, 1997).
Consumption characteristics Consumption_character Module	Consumption characteristics come from 24-hour recall data from CFPNS (1992–1995), which addressed the nutritional habits of non-institutionalized adults between 18 and 74 years old in Québec, Nova Scotia, Saskatchewan, Alberta and Prince Edward Island.
Prevalence and concentration Prevalence_and_conce Module	Prevalence and concentration module determines simulated distributions for the prevalence of *L. monocytogenes* in food and the concentration in contaminated food, nominally at retail.
Storage characteristics Storage_characterist Module	Storage characteristics module determines simulated distributions for the refrigerator temperature that the consumer stores the food at, and the length of time, measured as if from retail purchase, to the time of consumption.
Growth Growth2 Module	Growth module determines growth characteristics for *L. monocytogenes* in the food. Growth is characterized by exponential growth rates and stationary phase population size for each foodstuff.
Food amount eaten Serving_size1 (g) Chance node	Food amount eaten is the simulated distribution of the daily serving size for individuals from the non-susceptible group and for individuals from the susceptible group. Food amount eaten comes directly from the consumption amounts generated in the Consumption characteristics module.
Annual meals Annual_meals Objective node	The number of Annual meals for an individual is calculated from the Food frequency probability of consumption on a given day, by implementing Binomial sampling. The number of meals (population days with consumption) is calculated in the Consumption characteristics module, rounded here for display Table(Annualmealsreporting)(Round(Binomial(365, Mealfrequency[Annualmealsreporting='Individual'])), Round(Mealfrequency[Annualmealsreporting='Population']))
Prevalence in ingested food Prevalence_in_ingest Chance node	The Prevalence in ingested food node is the simulated distribution for how often a serving contains any L. monocytogenes contamination. For Icebox:=Refrigerator_studies Do For Person:=Risk_group_definitio Do Correlatedprevalence * (1-Exp(-10^Finalconcentration[Refrigerator_studies=Icebox] * Serving_size1[Risk_group_definitio=Person]))
Contaminated serving or not There_or_not_there Chance node	Contaminated serving or not is a simple accounting of whether a serving is contaminated or not. It is generated by sampling from the outcomes Not contaminated and Contaminated, with probabilities Not contaminated (1-Prevalence_in_ingest) Contaminated Prevalence_in_ingest

Title Identifier Structure	Description and Definition
Concentration in ingested food Finalconcentration (log_{10} CFU/g) Chance node	Concentration in ingested food is the simulated distribution of the concentration of *L. monocytogenes* in contaminated food at ingestion. Initial concentrations grow into final concentrations according to the growth determined in the Growth module. Final concentrations are restricted to theoretical maximum population densities or stationary phase densities. Using Calculatedfinal:= Initialconcentration + Unconstrgrowthamount Do (if Maximum_population >0 then (if Calculatedfinal <= logten(Maximum_population) then Calculatedfinal else logten(Maximum_population)) else Calculatedfinal)
Lm ingested given >0 Dose (CFU) Chance node	The Lm ingested given >0 node records the simulated distribution of the number of *L. monocytogenes* organisms in a serving of food, in those cases where the number of organisms is larger than 0. The Prevalence in ingested food node lets us derive how often the number of organisms is 0. Lm ingested given >0 generates non-zero observations by sampling on [1,∞) with Poisson probabilities. For Icebox:=Refrigerator_studies Do For Person:=Risk_group_definitio Do logten(Round(Conditional_poisson(10^Finalconcentration[Refrigerator_studies=Icebox] * Serving_size1[Risk_group_definitio=Person])))
Lm ingested Lm_ingested (CFU) Objective node	Lm ingested is one of the objective nodes for this exposure assessment. It is calculated by combining the simulated distributions for the Prevalence in ingested food and the Lm ingested given >0. The number of *L. monocytogenes* organisms ingested when the food is not contaminated is assumed to be 0. For temp:=Run Do if There_ot_not_there[Run=temp]='Not contaminated' then 0 else 10^Dose
Study indices Row_and_column_inde1 Library	Study indices is a collection of index information that structure the results.

A2.2.2 Non-susceptible and susceptible populations

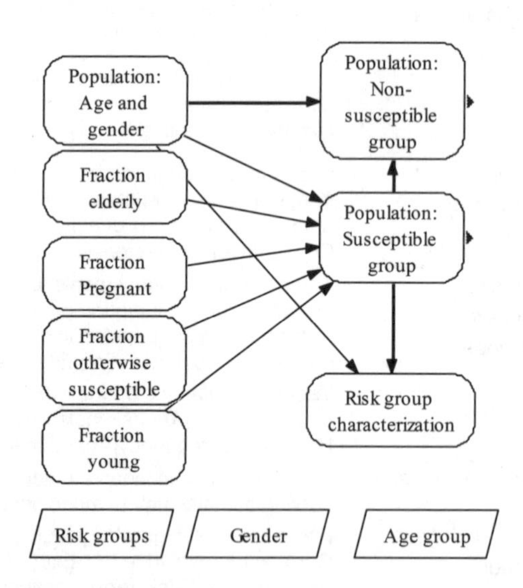

Figure A2.2 Influence diagram for non-susceptible and susceptible populations.

The susceptible population is determined by the fractions of persons who have one of the characteristics named: elderly, pregnant, otherwise susceptible, young. For the present implementation of the exposure assessment, where there are no consumption data for persons under 18 years old from Canadian data (CFPNS, 1992–1995), the fraction young is moot. In Figure A2.2, *Population age and gender* table holds domain estimates for the Age and Gender groups used. For Canadian consumption data, estimates represent the population counts, in the years of the surveys, in five provinces: Alberta, Nova Scotia, Prince Edward Island, Quebec and Saskatchewan, for which consumption information is available (CFPNS, 1992–1995). *Fraction elderly, Fraction pregnant, Fraction otherwise susceptible* and *Fraction young* are tables that describe the fraction of the population, in Gender × Age groups, who would be attributed to the susceptible group, for the named reason. For example, a fraction of Females, 18–34 and 35–49 would be attributed to the susceptible group in the *Fraction pregnant* table. All persons 65 and older, but no others, would be attributed to the susceptible group in the *Fraction elderly* table. Allocations were based on Miller, Whiting and Smith (1997), as applied to the Canadian population. The *Population susceptible group* table collects the fractions together to give the population size in the susceptible group, by Gender and by Age. The *Population non-susceptible group* table is derived by difference from the *Population age and gender* and *Population susceptible group* tables to give the population size in the non-susceptible group, by Gender and by Age. *Risk group characterization* is a useful summary of the attribution of individuals to the non-susceptible and susceptible groups. This module is also a natural holding place for parallelogram-shaped index nodes, *Risk groups*, *Gender* and *Age group*, that structure the results through many modules in the model. Details of the nodes are described in Table A2.2.

Table A2.2 Nodes for non-susceptible and susceptible populations module.

Title, Identifier, Structure	Description and Definition
Population Age and gender Popn_age_gender Module	Population age and gender table holds domain estimates for the Age and Gender groups used.
Fraction elderly Fraction_elderly Variable node	
Fraction Pregnant Fraction_pregnant Variable node	Fraction elderly, Fraction Pregnant, Fraction otherwise susceptible and Fraction young are tables that describe the fraction of the population, in Gender × Age groups, whom we would attribute to the susceptible group, for the named reason.
Fraction otherwise susceptible Fraction_otherwise_s Variable node	
Fraction young Fraction_young Variable node	
Population non-susceptible risk group Population_normal Variable node	Population non-susceptible group table is derived by difference from the Population Age and gender and Population Susceptible group tables to give the population size in the non-susceptible group, by Gender and by Age. Popn_age_gender — Population_high_risk
Population susceptible group Population_high_risk Variable node	Population susceptible group table collects the fractions together to give the population size in the susceptible group, by Gender and by Age. Popn_age_gender * (Fraction_elderly + Fraction_otherwise_s + Fraction_pregnant + Fraction_young)
Risk group characterization Riskgroupcharacteriz Variable node	Risk group characterization is a useful summary of the attribution of individuals to the risk groups.
Risk groups Risk_group_definitio Index node	Risk groups are defined as ['Non-susceptible', 'Susceptible']
Gender Gender_definition Index node	Gender is defined as ['Female', 'Male']
Age group Age_group_definition Index node	Age group is defined as ['18-34', '35-49', '50-64', '65-74']

A2.2.3 Consumption characteristics

Consumption characteristics are the amount of food eaten in a serving, the daily probability of consuming the food and the annual number of meals (days with consumption) in the population. The nodes *Food amount eaten* and *Food frequency* are collectors for the characteristics calculated in each of the food-specific modules: *Ice cream*; and *Fluid milk, pasteurized. Food amount eaten* is the distribution of amounts eaten (g). *Food frequency* is the probability of consuming the food on a given day. *Survey results* and *Ecdf columns* are index nodes that structure data tables in the *Fluid milk, pasteurized* and *Ice cream* modules (Table A2.3, Figure A2.3). Exposure assessment examples for pasteurized milk and ice cream were implemented using the modules described here.

Table A2.3 Nodes for Consumption characteristics module.

Title, Identifier, Structure	Description and Definition
Ice cream Consumption_char1 Module	The main consumption characteristics are the serving size and the frequency of consumption for ice cream.
Fluid milk, pasteurized Consumption_char3 Module	The main consumption characteristics are the serving size and the frequency of consumption for pasteurized fluid milk.
Food amount eaten Serving_size (g) Chance node	Food amount eaten is the distribution of amounts eaten (g). DetermTable(Food_groups)(Samp_cons_ice, Samp_cons_pmilk)
Food frequency Meal frequency Chance node	Food frequency is the probability of consuming the food on a given day and the annual meals (population consumption days) for population. Determtable(Food_groups, Annualmealsreporting)(Samp_freq_ice, Annualmealsicecream, Samp_freq_pmilk, Annualmealspmilk)
Survey results Survey_results Index node	Survey results structures data tables in Fluid milk, pasteurized and Ice cream modules. ['Respondents', 'Consumers']
Ecdf columns Ecdf_columns Index node	Ecdf columns structures data tables in Fluid milk, pasteurized and Ice cream modules. ['Amount', Fraction']

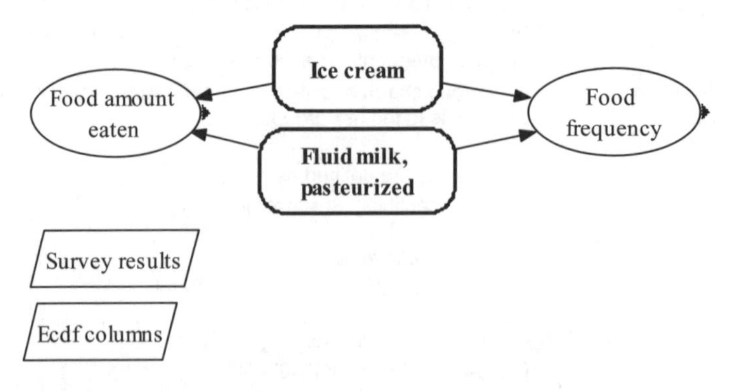

Figure A2.3 Influence diagram for *Consumption characteristics* module.

A2.2.4 *Ice cream* and *Fluid milk, pasteurized* modules

Consumption characteristics modules are specific to the RTE food considered. The structure of consumption characteristics modules for *Ice cream* and for *Fluid milk, pasteurized* are identical. Both are described together, referring to the *Fluid milk, pasteurized* example (Table A2.4, Figure A2.4).

Table A2.4 Nodes for *Ice cream* and for *Fluid milk, pasteurized* modules.

Title, Identifier, Structure	Description and Definition
Gender, age consumption frequency Gender_age_freq6 Variable node	Gender, age consumption frequency is a table indexed by Age group and Gender, holding point estimates of daily consumption probabilities for Pasteurized milk. Other nodes in the module use the point estimates to determine what proportions of Age and Gender group characteristics to include in non-susceptible and susceptible populations.
Susceptible group proportions Gender_age_pro_high6 Variable node	Susceptible group proportions is a table that holds proportions of total eating episodes assigned to Gender and Age group for individuals in the susceptible group. We have adjusted Gender, age consumption frequency proportions to reflect membership in susceptible groups, Population: susceptible group.
Susceptible group gender Sampled_high_risk_g6 Chance node	Susceptible group gender and Susceptible group age are stochastic nodes that hold a Gender and an Age Group, sampled so that Gender × Age group proportions among consumers in the susceptible population are respected.
Susceptible group age Sampled_high_risk_a6 Chance node	
Non-susceptible group proportions Normal_intake_pr6 Variable node	Non-susceptible group proportions is a table that holds proportions of total eating episodes assigned to Gender and Age group for individuals in the non-susceptible group. We have adjusted Gender, age consumption frequency proportions to reflect membership in non-susceptible groups, Population: Non-susceptible group.
Non-susceptible group gender Sampled_gender6 Chance node	Non-susceptible group gender and Non-susceptible group age are stochastic nodes that hold a Gender and an Age Group, sampled so that Gender × Age group proportions among consumers in the non-susceptible population are respected.
Non-susceptible group age Sampled_age_group6 Chance node	
Nutrition survey results, milks Nutrition_survey_re6 Variable node	Nutrition survey results, milks is a table that holds inferential statistics from the nutrition surveys: the number of survey respondents and the number who reported consuming Pasteurized milk on a given day.
Pasteurized milk amounts Pmilk_amount Variable node	Pasteurized milk amounts is a table that holds empirical daily Pasteurized milk amounts collected from the nutrition surveys. The table has columns Amount, an amount consumed (g) and Fraction, the inverse of the design-based weights associated with Pasteurized milk consumers in the nutrition surveys. There is a separate table for each Gender × Age group.
Milk amount index Pmilk_amount_index Index node	Milk amount index structures the table of amounts, Pasteurized milk amounts. Range of sequence corresponds to a set of rows in an Excel spreadsheet. Sequence(5, 459, 1)
Beta, milks frequency Beta_frequency6 Chance node	Beta, milks frequency uses a Beta distribution to represent uncertainty or variability over a Gender × Age group, for consumers' consumption probability on a given day. Beta, milks frequency is assumed to be Beta(x+1, n-x+1), where n is the number of respondents, and x is the estimated number of Pasteurized milk consumers, $n\pi$.

Title, Identifier, Structure	Description and Definition
Annual meals, susceptible group Annualmealshighrisk6 Chance node	Annual meals, susceptible group simulates the annual meals (population days with consumption) in susceptible population. It incorporates the variability and uncertainty about the fraction of adults who consume Pasteurized milk on a given day. It samples binomially among population days (population × 365), but uses a Normal approximation. We truncate the Normal distribution at 0 and the total population days. Sum(Sum(Using Beta_value:= Beta_frequency6[Risk_group_definitio='High risk'] Do Using People_days:=Population_high_risk*365 Do Using Interim:= -Truncate(-(Truncate(Normal(People_days*Beta_value, People_days*Sqrt(Beta_value*(1-Beta_value))), 0)), -People_days) Do if Interim<=0 then 0 else if Interim>=People_days then People_days else Round(Interim), Gender_definition), Age_group_definition)

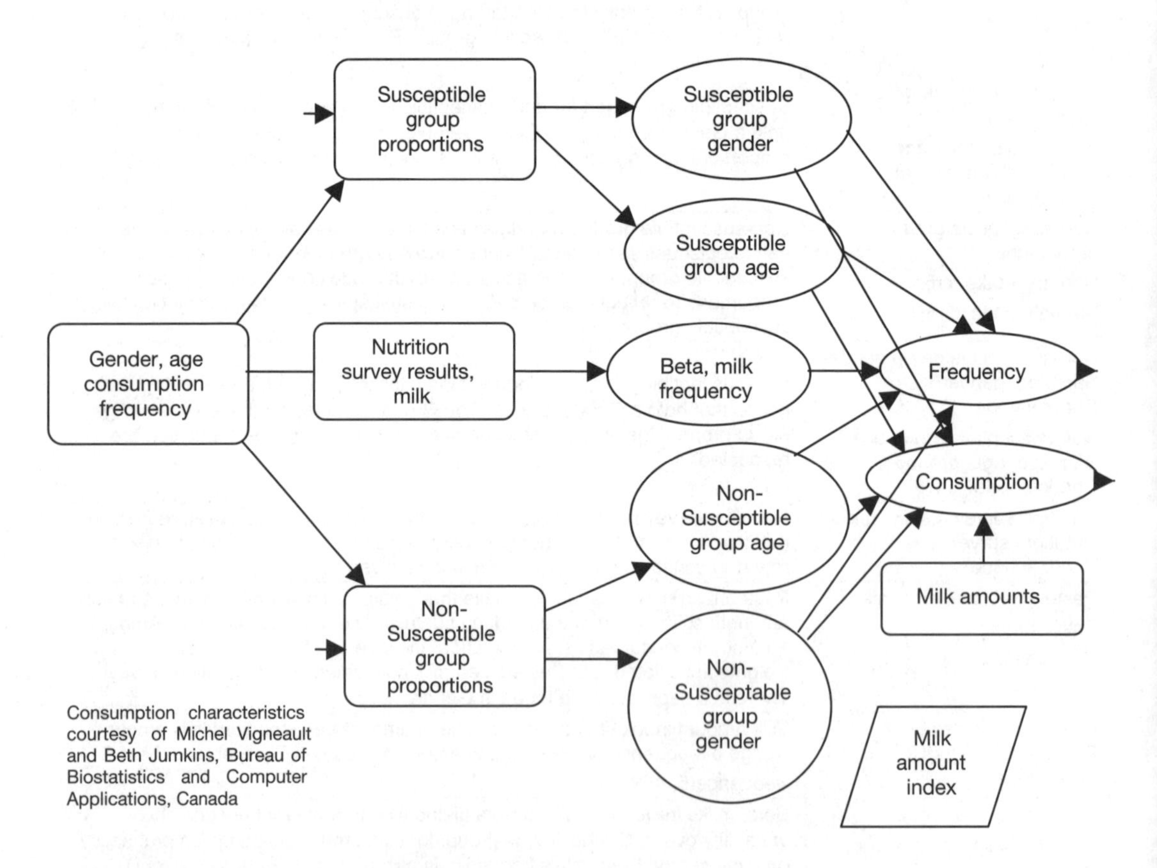

Figure A2.4 Influence diagram for *Pasteurized milk* consumption characteristics module. Except for changes to node identifiers, the *Ice cream* consumption characteristics module is identical.

A2.2.5 Prevalence and concentration

The *Prevalence and concentration* module simulates the *Prevalence characteristics* – nominally prevalence of *L. monocytogenes* contamination in foods at retail or source for the consumer – and simulates the *Concentration characteristics* – nominally the *L. monocytogenes* concentration in the food at that point. The module lets one specify a rank correlation coefficient between the prevalence and concentration. Last, it simulates the *Prevalence in ingested food* and the *Concentration in ingested food*, when the concentration is larger than 0 (Figure A2.5, Table A2.5).

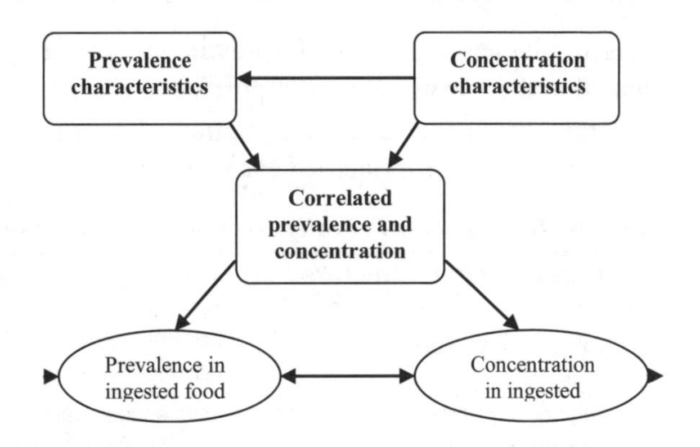

Figure A2.5 Influence diagram for *Prevalence and concentration* characteristics module.

Table A2.5 Nodes for *Prevalence and concentration* module.

Title, Identifier, Structure	Description and Definition
Prevalence characteristics Prevalence_character Module	Prevalence characteristics simulates the prevalence of *L. monocytogenes* as measured at retail, in the packages or units that the consumer would purchase.
Concentration characteristics Concentration_charac Module	Concentration characteristics simulates the *L. monocytogenes* concentration as measured at retail or source, in the packages or units that the consumer would purchase.
Correlated prevalence & concentration Correlated_prevalenc Module	Correlated prevalence & concentration is the means to specify the rank correlation coefficient between the prevalence and concentration.
Prevalence in ingested food Prevalence_in_ingest Chance node	The Prevalence in ingested food node is the simulated distribution for how often a serving contains any *L. monocytogenes* contamination. Some servings from a contaminated package will carry no organisms (exp(-mμ), where the serving size is mg and the concentration is μg^{-1}). Those probabilities adjust the prevalence estimates that emerge from the Prevalence and concentration characteristics module. Prevalence in ingested food is calculated as follows. For Icebox:=Refrigerator_studies Do For Person:=Risk_group_definitio Do Correlatedprevalence * (1-Exp(-10^Finalconcentration[Refrigerator_studies=Icebox] * Serving_size1[Risk_group_definitio=Person]))
Concentration in ingested food Finalconcentration ($\log_{10}$ CFU/g) Chance node	Concentration in ingested food is the simulated distribution of the concentration of *L. monocytogenes* in contaminated food at ingestion. Initial concentrations grow into final concentrations according to the growth determined in the Growth module. Final concentrations are restricted to theoretical maximum population densities or stationary phase densities. Using Calculatedfinal:= Initialconcentration + Unconstrgrowthamount Do (if Maximum_population >0 then (if Calculatedfinal <= logten(Maximum_population) then Calculatedfinal else logten(Maximum_population)) else Calculatedfinal)

A2.2.6 Prevalence characteristics

This implementation models prevalence as measured at retail, in the product packages or units that the consumer would purchase. The *Prevalence parameters* table specifies the α and β parameters of a Beta distribution. *Prevalence in packages* is defined as Beta (α, β) parameters as appropriate for each food group (Table A2.6).

Table A2.6 ***Nodes for Prevalence characteristics module.***

Title, Identifier, Structure	Description and Definition
Prevalence parameters Prevalenceparameters Variable node	The Beta distribution, which has support on [0, 1], is a common way to characterize the heterogeneity in the prevalence. Determtable(Food_groups,Prevdistrnparameters)(0.424, 0.55, 155.47)
Prevalence in packages Prevalenceinpackage Chance node	Prevalence in packages samples from the Beta distribution specified by Prevalence parameters and makes the Packaging adjustment required Packaging_adjustment * (Using localalpha:=Prevalenceparameters[Prevdistrnparameters='alpha'] Do Using localbeta:=Prevalenceparameters[Prevdistrnparameters='beta'] Do Beta(localalpha, localbeta))
Prevalence distribution parameters Prevdistrnparameters Index node	Prevalence distribution parameters indexes the columns of the Prevalence parameters table. ['alpha', 'beta']

A2.2.7 Concentration characteristics

Concentration distributions were derived from published studies for two groups of RTE foods. In Figure A2.6, rounded rectangular shapes are variables holding a table of data that describes the empirical distribution function or a set of quantiles, or parameters for a distribution function for the concentrations .

At each iteration, a value is sampled from the distribution and collected into one of the elliptical nodes. The *Initial concentration* node collects all results, still separate, together in the same place. The parallelogram at the bottom of the Figure A2.6, *Concentration table columns*, lists the columns that appear in concentrations tables: Concentration, and Quantile. Data collection and organization from referenced studies provide concentration distributions that represent levels of concentrations in recognizable packages or units of products (Table A2.7).

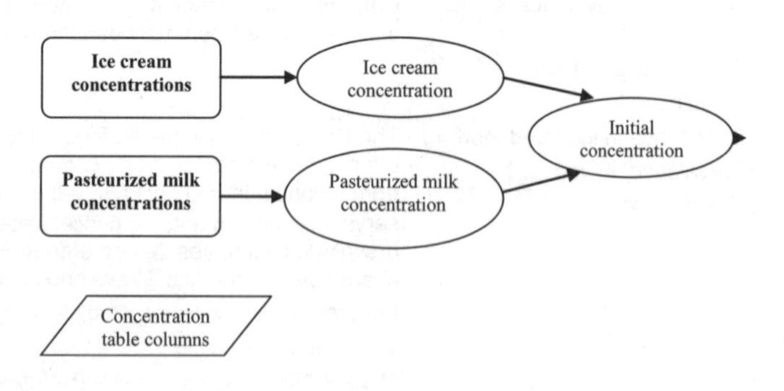

Figure A2.6 Influence diagram for *Concentration characteristics* module.

Table A2.7 Nodes for *Concentration characteristics* module.

Title, Identifier, Structure	Description and Definition
Ice cream concentrations Icecreamconctable Variable node	Ice cream concentrations holds quantile distribution for *L. monocytogenes* concentration in ice cream.
Ice cream concentration Icecreamconc Chance node	Ice cream concentration samples from the cumulative distribution that is specified in Ice cream concentrations. Cumdist(Icecreamconctable[Concentration_tables='Quantile'], Icecreamconctable[Concentration_tables='Concentration'])
Pasteurized milk concentrations Pastmilkconctable Variable node	Pasteurized milk concentrations holds quantile distribution for *L. monocytogenes* concentration in pasteurized milk
Pasteurized milk concentration Pastmilkconc Chance node	Pasteurized milk concentration samples from the cumulative distribution that is specified in Pasteurized milk concentrations. Cumdist(Pastmilkconctable[Concentration_tables='Quantile'], Pastmilkconctable[Concentration_tables='Concentration'])
Concentration table columns Concentration_tables Index node	Concentration table columns structures the concentration tables. ['Concentration', 'Quantile']
Initial concentration Initialconcentration Chance node	Initial concentration node collects all results, still separate, together in the same place. Determtable(Food_groups)(Icecreamconc, Pastmilkconc)

A2.2.8 Correlated prevalence and concentration

The Analytica™2.0.1 and 2.0.5 software does not directly implement built-in methods for generating random variables with a desired correlation structure. The installation does provide a Library module, *Correlated Distributions*, which provides the mechanics to achieve the desired result. The *Correlated Distributions* library module implements the method of Iman and Conover (1982), which makes the rank correlation between specified variables meet the desired result. So, *Prevalence in packages*, from the *Prevalence characteristics* module and *Initial concentration*, from the *Concentration characteristics* module, are re-ordered to produce rank-correlated *Correlated prevalence* and *Correlated concentration*.

A2.2.9 Storage characteristics

Storage temperature and *Storage time* characterize storage conditions in this exposure assessment. Storage temperature is intended to represent storage in the consumer's refrigerator, after purchase of the food product from retail. Storage time is intended to represent the length of time that the food product is stored at that temperature, measured from retail purchase until the consumer eats a portion. A simple implementation assumes constant temperature. More complicated implementations that depend on quantitative data lacking here could incorporate time and temperature integration.

Storage temperature comes from four separate sources in different countries, which were included through the whole exposure to examine the effects of different assumptions about temperatures, or different distributions of temperatures. It is assumed that refrigerator storage temperatures are the same for any food product – unrealistic, but simplifying (Table A2.8).

Table A2.8 Quantiles for refrigerator temperature (°C) distributions, showing point estimates for cumulative probabilities from four studies.

Audits International (2000)		Johnson et al. (1998)		Sergelidis et al. (1997)		O'Brien (1997)	
°C	Cum. Prob.	°C	Cum. Prob.	°C	Cum. Prob.	°C	Cum. Prob.
0	0	-2.5	0	0	0	0	0
0.14	0.03	-2	0.002	9	0.45	4	0.40
0.83	0.06	-1	0.002	10	0.75	11	1
1.94	0.15	0	0.01	13	1		
3.05	0.37	1	0.02				
4.16	0.74	2	0.07				
5.28	0.80	3	0.11				
6.39	0.91	4	0.18				
7.50	0.98	5	0.30				
8.61	0.99	6	0.44				
9.72	1.00	7	0.76				
10.28	1	8	0.92				
		9	0.96				
		10	0.99				
		11	0.997				
		12	0.998				
		13	1				

Storage time represents the length of time that the consumer stores the product before eating a serving from it. Storage time distributions are modelled as specific to the food product under consideration. Following FDA/FSIS (2001), *Minimum time*, *Mode time* and *Maximum time* parameterize storage time distributions, via Triangular(Minimum time, Mode time, Maximum time). Minimum time is set to a constant 0.5 days for all products, but Mode and Maximum are intended to depend on the food. Mode time and Maximum time are allowed to be stochastic. Mode time varies as Uniform(±20% nominal). Maximum time varies as Uniform(±50% nominal). Nominal values are listed in Table A2.9. Mode time and Maximum time are strictly related, so that nonsensical values are not generated. Consequently, this implementation calculates an *Indep. Storage time* that aligns the smallest *Mode time* with the smallest *Indep. Maximum* (Table 2.10).

Table A2.9

Post-retail storage times (days).

	Min	Mode	Max
Ice cream	0.5	7	30
Fluid milk, pasteurized	1	5	12

Storage life for pasteurized milk depends on the growth of spoilage bacteria, which depends on temperatures. The effect would be to truncate the time distribution differently at different temperature values. General tendencies would be the same. Distribution shapes would change. The storage life for pasteurized milk is assumed to be 12 days at 4°C, with storage life at other temperatures determined by the relationship $Life(T) = 12 \times \left[\frac{4+7.7}{T+7.7}\right]$ in Neumeyer, Ross and McMeekin (1997) and Neumeyer et al. (1997). The influence diagram is shown in Figure A2.7.

Table A2.10 Nodes for Storage characteristics module.

Title, Identifier, Structure	Description and Definition
Audit International 2000 Audit2000 Variable node	Audits International 2000 is 1 of 4 sources of refrigerator storage temperatures (Table A2.8).
Refrigerator characteristics Refrigerator Index node	Refrigerator characteristics node structures the data in the refrigerator temperature table. ['Temperature', 'Frequency', 'Cumulative frequency']
Refrigerator studies Refrigerator_studies Index node	Refrigerator studies node maintains a consistent list of the 4 refrigerator temperature source studies. ['Audits International 2000', 'Johnson et al., 1998', 'Sergelidis et al., 1997', 'O'Brien 1997']
Storage temperature Storage_temperature Chance node	Sstorage temperature is sampled from the information in the four refrigerator temperature studies. Cumdist(Audit2000[Refrigerator='Cumulative frequency'], Audit2000[Refrigerator='Temperature'])
Post-retail storage time Post_retail_storage_ Variable node	Post-retail storage time holds minimum, mode and maximum storage time for each Food groups label and for each Updates label. Storage time represents the length of time that the consumer stores the product before eating a serving from it. Storage time distributions are modelled as specific to the food product under consideration. Following FDA/FSIS (2001), Minimum time, Mode time and Maximum time parameterize storage time distributions, via Triangular(Minimum time, Mode time, Maximum time) (Table A2. 9).
Minimum time Minimumtime Chance node	Minimum time extracts the minimum time from Post-retail storage time appropriate to the selected Food groups. Post_retail_storage_[Post_retail_storage_='Minimum']
Mode time Modetime Chance node	Mode time extracts the nominal mode time from Post-retail storage time appropriate to the selected Food groups. Mode time varies as Uniform (±20% nominal). Using Temp:=Post_retail_storage_[Post_retail_storage_='Mode'] Do Uniform(0.8*Temp, 1.2*Temp)
Indep. maximum Maximumindeptime Chance node	Indep. maximum extracts the nominal maximum time from Post-retail storage time appropriate to the selected Food groups. Maximum time varies as Uniform(±50% nominal). Using Temp:=Post_retail_storage_[Post_retail_storage_='Maximum'] Do Uniform(0.5*Temp, 1.5*Temp)
Corr. Maximum Maximumtime Chance node	The Mode time is assumed to follow a Uniform(0.8*mode, 1.2*mode) and the Indep. maximum to follow a Uniform(0.5*maximum, 1.5*maximum). To avoid nonsensical parameter combinations, and to represent what would to be a sensible set of conditions, the random mode and random maximum have a correlation coefficient of 1. For Onebyone:=Updates Do Using Another:=Rank(Maximumindeptime[Updates=Onebyone],Run) Do Using Sortedmaximum:=Maximumindeptime[Updates=Onebyone, Run=Sortindex(Another,Run)] Do Sortedmaximum[Run=Rank(Modetime[Updates=Onebyone], Run)]
Storage time Preliminarytime Chance node	Storage time is the storage time before acting to make the time and temperature related. Triangular(Minimumtime, Modetime, Maximumtime)
Truncated storage time Storage_time Chance node	The storage life for pasteurized milk is assumed to be 12 days at 4°C, with storage life at other temperatures determined by the relationship in Neumeyer, Ross and McMeekin (1997) and Neumeyer et al. (1997). (Using local1 := (1643/((Storagetemperature+7.7)^2)) Do Using local2 := (Storage_time1>local1) Do ((Storage_time1*(1-local2))+(local2*For local3 := Run Do (If local2[Run=local3] Then (-Truncate((-Storage_time1),(-local1[Run=local3]))) Else 0))))

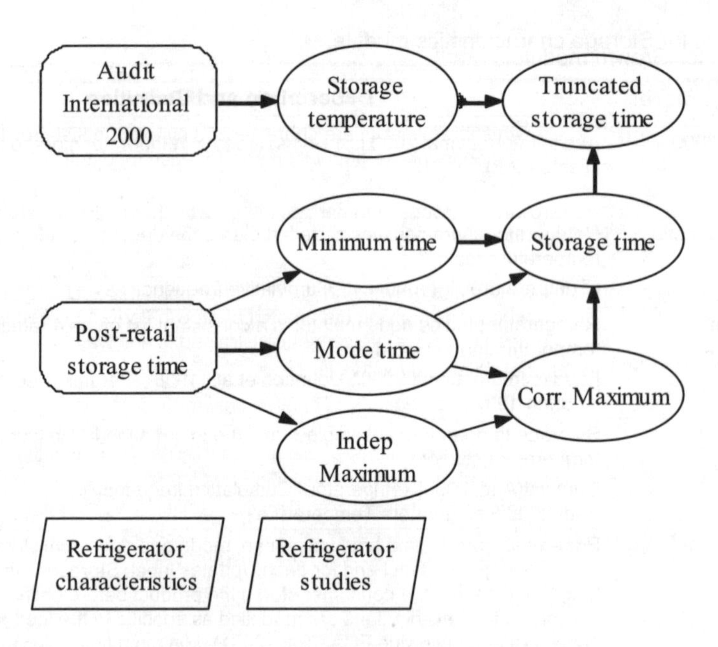

Figure A2.7 Influence diagram for *Storage characteristics* module.

A2.2.10 Growth

The *Growth* module uses simulated *Growth characteristics* to determine the *Growth per day* If it is assumed that every day, or every part day, has the same *Storage conditions* and the same *Growth characteristics*, then the *Unconstrained growth and die off* can be determined in a straightforward manner, constraining that *Unconstrained growth* by the stationary phase maximum population density. The influence diagram is shown in Figure A2.8 and the noted are described in Table A2.11.

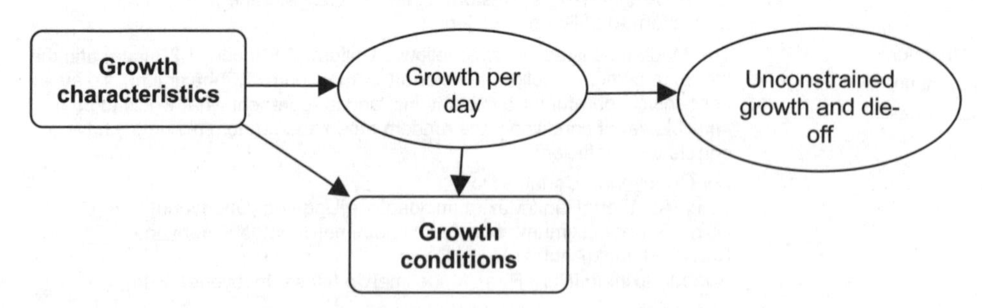

Figure A2.8 Influence diagram for *Growth* module.

Table A2.11 Nodes for *Growth* module.

Title, Identifier, Structure	Description and Definition
Growth characteristics Growth_characteristi Module	Growth characteristics are characterized by exponential growth rates at 5°C and stationary phase population size for each foodstuff.
Growth conditions Growth_conditions Module	Growth conditions summarizes the growth characteristics and growth conditions.
Unconstrained growth and die off Unconstrgrowthamount ($\log_{10}$ CFU/g) Chance node	Unconstrained growth and die off is the simple product of Storage time and Growth per day, giving the amount of growth ($\log_{10}$ CFU/g) over the whole storage time, were growth not constrained in any way. Storage_time * Growthdaily
Growth per day Growthdaily ($\log_{10}$ CFU/g/day) Chance node	Growth per day adjusts calculated growth (5°C) to Storage temperature. Convert to an EGR at some other temperature via McMeekin et al. (1993). In no growth conditions – zero growth rate or Storage temperature below minimum growth temperature – the zero growth rate remains as is. For Icebox:=Refrigerator_studies Do Using localtemperature:= Storage_temperature[Refrigerator_studies=Icebox] Do if localtemperature <= Minimum_growth_tempe then (if Growth_rate<=0 then Growth_rate else 0) else if Growth_rate<=0 then Growth_rate else Growth_rate * (localtemperature-Minimum_growth_tempe)^2/(5-Minimum_growth_tempe)^2

A2.2.11 Growth characteristics

Growth characteristics are the exponential *Growth rates*, the *Minimum Growth Temperature* and the *Stationary Phase Population*. The influence diagram is shown in Figure A2.9. The Growth rates node is a table of means and standard deviations for the growth rate, $\log_{10}$/day, at 5°C, gleaned from versions of FDA/FSIS (2001). Storage temperature is explicitly accounted for as a dependent condition for growth. It is assumed that the range of growth rates (FDA/FSIS, 2001) samples among the other dependent conditions (a_W, pH, NaCl, NO_3). Values are shown in Table A2.12. Growth rate selects values according to Normal(mean, standard deviation) for this exposure assessment. FDA/FSIS (2001) provides maximum Stationary phase population values that change with temperature. Minimum growth temperature is implemented as Triangular(1°C, 1.1°C, 2°C) for this exposure assessment.

A2.2.12 Growth conditions

The *Growth conditions* module gives a summary of growth and survival. For example, it converts the *Growth rate* simulated for 5°C into a *Generation time*, via $\log_{10}(2)$/Growth rate. Also, it summarizes the growth situations, by tabulating from the simulated growth, to report the fraction of cases where the conditions jointly point to growth, no growth and die-off of the population. Nodes on the left-hand side of the diagram (Figure A2.10) are defined elsewhere, but are displayed here for continuity. Storage temperature and Storage time are defined in the Storage conditions module. Growth rate and Minimum growth temperature are defined in the Growth characteristics module. Growth per day is defined in the Growth module.Details of the notes are presented in Table A2.13.

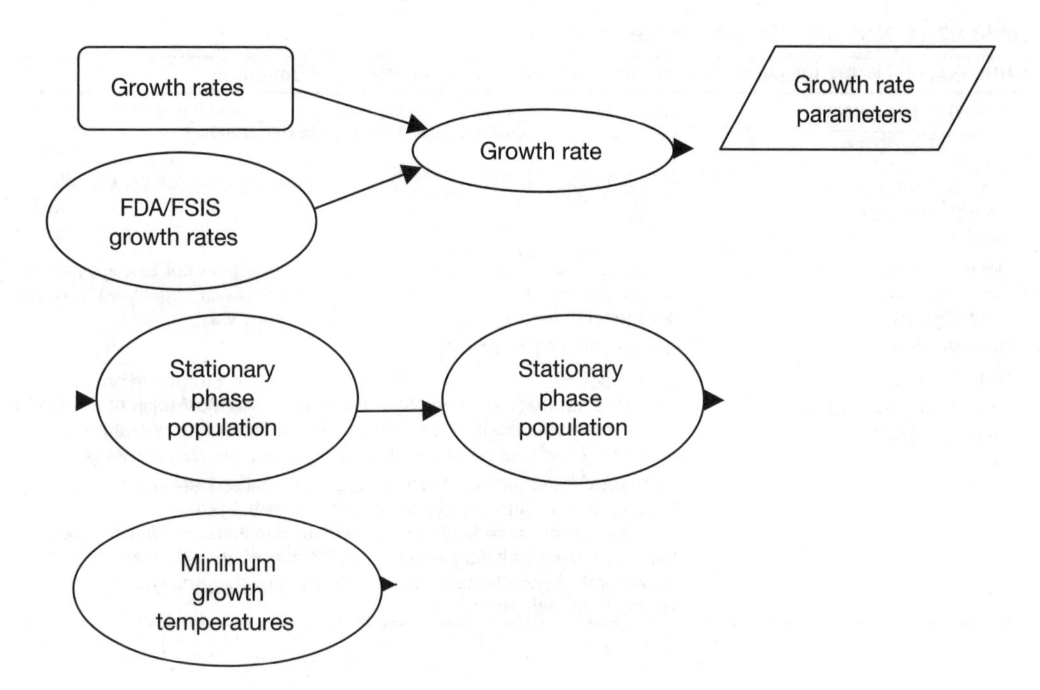

Figure A2.9 Influence diagram for *Growth characteristics* module.

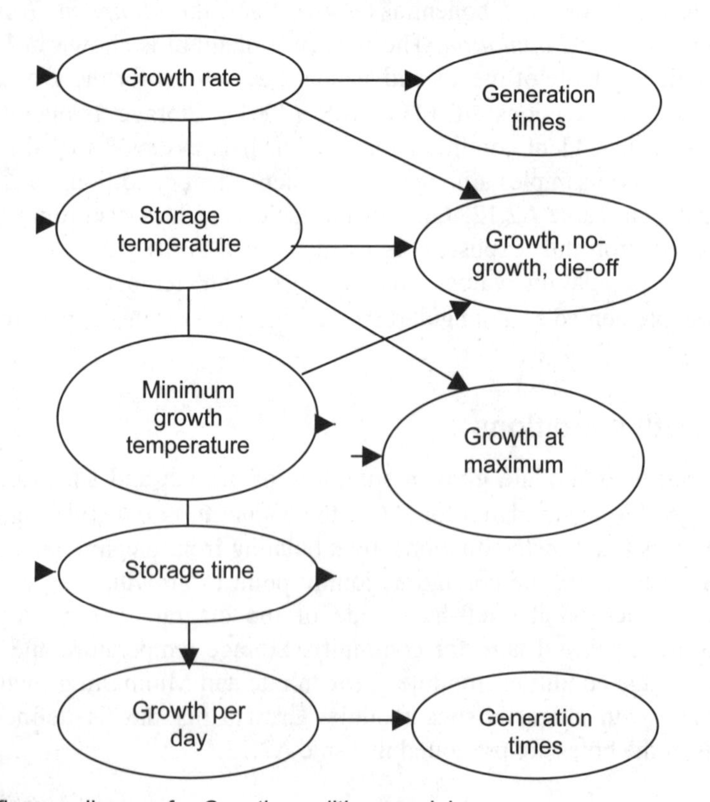

Figure A2.10 Influence diagram for *Growth conditions* module.

Table A2.12 Nodes for *Growth characteristics* module.

Title, Identifier, Structure	Description and Definition
Growth rates Growth_rates Variable node	FDA/FSIS 2001 suggests these Growth rates. We summarize the growth rates by the mean and standard deviation. Other summaries of growth rates appear in such as Farber and Peterkin (2000: Table 44-10). In order by Food groups, the definition specifies the mean, variance and number of studies. Determtable(Food_groups,Growthtableparameter)(0,Uniform(0.092, 0.434))
FDA/FSIS growth rates Fda_fsis_growth_rate Chance node	FDA/FSIS growth rates Determtable(Food_groups) (0,Uniform(0.092, 0.434))
Stationary phase population intermediate Stationary_phase_int Chance node	Stationary phase population intermediate is modelled as different for milks and for other foods of interest. Also, it varies with a range of Storage temperature. Using Allotherfoods:=(if Storage_temperature<5 then 10^5 else if Storage_temperature>7 then 10^8 else 10^6.5) Do Using Milks:=(if Storage_temperature<5 then 10^7 else if Storage_temperature>7 then 10^8 else 10^7.5) Do Table(Food_groups) (Allotherfoods, Allotherfoods, Milks, Milks, Allotherfoods, Allotherfoods, Allotherfoods)
Minimum growth temperature Minimum_growth_tempe Chance node	Farber and Peterkin (2000: Table 44-9) and its references suggest Minimum growth temperature between 1°C and 2°C for foods. The structure of the definition leaves room for the Minimum growth temperature to be different for each Food group, but leaves it the same, regardless of the set of Growth rates. Determtable(Food_groups,Updates)(Triangular(1, 1.1, 2), -1.18, Triangular(1, 1.1, 2), -1.18)
Growth rate Growth_rate Chance node	Though indexed by Updates (WHO/FAO 2000.06.17, FDA/FSIS 2000.05.19), Growth rate uses the FDA/FSIS growth rates for both. Growth rate has the simulated distribution of growth rate, for the food of interest, at 5°C. Table(Updates) (Fda_fsis_growth_rate, Fda_fsis_growth_rate)
Stationary phase population Maximum_population Chance node	Stationary phase population selects only the maximum density from Stationary phase population intermediate, for the selected Food groups. Stationary_phase_int[Food_groups=Food_groups]
Growth table parameters Growthtableparameter Index node	Growth table parameters structures the Growth rates table. ['mean', 'std. dev.', '# studies']

Table A2.13 Nodes for *Growth conditions* module.

Title, Identifier, Structure	Description and Definition
Generation times Generation_times (h) Chance node	Generation times calculated at the exponential growth rates at 5°C. 24 * logten(2)/Growth_rate
Growth, no growth, die-off Growth1 Chance node	Growth, no growth, die-off is a simple summary. if Growth_rate<0 then 'Die-off' else if Growth_rate=0 Or Storage_temperature<Minimum_growth_tempe then 'No growth' else 'Growth'
Growth at maximum Growth_at_maximum Chance node	Growth at maximum tabulates how the maximum population density constrains growth. Sometimes, simulated growth is "Below maximum". Sometimes, simulated growth is constrained by the maximum population density set by the lower Storage temperature range ('At Lower'), set by the middle Storage temperature range ('At Mid'), and set by the upper Storage temperature range ('At Higher'). if 10^Finalconcentration < Maximum_population then 'Below maximum' else if Storage_temperature<5 then 'At Lower' else if Storage_temperature>7 then 'At Higher' else 'At Mid'
Generation times Generation_times1 Chance node	Generation times

A2.2.13 Study indices

The *Study indices* library module stores 4 index nodes that structure results from the exposure assessment (Table A2.14).

Table A2.14 Nodes for *Study indices* module.

Title, Identifier, Structure	Description and Definition
Food groups Food_groups Index node	Food groups lists the food commodities that the exposure assessment addresses. ['Ice cream', 'Fluid milk, pasteurized']
Updates Updates Index node	Updates lets the exposure assessment address different sets of storage, time and growth conditions. ['WHO/FAO 2000.06.17', 'FDA/FSIS 2000.05.19']
Annual meals reporting Annualmealsreporting Index node	Annual meals reporting indexes the Annual meals objective node ['Individual', 'Population']
Contaminated or not Contaminated_or_not Index node	Contaminated or not defines the domain of the Chance node Contaminated or not. ['Not contaminated', 'Contaminated']

A2.3 CONSUMPTION CHARACTERISTICS

A2.3.1 Overview

The exposure assessments characterize consumption by meal size and meal frequency, noting and reporting differences in consumption patterns in the population sub-groups with different susceptibility. The meal or serving size is the estimated portion that people eat and has a distribution estimated from survey data. Similar surveys derive the frequency of eating specific RTE foods. Sources of consumption data are discussed earlier in Sections 3.1, 4.1.2 and 4.1.4.3 of this rport. For the pasteurized milk and ice cream assessments, data describe the consumption characteristics of adult Canadians.

A2.3.2 The data

Consumption characteristics are derived from 24-hour recall data from Canadian Federal-Provincial Nutrition Surveys (CFPNS, 1992–1995), which addressed the nutritional habits of non-institutionalized adults between 18 and 74 years old in the Provinces of Québec, Nova Scotia, Saskatchewan, Alberta and Prince Edward Island. Results are based on data from a total of 10 162 individual respondents from the 1990 Nova Scotia Nutrition Survey (2212 respondents), the 1990 Québec Survey (2118 respondents), the 1993/94 Saskatchewan Survey (1798 respondents), the 1994 Alberta Survey (2039 respondents) and the 1995 Prince Edward Island Survey (1995 respondents). Detailed one-day 24-hour recall data were used to examine consumption of various foods that would help to describe the consumption frequency and amounts eaten for the food groups relevant to this exposure assessment. All survey estimates are weighted to adjust for the sample design, and balance the ages and provinces according to their representation in the populations of those provinces. It is assumed that the remainder of the Canadian adult population eats like this group.

By using single occasion or daily consumption, estimates represent the fraction of the total population consuming the selected food on a given day, essentially a day at random. Food intakes are subject to day-to-day variation among individuals. Thus, the estimates are not indicative of "usual" intake, but are more indicative of the episodic intake with which would be associated foodborne illness. Distributions of "usual" intakes are unobservable in the 24-hour, one-day recall data that the Nutrition Surveys provide. Bureau of Biostatistics and Computer Applications has developed methods to remove the day-to-day within person variability (Junkins and Laffey, 2000: Junkins, Laffey and Weston, 2001; Hayward, [2001]). Those synthesized distributions of "usual" intakes are less heavily tailed than distributions of intakes that retain inter- and intra-person variability as is appropriate for the consumption distributions for these exposure assessments.

There is some uncertainty due to extrapolation of the results to 365 days' experience, when simulating factors such as annual consumption in the population, or to any reference period. The consumption of milk or ice cream were represented by reference to consumption of any of several foodcode categories, a classification system that the surveys employed. Selection of foodcodes from the nutrition surveys' databases was intended to reflect both consumption frequency and amount consumed on eating occasions. The information from individual all-eating episodes that included the food was used, except when the eating episode involved preparation such as cooking. When the food was an ingredient in the serving, an appropriate amount of the food to include was derived or estimated. There are uncertainties associated with this representation of intended foods by particular foods identified in the surveys. Additionally, there might be underreporting or overreporting errors associated with respondent errors and misclassification errors. Trained interviewers estimated amounts consumed on respondents' eating occasions. This is methodologically preferable to a practice that lets a respondent estimate the amounts consumed. However, it is recognized that the amounts recorded contain reporting errors and variability due to the interviewers' estimation methods.

It was difficult to adequately identify how to appropriately aggregate the sometimes many individual foods into the same eating occasion within the respondent's reference day. Therefore, all eating episodes for a food on the same day were aggregated into a daily consumption amount. That practice loses the distinction that one might wish to strike among occasions when the food was consumed alone, as an ingredient in a recipe or as only one element among several elements in a meal. Although the Nutrition Surveys distinguish consumption at an individual's home from consumption at another establishment outside the home, the distinction has been ignored for this exposure assessment. Past work suggests that consumption frequency differences and consumption amounts differences do exist between home and away consumption. Consequently, the consumption that is incorporated is assumed to represent a combination of all eating occasions. Combined independently with foodborne contamination, it is implicitly assumed that there are no differences in contamination rates and concentrations between food consumed at home and food consumed away from home (E.A. Junkins, pers. comm., 2000; M. Vigneault, pers. comm., 2000).

Gender and Age groups

The Nutrition Surveys do not classify respondents into groups to which might be attributed characteristics like higher susceptibility to foodborne illness. Rather, membership in non-susceptible and susceptible groups is imputed from Gender and Age attributes. Consumption characteristics of susceptible and non-susceptible groups of individuals, then, are different only in the manner that constituent Gender and Age characteristics are present in those groups. A susceptible group that is represented by elderly consumers would therefore possess consumption characteristics that differ from the non-susceptible group, solely because of differences between the consumption characteristics of elderly consumers and other consumers.

Simulating consumption amounts

To make it easier to specify the consumption distribution, some conventions were followed. Distributions were described by sampling in the correct proportions, from distributions that describe consumption in 4 Age × 2 Gender ranges, both for frequency of consumption and consumption amount. Eating episodes, both at home and away from home, were combined into the same distribution, capturing some variability but not distinguishing their separate influences. Non-susceptible and susceptible populations were defined by assuming that some fraction of each Gender × Age group is more susceptible. It is assumed also that the consumption characteristics of all persons of the same age and gender are the same, whether the person is in the susceptible or non-

Table A2.15 Fraction of population in Non-susceptible and Susceptible risk groups attributed to Gender × Age groups.

	Susceptible group		
Age	Female	Male	Total
18–34	0.16	0.04	0.20
35–49	0.08	0.03	0.12
50–64	0.02	0.02	0.04
65–74	0.35	0.29	0.64
	0.61	0.39	1.00
	Non-susceptible group		
Age	Female	Male	Total
18–34	0.20	0.23	0.43
35–49	0.17	0.18	0.35
50–64	0.11	0.11	0.22
65–74	0.00	0.00	0.00
	0.48	0.52	1.00

susceptible group. So, consumption in the susceptible and non-susceptible groups can be correctly simulated by sampling in the correct proportions from consumption characteristics captured for the Gender × Age groups. However, differences in the distributions of consumption for persons from the non-susceptible and susceptible groups are attributable only to the different Gender × Age group make-up of the groups. The values used are presented in Tables A2.15 and A2.16.

Table A2.16 Fraction of population in Gender × Age group attributed to Susceptible group.

Age	Female	Male
18–34	0.12	0.03
35–49	0.08	0.03
50–64	0.03	0.03
65–74	1.00	1.00

A2.4 NON-SUSCEPTIBLE AND SUSCEPTIBLE POPULATIONS

Among adults, susceptible groups are defined to include all adults 65 and older, pregnant women (1.3% of the population) and individuals with suppressed immune systems and certain medical conditions, such as cancer and recent organ transplantation (3.3% of the population) (Miller, Whiting and Smith, 1997). Different fractions of the population in Age and Gender groups are attributed to the susceptible group (Table A2.15), so non-susceptible and susceptible groups include gender and age groups in different fractions (Table A2.16). Fractions attributed to a Gender × Age group depend on the population size for that group, and are criteria for attributing risk categorization to that group. When individual food products are considered, the fraction of individuals who are susceptible depends, too, on consumption characteristics for the population. Susceptible groups would include also all children under 6 months, or perhaps 0-4 years old (J.M. Farber, pers. comm., 2000), some fraction of individuals under 18 years old, and all persons older than 74 years. Among adults aged 18–74, 15% would fit into the susceptible group and 85% would fit into the non-susceptible group when the definition described here is applied (Tables A2.15, A2.16 and A2.17).

An alternative approach follows one suggested in FDA/FSIS (2001). Observations from the FoodNet database describe listeriosis in the United States of America by age group. One might hypothesize that the incidence in an age group depends on, particularly, susceptibility, consumption characteristics and population representation. United States of America incidence data were used and combined with Australian populations in different age groups. Figure A2.11 scales those incidence data so that the incidence in the 10–19-year-old age group corresponds to 1. The relative incidence in the <30 days age group is 300 times the incidence in the 10–19 age group. Similarly, consumption characteristics among populations could be used to account for consumption differences. The remaining differences would affect the different age groups' susceptibility to listeriosis, perhaps forming a surrogate representation for the contrast between susceptible and non-susceptible populations.

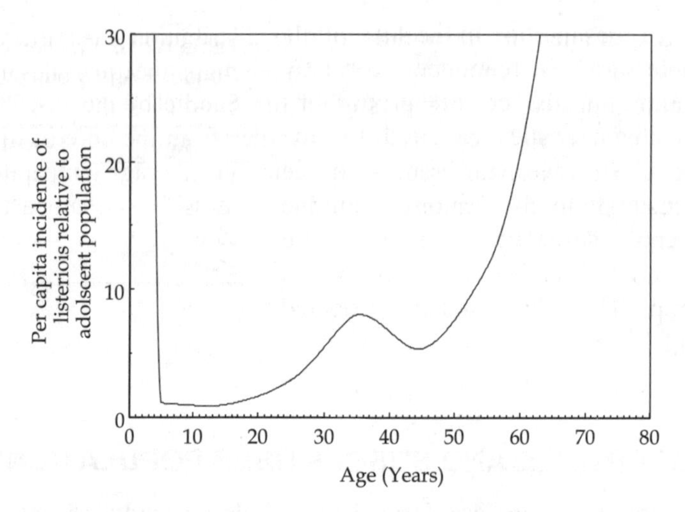

Figure A2.11 Relative per capita incidence of listeriosis (See text for details of calculation).

Table A2.17 Population of Canada allocated to susceptible and non-susceptible groups, using factors from Table A2.15.

Population of Canada				Susceptible population size				Non-susceptible population size			
Age	Male	Female	Total	Age	Male	Female	Total	Age	Male	Female	Total
0-4	911 028	866 302	1 177 330	0-4	911 028	866 302	1 777 330	0-4	0	0	0
5-17	2 738 162	2 598 365	5 336 526	5-17	82 145	77 951	160 096	5-17	2 656 017	2 520 414	5 176 431
18-34	3 711 154	3 591 613	7 302 768	18-34	111 335	430 994	542 328	18-34	3 599 820	3 160 620	6 760 439
35-49	3 823 789	3 802 863	7 626 652	35-49	114 714	304 229	418 943	35-49	3 709 075	3 498 634	7 207 709
50-64	2 403 311	2 453 603	4 856 914	50-64	72 099	73 608	145 707	50-64	2 331 212	2 379 995	4 711 207
65-74	1 000 723	1 134 443	2 135 166	65-74	1 000 723	1 134 443	2 135 166	65-74	0	0	0
74+	644 742	1 069 989	1 714 731	74+	644 742	1 069 989	1 714 731	74+	0	0	0
	15 232 909	15 517 178	30 750 087		2 936 785	3 957 516	6 894 301		12 296 124	11 559 662	23 855 786

SOURCE: Adapted from: Statistics Canada, www.statcan.ca/english/Pgdb/People/Population/demo10a.htm (November 2000), except for 15–19 years age group prorated into 5–17 years and 18–34 years age groups in table above.

A2.5 HOME STORAGE CHARACTERISTICS

A2.5.1 Home refrigeration temperatures

Four studies[4] contributed information about the distribution of refrigeration temperatures, important as one of the main determinants of growth of *L. monocytogenes* during storage. Audits International (2000) surveyed homes in the United States. Johnson et al. (1998) surveyed persons 65 years and older in the United Kingdom. Sergelidis et al. (1997) published results from a survey of homes in Athens, Greece. O'Brien (1997) also considered homes in the United States of America (Figure A2.12). Quantiles from Johnson et al. (1998)

4. A comment made on a late draft of this report pointed to two other references. Notermans et al. (1997) report refrigerator temperatures for households in the Netherlands, for pasteurized milk. Willocx, Hendrickx and Tobback (1993) report, *inter alia*, refrigerator temperatures for Belgian residences.

are approximately 3°C higher in the middle of the distribution than the ones in Audits International (2000). Based on limited presentation, though, the quantiles from Sergelidis et al. (1997) are approximately 3°C higher still in the middle of the distribution. O'Brien (1997) and Sergelidis et al. (1997) report 1 or 2 quantiles from the core of their temperature distributions (Table A2.8). It has been assumed that the same storage temperature distribution is appropriate for all RTE foods of interest. Further, it has been assumed that the food is stored at the same temperature throughout its shelf life.

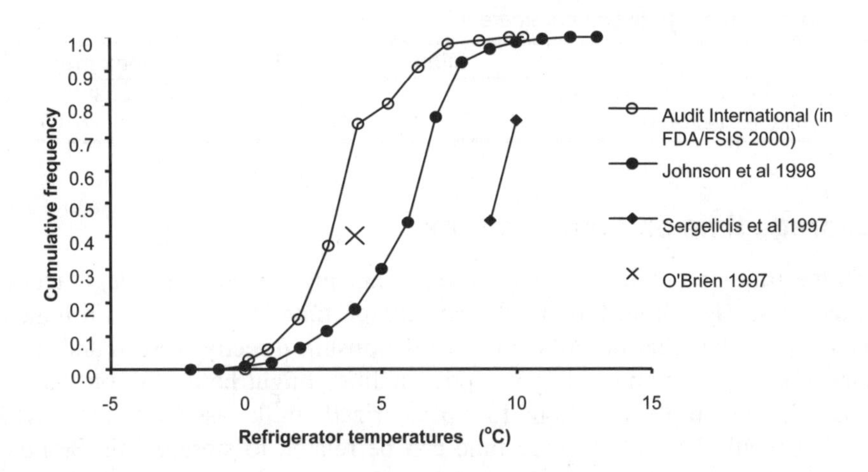

Figure A2.12 Four storage temperature distributions.

A2.5.2 Home storage times

No specific references that describe the length of time that consumers store foods in the home before eating were found[5]. Several characteristics that simulated results should try to emulate, at least qualitatively, might be considered. First, there should be at least a short, minimum storage time associated with all food consumed, representing, at least, the time from retail purchase to the individual's home. Second, storage time distributions that describe variability should have some maximum time that should be constrained by when the consumer would no longer accept the product. The maximum time would be related to the product's shelf-life, but should also reflect variability among individuals' practices of choosing whether to consume foods that have been stored beyond recommended limits. Some authors have studied the relationship between food spoilage, as represented by growth of some spoilage bacteria to high concentrations, and the organoleptic qualities of the food – qualities that help individuals to decide whether to eat a food or not (Priepke, Wei and Nelson, 1976; King, Henderson and Lill, 1986; Garcia-Gimeno and Zurera-Cosana, 1997). Several countries use 10^6 CFU/g concentrations of mesophiles as a guideline for acceptability.

5. A comment on a late draft of this report pointed to Notermans et al. (1997), which reports summary results for the storage time (after pasteurization of the milk) for pasteurized milk in households in the Netherlands.

In another manner, FDA/FSIS (2001) provides some useful information about storage time to consumption among individuals (Table A2.18). For the present, this exposure assessment uses those same criteria. FDA/FSIS (2001) accounts for individual variability in the storage time by describing a distribution specified by Triangular(Minimum, Mode, Maximum). That work also introduces variability and uncertainty in their representation by varying the Mode uniformly in the interval Mode ±20% and the Maximum in the interval Maximum ±50%.

Table A2.18 Storage time distribution parameters.

	Minimum	Mode	Maximum
Ice cream	0.5	7	30
Fluid milk, pasteurized	1	5	12

A2.5.3 Storage time and temperature

Clearly, storage time and storage temperature are not independent. Spoilage actions would severely truncate storage time, forcing shorter storage times to happen with higher storage temperatures. No studies that describe such a relationship directly were found. Studies that directly relate spoilage bacteria to organoleptic qualities might be useful, but have not been explored. A simple implementation for pasteurized milk assumes that organoleptic preferences that would truncate storage time can be related to storage life. Storage life for pasteurized milk depends on the growth of spoilage bacteria, which depends on temperatures. The storage life for pasteurized milk is assumed to be 12 days at 4°C, with storage life at other temperatures determined by the relationship $Life(T) = 12 \times \left[\frac{4+7.7}{T+7.7}\right]$ (Neumeyer, Ross and McMeekin, 1997; Neumeyer et al., 1997). To account for variability among individuals, the relationship time $\propto$ temperature^{-1} would not be deterministic and this relationship would be directed only to constrain the most extreme storage length at a given temperature.

A2.6 GROWTH CHARACTERISTICS

A2.6.1 Introduction

Extensive discussions about the microbial ecology of *L. monocytogenes*, predictive microbiology and growth characteristics are discussed elsewhere in this report. For the assessment examples, it has been assumed that the *L. monocytogenes* organisms were in the food sufficiently long for the lag phase to have passed. With that assumption, growth dynamics can be described with the exponential growth rate alone. It is assumed, further, that growth characteristics that must be explicitly accounted for are: the exponential growth rates, the minimum growth temperature and the stationary phase population.

A2.6.2 Growth rates

It is assumed that there is no growth or decline of *L. monocytogenes* populations in contaminated ice cream. For the milk exposure assessment example, growth rates reported in FDA/FSIS (2001) have been used to capture level and variability in growth of *L. monocytogenes*. Specific parameterization is given in Table A2.19, where growth rates refer to growth at 5°C. FDA/FSIS (2001) reports that in their referenced studies, study-specific growth rates were converted from the growth rate at a specific study temperature other than 5°C using relationships from McMeekin et al., 1993. At a temperature T°C other than 5°C, growth is calculated using the relationship $\sqrt{\mu_T} = \sqrt{\mu_5}(T - T_{\min})(5 - T_{\min})^{-1}$, where μ is the growth rate and T_{min} is the minimum growth temperature. Storage temperature as a dependent condition for the growth rate has been explicitly accounted for. However, it is assumed that the assumed distribution of growth rates effectively samples among the other dependent conditions (a_W, pH, NaCl, NO_3) in the same proportions that would occur in real environments. Bovill et al. (2000) note that competitive flora in the growth environment and the physiological state of the *L. monocytogenes* organisms might also be considered to be growth conditions. There is additional uncertainty in the estimated growth rates, not explicitly accounted for. Study methods and measurements contribute generally random effects that increase variability in replicated results for a given set of conditions, and therefore contribute uncertainty regarding what the true rate would be at those conditions.

Growth rates in Table A2.19 have been converted to refer to growth at 5°C, from the growth rate at a specific study temperature, using the square root relationship (McMeekin et al., 1993). So, growth rates at 5°C, as a baseline, do not explicitly account for variability and uncertainty in model extrapolation (or interpolation) from a study temperature back to 5°C.

A2.6.3 Stationary phase population

This implementation of maximum population densities is straightforward. FDA/FSIS (2001) reports stationary phase population values that change with storage temperature (Table A2.19). The stationary phase population is viewed as one of many constraints on the growth of *L. monocytogenes*. Other characteristics include competition with other microorganisms and growth of total spoilage bacteria populations to the extent that the food is not organoleptically acceptable, but these other characteristics are not accounted for.

A2.6.4 Minimum growth temperature

Based on Farber and Peterkin (2000), minimum growth temperature is implemented as Triangular(1°C, 1.1°C, 2°C) for this exposure assessment. Alternatives, such as -1.18°C (FDA/FSIS, 2001), set lower minimum growth temperature for *L. monocytogenes* than are implemented here. It is assumed also that minimum growth temperature is the same for the example foods to which it is applied.

A2.6.5 Implementation of microbial growth

The amount of growth using *daily growth × days storage* is calculated and applied to initial concentrations using $Concentration_{Final} = Concentration_{Initial} + Growth$ to get final concentrations, when quantities are expressed on a $\log_{10}$ scale. Growth rates at a stochastic storage temperature are adjusted for, using the relationship $\sqrt{\mu_T} = \sqrt{\mu_5}(T - T_{\min})(5 - T_{\min})^{-1}$ (McMeekin et al., 1993). Doing so incorporates variability associated with storage temperature, but does not explicitly incorporate uncertainty in extrapolating from a growth rate at 5°C to a growth rate at another temperature. The amount of growth until consumption of a portion is the simple product of the daily growth rate and the length of the storage time. This incorporates variability associated with the storage time, but assumes constant growth rate over the whole storage time. Growth is constrained so that final concentrations cannot be simulated to exceed the maximum population density.

Table A2.19 Population growth characteristics for *L. monocytogenes*, giving growth rates at 5°C and stationary phase populations at various temperatures.

	Growth rate distribution		Stationary population		
			<5°C	5°C–7°C	>7°C
Ice cream	0				
Milks	Uniform(0.092, 0.434)	Milks	10^7	$10^{7.5}$	10^8

SOURCE: FDA/FSIS, 2001.

A2.7 PREVALENCE AND CONCENTRATION

This exposure assessment model copies the practice that separates prevalence of contaminated servings and the concentration of *L. monocytogenes* in contaminated servings. This practice is similar to some published quantitative risk assessments (Cassin, Paoli and Lammerding, 1998; Lindqvist and Westöö, 2000), but differs from others (Bemrah et al., 1998, in part; FDA/FSIS, 2001). The practice separates concentration zeros (non-prevalence) from concentration non-zeros. First, the literature presents large data sets that count qualitatively positive and qualitatively negative samples. Concentrations, when presented, come from the typically small number of qualitatively positive samples. Second, it makes the simulation more efficient. The same 10^k iterations can define the probability of contaminated product and the distribution of concentrations, given contaminated product. Were both zero and non-zero concentrations combined, then that 10^k simulated observations would generate $\sim 10^{k-m}$, with m generally smaller than 1, number of zeros and only $\sim 10^m$ non-zeros, reducing the amount of precision that the simulation generates about the concentration distribution.

This implementation acts as if the declarations that positive and negative samples make are exact. Hence it calls the concentration in qualitatively negative samples exactly 0 CFU/g. The concentration in qualitatively negative samples should be modelled as random variables on $[0, \infty)$.

A2.8 COMBINING INDEPENDENT PREVALENCE ESTIMATES

A2.8.1 Introduction

Prevalence estimates for the presence of *L. monocytogenes* in foods come from surveys that typically provide summary information that includes the number of samples detected positive and the total number of samples tested. In some cases, though rarely, a detailed study design is also provided. Most studies give some context for the source of the samples – geography, food types or textures, points of origin, raw materials used or motivation. Most studies describe the methods used to test for *L. monocytogenes* presence. Most often, research has come via the microbiological literature. Some research has come from reports issued by national agencies. The food industry has also provided data sets. One could include prevalence estimates whose source is a modelled estimate, as is common in a quantitative risk assessment. When several studies are available, it is useful to take advantage of the observed variability between study estimates to provide a proxy model for including uncertainty and variability in a probabilistic risk assessment. To this end, it is assumed that k *combinable* studies are available, providing summary data $\{(Y_i, n_i); i = 1, \ldots, k\}$, where Y is the number of samples positive for *L. monocytogenes* and n is the number of samples. This assumes that, within studies, sample designs behave as simple random samples, that samples are independent, and that there is constant probability of a positive sample.

A2.8.2 Beta-binomial model for combining prevalence estimates

A simple assumption about the stochastic structure of a collection of studies gives binomial variability to the individual study estimates and a Beta distribution to the between-study variability of the true study prevalences. Also, this assumes that there are no overriding factors that are present that would group the studies into subsets, part of the assumption that the studies can be combined with a simple mixing distribution. More formally, the following two-stage model is assumed: $Y_i|n_i,\pi_i \sim \text{Binomial}(n_i,\pi_i)$, $i = 1, \ldots, k$ and $\pi_i \sim \text{Beta}(\alpha,\beta)$. For risk assessments, the mixing distribution is of importance. The role played by the distribution of the true study prevalence values is understood in the following way. If the mixing distribution is primarily a description of uncertainty, with a common fixed underlying prevalence value, then information from these several studies could be simply combined to give a more precise estimate of that single, fixed, true prevalence than the individual studies give. In this case, the Beta distribution plays the role of a prior density on the prevalence parameter. However, if the distribution of true study prevalence values also reflects variability in the prevalence value, then increasing the number of observations does not reduce this variability, though it can improve knowledge of the underlying distribution. In this case, the Beta distribution is an intrinsic component of the variability of the phenomenon under consideration among circumstances, situations or scenarios. Information from the studies can be appropriately combined to estimate the unknown parameters of that Beta distribution (Ross, pers. comm., 2000). There are a number of approaches available for estimating the parameters α and β of the Beta mixing distribution (Vose, 2000).

A2.8.3 Other alternatives appropriate to some circumstances

Alternatives can be found appropriate to some circumstances that Lindqvist and Westöö (2000) and Vose (2000) illustrate. Lindqvist and Westöö (2000) present prevalence data that are proportions $\{p_j, j = 1, \ldots, k\}$, where the sample sizes and the numbers of positive samples are either ignored or not reported. Those authors pool the observed data, treating them as independently and identically distributed from a distribution that they describe by the quantiles of that pooled sample. There, quantiles are defined by associating the *j*th largest observed fraction p_j with the $j(k+1)^{-1}$ th point of the distribution. Such a derivation is appropriate when the sample sizes used to estimate the individual fractions are the same or nearly the same, so that they are ignorable. Retaining the sample sizes, nevertheless, is useful to properly account for the uncertainty that one would associate with the fractions as estimates of true fraction values drawn from that empirically defined distribution. The true fraction values play the same role as described above. They can be understood as describing the variability among the true values of the fraction obtained under the conditions that the pooled sample describes. Alternatively, they can be understood as an expression of the uncertainty about the single true prevalence for that same population, from which the samples form independent observations. Vose (2000) discusses several methods to use with an example data set of fractions $\{p_j, j = 1, \ldots, k\}$, where the sample sizes and the numbers of positive samples are not reported. He specifies a Beta mixing distribution for the true values that the data set observes and describes how to estimate the parameters, α and β, by maximum likelihood methods and by the method of moments. He also describes in brief a procedure similar to the one that Lindqvist and Westöö (2000) carry out with their pooled data.

Alternatively, some knowledge of the population might be constructed by considering how to appropriately mix the various conditions of the studies sampled from, rather than basing the mixing distribution for the variability or uncertainty about the true prevalence values on the Beta distribution. One might consider mixing distributions for recognizable parts of a food supply: geographical, food type, point of origin, or raw materials used.

A2.9 DISTRIBUTIONS FOR *L. MONOCYTOGENES* CONCENTRATIONS IN FOODS

A2.9.1 Empirical distribution functions and fitted distributions

Small samples of observations for *L. monocytogenes* concentration in contaminated samples capture the distribution only with some uncertainty, both in the centre of the distribution and in the tails. Of particular concern is the upper tail of the distribution, where large concentrations sit. Studies that were reviewed seldom record high concentrations, or only under exceptional circumstances, making it difficult to model the thickness of the tail. Theoretical constraints on the length of the tail probably can be derived from predictive microbiology, but these require knowledge of growth conditions such as temperature and medium, and might be so much larger than empirical data produce that they would be somewhat unrealistic for practical use. Empirical distributions, too, are somewhat limited in their ability to capture the distribution very precisely in the upper tails. Confidence intervals can capture some notion of uncertainty, but will be a constant width in the tails, above the largest recorded observation. Uncertainty about the whole distribution can be captured non-parametrically by determining confidence intervals about the empirical density function or the empirical distribution function, or as a summary of the empirical distribution at selected quantiles. Last, given some assumptions, one can capture the shape of the distribution by fitting parametric distributions to the data. Uncertainty can be captured by varying the parameters among a confidence set, encompassing all combinations of the parameters that produce distributions that are consistent with the data. The distributions themselves capture variability among *L. monocytogenes* concentrations in different conditions. Parameter uncertainty and confidence intervals may be considered to describe some combination of that variability and uncertainty about that variability.

A2.9.2 Families of distributions

One alternative is to fit an analytical distribution to the data. The families of distributions considered as candidates for describing the concentration distribution should, first, respect the domain of the distribution. As used here, concentrations in contaminated foods have support on $(0, \infty)$ or on a subset, truncating $(0, \infty)$ at a minimum and at a maximum value. Second, consideration of candidate probability distributions would be restricted to ones that refer to a continuous random variable, and not a discrete random variable. FDA/FSIS (2001), for example, considered candidates like the lognormal distribution, the logistic distribution (folded or half-logistic, since the logistic distribution is defined on $(-\infty, \infty)$) and the Beta distribution. Method of moments could be used to estimate parameters for the analytical distributions. Preferable would be to use maximum likelihood methods. Nevertheless, some attention to the fit that the analytical distribution provides for the data, and in which parts of the domain, and goodness-of-fit criteria for the fit over the whole range of the data, should be considered. Distributions selected should also represent what is known about the sampling methods. That is, a point estimate is made about the concentration in an amount of product based on a small sample.

Minimum and maximum concentrations

Setting limits on the length of upper and lower tails can be straightforward and heuristic. When working with concentration distributions in this exposure assessment, the extent of the lower limit of contamination has been set to 0.04 CFU/g (1 CFU per 25 g), a lower detection limit, in effect, for every foodstuff. Upper limits are often set based on authors' suggestions, or set a judged limit larger than the largest observation. In some studies, the largest observed concentration stands as the upper limit, though this might be considered to be unrealistic. A more rigorous approach to setting maxima would consider the operating characteristic curve that is associated with the sample size and sample design of the studies that form the data sets. Minimum and maximum concentrations might also be used in conjunction with fitted distribution functions. The distribution function would define the shape of the distribution; the limits would define the domain of the distribution.

Heterogeneity of the organism in the package

Data collection and organization from referenced studies provide concentration distributions that represent levels of concentrations in recognizable packages or units of products, or give measurements from which one makes an inference about the concentration in the package or unit of product.

A2.10 REFERENCES CITED IN APPENDIX 2

Audits International. 2000. 1999 U.S. Food Temperature Evaluation. Design and Summary Pages. Audits International and U.S. Food and Drug Administration. 13p.

Bemrah, N., Sana, M., Cassin, M.H., Griffiths, M.W. & Cerf, O. 1998. Quantitative risk assessment of human listeriosis from consumption of soft cheese made from raw milk. *Preventative Veterinary Medicine*, **37**: 129–145.

Bovill, R., Bew, J., Cook, N., D'Agostino, M., Wilkinson, N. & Baranyi, J. 2000. Predictions of growth for *Listeria monocytogenes* and *Salmonella* during fluctuating temperature. *International Journal of Food Microbiology,* **59**: 157–165.

Cassin, M.H., Paoli, G.M. & Lammerding, A.M. 1998. Simulation modeling for microbial risk assessment. *Journal of Food Protection,* **61**(11): 1560-1566.

CFPNS [Canadian Federal-Provincial Nutrition Surveys]. 1992–1995. Bureau of Biostatistics and Computer Applications, Food Directorate, Health Canada. See also: Karpinski & Nargundkar, 1992; Junkins & Karpinski, 1994; Junkins, 1994; Junkins & Laffey, 2000; Junkins, Laffey & Weston, 2001.

Farber, J. M., & Peterkin, P.I. 2000. *Listeria monocytogenes*. pp. 1178–1232, *in:* B.M. Lund, T.C. Baird-Parker and G.W. Gould (eds). *The Microbiological Safety of Food.* Vol. 2. Gaithersburg, Maryland: Aspen Publishers.

FDA/FSIS [U.S. Food and Drug Administration/Food Safety and Inspection Agency (USDA)]. 2001. Draft Assessment of the relative risk to public health from foodborne *Listeria monocytogenes* among selected categories of ready-to-eat foods. Center for Food Safety and Applied Nutrition (FDA) and Food Safety Inspection Service (USDA) (Available at: www.foodsafety.gov/~dms/lmrisk.html). [Report published September 2003 as: Quantitative assessment of the relative risk to public health from foodborne *Listeria monocytogenes* among selected categories of ready-to-eat foods. Available at: www.foodsafety.gov/~dms/lmr2-toc.html].

Garcia-Gimeno, R.M. & Zurera-Cosana, G. 1997. Determination of ready-to-eat vegetable salad shelf-life. *International Journal of Food Microbiology,* **36:** 31–38.

Hayward, S. [2001]. Multivariate adjustment of nutrient intakes. Draft Technical Report, Bureau of Biostatistics and Computer Applications, Food Directorate, Health Products and Food Branch, Health Canada, Ottawa.

Iman, R.L. & Conover, W.J. 1982. A distribution-free approach to inducing rank correlation among input variables. *Communications in Statistics,* B. **11**(3): 311–334.

Johnson, A.E., Donkin, A.J., Morgan, K., Lilley, J.M., Neale, R.J., Page, R.M. & Silburn, R. 1998. Food safety knowledge and practice among elderly people living at home. *Journal of Epidemiology and Community Health,* **52:** 745–748.

Junkins, E. 1994. Saskatchewan Nutrition Survey 1993/94. Methodology for estimating usual intake. BBCA Technical Report E451311-005. Bureau of Biostatistics and Computer Applications, Food Directorate, Health Canada.

Junkins, E. & Karpinski, K. 1994. Enquéte québécoise sur la nutrition. Méthodologie pour estimer l'apport habituel, les statistiques sommaires et les erreurs-types. Bureau of Biostatistics and Computer Applications, Food Directorate, Health Canada.

Junkins, E. & Laffey, P. 2000. Alberta Nutrition Survey 1994. Methodology for estimating usual intake. BBCA Technical Report E451311-006. Bureau of Biostatistics and Computer Applications, Food Directorate, Health Canada.

Junkins, E., Laffey, P. & Weston, T. 2001. Prince Edward Island Nutrition Survey 1995. Methodology for estimating usual intake. BBCA Technical Report E451311-007. Bureau of Biostatistics and Computer Applications, Food Directorate, Health Canada.

Karpinski, K. & Nargundkar, M. 1992. Nova Scotia Nutrition Survey. Methodology Report. BBCA Technical Report E451311-001. Bureau of Biostatistics and Computer Applications, Food Directorate, Health Canada.

King, G.A., Henderson, K.G. & Lill, R.E. 1986. Asparagus: effect of controlled atmosphere storage on shelf-life of four cultivars. *New Zealand Journal of Experimental Agriculture,* **14:** 421–424.

Lindqvist, R. & Westöö, A. 2000. Quantitative risk assessment for *Listeria monocytogenes* in smoked or gravad salmon/rainbow trout in Sweden. *International Journal of Food Microbiology,* **58**: 181–196.

McMeekin, T.A., Olley, J., Ross, T. & Ratkowsky, D.A. 1993. *Predictive Microbiology. Theory and Application*. Taunton, UK: Research Studies Press. 340p.

Miller, A.J., Whiting, R.C. & Smith, J.L. 1997. Use of risk assessment to reduce listeriosis incidence. *Food Technology*, **51**: 100–103.

Neumeyer, K., Ross, T. & McMeekin, T.A. 1997. Development of a predictive model to describe the effects of temperature and water activity on the growth of spoilage pseudomonads. *International Journal of Food Microbiology,* **38**: 45–54.

Neumeyer, K., Ross, T., Thompson, G. & McMeekin, T.A. 1997. Validation of a model describing the effects of temperature and water activity on the growth of psychotrophic pseudomonads. *International Journal of Food Microbiology,* **38**: 55–63.

Notermans, S., Dufrenne, J., Teunis, P., Beumer, R., te Giffel, M. & Weem, P. 1997. A risk assessment study of *Bacillus cereus* present in pasteurized milk. *Food Microbiology,* **14**: 143–151.

O'Brien, G.D. 1997. Domestic refrigerator air temperatures and the public's awareness of refrigerator use. *International Journal of Environmental Health Research*, **7:** 141–148.

Priepke, P.E., Wei L.S. & Nelson, A.I. 1976. Refrigerated storage of prepackaged salad vegetables. *Journal of Food Science*, **41:** 379–382.

Sergelidis, D., Abrahim, A., Sarimvei, A., Panoulis, C., Karaioannoglou, P. & Genigeorgis, C. 1997. Temperature distribution and prevalence of *Listeria* spp. in domestic, retail and industrial refrigerators in Greece. *International Journal of Food Microbiology*, **34**: 171–177.

Statistics Canada (2000), "Population by sex and age", http://www.statcan.ca/english/Pgdb/People/Population/demo10a.htm, (2000.November)

Vose, D. 2000. *Risk analysis: a quantitative guide*. 2nd edition. Chichester, United Kingdom: Wiley.

Willocx, F., Hendrickx, M. & Tobback, P. 1993. Temperatures in the distribution chain. pp. 80–99 [oral presentation], *in:* Proceedings of the First European "Sous vide" cooking symposium. Leuven, Belgium, 25–26 March 1993.

Appendix 3. Predictive microbiology: concepts, application and sources

Predictive microbiology involves the systematic study and quantification of microbial responses to environments in foods and may be considered as the application of research concerned with the quantitative microbial ecology of foods. It is based upon the premise that the responses of populations of microorganisms in a defined environment are reproducible. By characterizing environments in terms of those factors that most affect microbial growth and survival, it is possible from past observations and experience to predict the responses of those microorganisms in other, similar environments. Ideally, the patterns of microbial behaviour are integrated with knowledge of the physiology of microbes. This knowledge can be expressed very succinctly using the language of mathematics, in the form of mathematical models. Those models can be considered as "condensed knowledge".

Predictive microbiology models provide a way to estimate changes in *L. monocytogenes* levels in foods as the product moves through the production-to-consumption chain. To make those estimates, periods, temperatures, product composition and concentrations of *L. monocytogenes* at some other point in the chain are required.

This section provides practical guidance for the application of predictive microbiology models in exposure assessment. Predictive microbiology has been extensively reviewed (Farber, 1986; Ross and McMeekin, 1994; Buchanan and Whiting, 1997; Ross, Baranyi and McMeekin, 1999; McDonald and Sun, 1999). McMeekin et al. (1993) provide a good introduction to the concept and its practical application.

The information below is drawn largely from Ross, Baranyi and McMeekin (1999).

A3.1 SOURCES OF GROWTH RATE MODELS AND DATA

Many data (*see* ICMSF, 1996) and many models for prediction of the growth rate of *L monocytogenes* are available (see Table A3.1, at the end of this appendix). In general, *L. monocytogenes* responds to environmental factors with the same patterns of response as other vegetative microorganisms and can be described by the same forms of model that describe growth rate responses of other organisms (Ross, 1993; Wijtzes et al., 1993; Tienungoon, 1998). However, it is reported that the temperature–growth rate relationship of *L. monocytogenes* is not as well described by existing models as it is described for other organisms, particularly at low temperatures that cause slow growth rates (Bajard et al., 1996; Ross, 1999). Generation times of *L. monocytogenes* under a range of conditions can be estimated easily using models such as most of those listed in Table A3.1. One can easily incorporate a published model into spreadsheet software.

A3.2 PRACTICAL CONSIDERATIONS

Models used in risk assessment must adequately reflect reality. Thus, before predictive models are used in exposure assessment, their appropriateness to that exposure assessment and overall reliability should be assessed. Users of models must be aware of the predictive limits of models, both in terms of the range of conditions that a model's interpolation region encompasses (Baranyi et al., 1996) and the variables that the model considers. Completeness error arises in model predictions when the model does not explicitly consider the effect of factors in a food that will affect the growth response of the microorganism modelled. The models referred to in Table A3.1 were developed to test different modelling strategies or, in the later published models, to include the effect of specific variables not included in earlier models. Ideally, a single model could encompass all the variables of relevance in all foods and is the ultimate aim of the scientific approach to predictive microbiology as the basis of a quantitative understanding of the microbial ecology of foods. However, creating such a model and scientific framework is time consuming. Alternatively, an iterative approach for development of product-oriented models i.e. based on observations in a system closely related to the food of interest, may satisfy the current technological needs of the food industry (Dalgaard, 1997; Dalgaard, Mejlholm and Huss, 1997).

Where completely appropriate models are not available, the limitations of the models should be documented and the implications of those limitations discussed as sources of uncertainty.

This section will consider assessment of model performance and limits. The discussion will be presented under the following headings:

- limits to application (i.e. interpolation or extrapolation);
- sources of variability and uncertainty; and
- performance evaluation.

A3.3 INTERPOLATION OR EXTRAPOLATION

No predictive models currently in use have a sound basis in theory, i.e. they are empirical descriptions and summaries of observations. A simple rule of modelling is that models without theoretical bases cannot be used reliably to make predictions by extrapolation, but only by interpolation. Interpolation relates to prediction made "between" the observations that the model is based on, while extrapolation is when predictions are made for conditions outside the range of those studied in the development of the models. A common interpretation of the interpolation region is that any combination of variables (e.g. temperature, water activity, pH, phenol, nitrite, etc.) that falls within the respective ranges of variables tested in the development of the model is within the interpolation region.

Certainly, microbial growth or death in a food cannot be predicted reliably when the conditions are *outside* the range of any individual factor tested in the model. However, the interpolation region is usually smaller than the simple interpretation suggested above. Few models are based on full factorial experimental designs. Unfortunately, the regions with fewest observations are usually those at the extremes of the ranges, where growth is slowest or may not occur at all due to the interaction of inhibitory factors (this is considered further in the section below on growth/no-growth models). However, these regions are often of most relevance when modelling because they are the conditions normally used to extend the shelf-life and safety of foods. As a result, users of models may inadvertently make predictions by extrapolation, particularly for conditions under which growth may be slow but of direct relevance in determining exposure to *L. monocytogenes* in RTE foods.

The determination of the true interpolation region, and the consequences of extrapolation, were discussed by Baranyi et al. (1996). Those authors concluded that models using a large number of parameters, e.g. higher order polynomial models, were more prone to unreliability resulting from inadvertent extrapolation because the predictions of the model often changed dramatically near the limits of the interpolation region.

Inadvertent extrapolation can also occur when using stochastic modelling techniques to describe effects of fluctuating temperature. Inadvertent extrapolation may also occur for other factors, but temperature is the factor most likely to fluctuate. Distributions can have infinitely long "tails", so it is important that the tails of the distributions used to model temperatures are truncated to match the interpolation range of the predictive microbiology model used.

A3.4 GROWTH/NO-GROWTH MODELS

Growth/no-growth models are a relatively new area of predictive microbiology. They aim to define the sets of combinations of factors that permit the growth of a modelled organism and those that do not. While there are absolute limits to growth of *L. monocytogenes* (see Table 3.1 in the main report) combinations of inhibitory factors can also prevent growth under milder conditions of each factor, a phenomenon widely employed in the food industry. These combinations of growth-preventing factors form a smooth surface in multi-dimensional space, or a smooth curve if one considers the interaction of two factors at a time as shown in Figure A3.2. There are relatively few growth/no-growth models currently available (Table A3.1). On the growth side of the interface, models can predict growth. On the no-growth side of the boundary, death occurs. Thus, growth/no-growth models provide additional information on the interpolation region of models.

A3.5 SOURCES OF VARIABILITY AND UNCERTAINTY

Model predictions can never perfectly match observations. Each step in the model construction process introduces some error, as outlined below, and presented in order of the magnitude of their contribution to the overall error in the models predictions.

- **Homogeneity error** arises because either some foods are clearly not homogeneous, or, at the scale of a microorganism, foods of apparently uniform consistency may comprise many different microenvironments. Current predictive models do not account for this non-homogeneity of foods.
- **Completeness error** arises because the model is a simplification, i.e., in practice, not all relevant factors can be included in the model.
- **Model function error** is similar to completeness error, and arises mainly from the compromise made when using empirical models, i.e. that the model is only an approximation to reality.
- **Measurement error** originates from inaccuracy in the raw data used to estimate the parameters of a model, i.e. due to methodological limitations in our ability to measure accurately the environment and the microbial response.

- **Numerical procedure error** includes all errors that are the consequences of the numerical procedures used for model fitting and evaluation, some of which are methods of approximation only. Generally, numerical procedure errors are negligible in comparison with the other types of errors.

The error in the estimate of maximum specific growth rate (or doubling time) of an organism determined from measurement of growth in laboratory media is ~10% per independent variable. As a rule of thumb, each additional environmental factor (pH, a_w, etc.) adds at least another 10% relative error to the model, assuming that the interpolation region of the model is comparable to the whole growth region. (Models with a small interpolation region have smaller error). An example of the interaction of factors limiting the growth of *L. monocytogenes* and the use of a model to predict those interactions is presented in Figure A3.1. Thus, the best performance that might be expected from a kinetic model encompassing the effect of three environmental factors on growth rate is ~30%.

A3.6 DISTRIBUTION OF RESPONSE TIMES

It is recognized that there is variation in the ecology of strains of *L. monocytogenes.* Begot, Lebert and Lebert (1997) reported variability in the growth rate responses of 58 strains. Peleg and Cole (1998) hypothesized that non-linear inactivation curves result from the natural variability that exists in microbial populations.

There has been lively discussion in the literature concerning the variability of bacterial growth rates. Using the limited amount of replicated published data concerning growth rate estimates under varying environmental conditions, Ratkowsky et al. (1991) concluded that growth rate responses became increasingly variable at slower growth rates, an observation confirmed by others (Fehlhaber and Krüger, 1998).

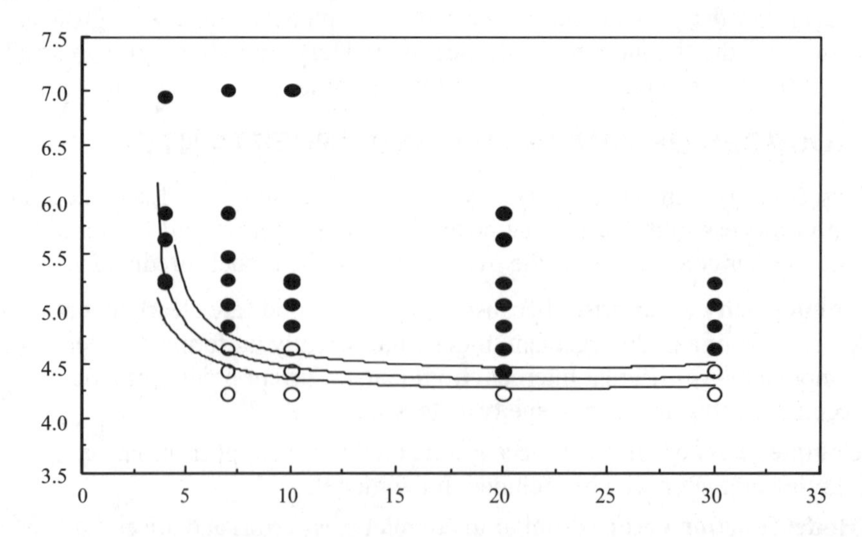

Figure A3.1 An example of the interaction of factors limiting the growth of *Listeria monocytogenes* and the use of a model to predict those interactions. The smooth lines are predictions of a model for the probability of growth of *Listeria monocytogenes* L5 (Tienungoon et al., 2000). The symbols (filled circles: growth observed; open circles, no growth observed) are the data of George, Lund and Brocklehurst (1988) for the effect of temperature and pH on the growth of *L. monocytogenes* NCTC 10357. In the figure, the lines indicate the predicted limits of growth at various levels of confidence (P = 0.9: lower curve; P = 0.5: middle curve; P = 0.1: lower curve).

In the data presented by Ratkowsky et al. (1991) variance in the square root of growth rate (var$\sqrt{rate}$) is constant, regardless of the magnitude of *rate*. Alber and Schaffner (1992) showed that for a strain of *Yersinia enterocolitica* (serotype 08), a logarithmic transformation of rate better "homogenizes" or "stabilizes" the variance. Dalgaard et al. (1994) reported that a transformation intermediate between the square root of rate and the logarithm of rate was required to normalize the variability in growth rate responses. Ratkowsky et al. (1996) reported similar observations, depending on the data used.

Ratkowsky (1992) presented the following general relationship between the variance in growth response times and the mean of those responses for a range of possible distribution types:

$$V = c\mu^n$$

where μ is the mean of the probability distribution, V is the variance of the probability distribution, n is an integer exponent having values 0, 1, 2 or 3, corresponding to the normal, Poisson, Gamma (logarithm of rate) and Inverse Gaussian (square root of rate) distributions, respectively, and c is a constant.

It is important to characterize the variability in responses, and to recognize that those responses are not normally distributed if that information is to be used within stochastic models for risk assessment.

A3.7 EVALUATION OF MODEL PERFORMANCE

A number of authors (Buchanan and Phillips, 1990; Wijtzes et al., 1993; George, Richardson and Peck, 1996; Fernandez, George and Peck, 1997; Walls and Scott, 1997; McClure et al., 1997; te Giffel and Zwietering, 1999) have evaluated the reliability of predictive microbiology models for *L. monocytogenes* growth rate and many have concluded that the models perform satisfactorily. However, in most of those assessments, no objective criterion for "satisfactory" was given.

Evaluation of model performance typically involves the comparison of model predictions to analogous observations not used to develop the model. Various measures have been used. Wijtzes et al. (1993) plotted literature values for the generation time of *L. monocytogenes* against the corresponding predictions of a model derived from studies in laboratory broth. From this plot, predictions that would be unsafe in practice could be visualized readily, and the overall reliability of the model assessed. Duh and Schaffner (1993) developed predictive equations for *Listeria* growth rate based on measurements in brain-heart infusion broth. Complementary literature values for the growth of the organism in food were then added to the data set and regression analysis of the supplemented dataset performed. The close similarity in mean square error (MSE) and correlation coefficient (r^2) values of the equations fitted to either data set were taken as an indication of the reliability of the models when applied to foods. Another measure of the accuracy of predictive equations was introduced by McClure, Zwietering and Roberts (1993), who compared their models on the basis of the sum of the squares of the differences of the natural logarithm of observed and predicted values. Zwietering et al. (1994) introduced the use of the *F*-ratio test. In this method the MSE of the models when assessed against data that are not used to generate the model was compared to the measurement error of the model itself, i.e. the model compared to the data used to generate it. If the measurement error is not significantly different from the prediction error, the model is considered to be satisfactory. te Giffel and Zwietering (1999) reviewed these measures in greater detail.

Two additional complementary measures of model performance can be used to assess the "validity" of models and are claimed to have the advantage of being readily interpretable (Ross, 1996), namely a bias factor and an accuracy factor.

The "bias factor" (B_f) is a multiplicative factor by which the model, on average, over- or under-predicts the response time. Thus, a bias factor of 1.1 indicates not only that a growth model is "fail-dangerous" because it predicts longer generation times than are observed, but also that the predictions exceed the observations, on average, by 10%. Conversely, a bias factor less than unity indicates that a model is, in general, "fail-safe", but a bias factor of 0.5 indicates a poor model that is overly conservative because it predicts generation times, on average, half of that actually observed. Perfect agreement between predictions and observations would lead to a bias factor of 1.

The "accuracy factor" (A_f) is also a simple multiplicative factor indicating the *spread* of observations about the model's predictions. An accuracy factor of two, for example, indicates that the prediction is, on average, a factor of two different from the observed value, i.e. either half as large or twice as large. The bias and accuracy factors can equally well be used for any time-based response, e.g. lag time, time to an *n*-fold increase, death rate, D value, etc. Modifications to the factors were proposed by Baranyi, Pin and Ross (1999).

Ideally, predictive models would have $A_f = B_f = 1$, but, typically, the accuracy factor will increase by 0.10–0.15 for every variable in the model. Thus, an acceptable model that predicts the effect of temperature, pH and water activity on *Listeria* growth rate could be expected to have $A_f = 1.3–1.5$. Satisfactory B_f limits are more difficult to specify because limits of acceptability are related to the specific application of the model. B_f is a measure of the extent of under- or over-prediction of the observed response rates by the model. Thus, a bias factor of 1.1 indicates not only that a generation time model is "fail-dangerous" not only because it predicts longer generation times than are observed, but also because the observations exceed the predictions, on average, by 10% in terms of $\log_{10}$ CFU. Conversely, B_f <1 indicates that a model is, in general, "fail-safe". Note, however, that when applied to rate-based data, $B_f > 1$ indicates the model *under*-predicts the observed rate, potentially leading to "fail-dangerous" predictions.

Armas, Wynne and Sutherland (1996) considered that B_f values in the range 0.6–3.99 were acceptable for the growth rates of pathogens and spoilage organisms when compared with independently published data. te Giffel and Zwietering (1999) assessed the performance of many models for *L. monocytogenes* against seven datasets, and found Bias factors in the range 2–4, which they considered to be acceptable, allowing predictions of the order of magnitude of changes to be made.

Other workers have adopted higher standards. Dalgaard (2000) suggested that B_f values for successful validations of seafood *spoilage* models should be in the range 0.8–1.3. Ross (1999) considered that, for pathogens, less tolerance should be allowed for $B_f > 1$ because that corresponds to under-predictions of the extent of growth and could lead to "fail-dangerous" predictions. Thus, Ross (1999) recommended that for models describing *pathogen* growth rate, B_f in the range 0.9–1.05 could be considered good, in the range 0.7–0.9 or 1.06–1.15 considered acceptable, and less than ~0.7 or greater than 1.15 considered unacceptable.

A3.8 SPECIFIC MODELS VERSUS GENERAL MODELS

The results of te Giffel and Zwietering (1999) and Ross (1999) showed that model performance is dependent on the data used to assess them. Differences in the performance of individual models were observed when the test datasets were disaggregated into food groups, or into ranges of growth rates. Some of these differences stem from the quality of the data used to assess the models, and the shortcomings of assessing models against data derived from the published literature have been commented on in several studies (Sutherland, Bayliss and Roberts, 1994; Ross, 1996; Walls and Scott, 1997; te Giffel and Zwietering, 1999). A second reason for poor performance may stem from completeness error. While te Giffel and Zwietering (1999) endorsed the performance of general models, Dalgaard (1997) and Dalgaard, Mejlholm and Huss (1997) proposed that strategies for model development based on observations in a system closely related to the food of interest will provide better performance for that specific product.

A3.9 PRACTICAL MICROBIAL ECOLOGY MODELLING IN RISK ASSESSMENTS

A3.9.1 Temperature distributions

Foods are rarely held under completely controlled temperature during their entire shelf-life. A common technique is to model the average temperature, based on temperature records obtained from surveys. The growth rate response of bacteria to temperature is complex and is not directly proportional to temperature. As noted by Cassin et al. (1998) the question arises whether the use of the average temperature over a time interval systematically biases the estimate of growth. This issue was addressed by Ross (1999) who used 246 temperature histories obtained using electronic temperature data-loggers for meat processing, transport and storage in Australia. Typically, the time interval between temperature recordings was a few minutes long.

Three methods were used to calculate the amount of microbial growth for each temperature history. In the first, the estimate of growth was based on the average temperature of all the temperatures recorded over the monitoring period. In addition, estimates were also generated for the *worst* case, i.e. based on the highest temperature recorded in each temperature record. The average and highest temperature values were substituted into models to predict the number of generations of pseudomonads and *E. coli* for each temperature history, respectively, by the two approaches. In the third method, the growth was determined using "time temperature function integration". For each time interval in the temperature history the growth rate of both pseudomonads and *E. coli* at the beginning and end of each time interval was calculated. The average of those growth rates was substituted into predictive models to calculate the number of generations over each recording interval, and the calculated number of generations for each time interval added to estimate the growth (i.e. number of generations) over the *entire* time monitored for each of the 246 temperature histories used.

In all methods, any temperature outside the ranges specified for each model were calculated to correspond to no growth, whether based on the average temperature over the interval, or full time-temperature integration.

For each organism-sector combination, the histograms of the distributions predicted by each method were plotted on a single graph to enable direct comparison of the effect of the three calculation methods. Representative plots of pseudomonad growth are shown in Figure A3.2.

The relationship between specific sets of predictions is lost in the preparation and presentation of the frequency distribution graph.

Ross (1999) showed mathematically that the average rate of growth at two temperatures *in the sub-optimal temperature region* is always greater than or equal to the growth rate at the average of two temperatures and that the difference between the two calculation methods is a function of the magnitude of the difference between the two temperatures. Using the dataset described the results indicated that *in practice* the difference between the two estimation methods is typically of the order of -0.1 to 0.2 $\log_{10}$ CFU, presumably because in most cases the range of temperatures experienced is small. This is a very small difference, particularly bearing in mind that the limits of accuracy of current microbial enumeration methods is approximately 0.3 $\log_{10}$ CFU (Jarvis, 1989).

However, there are certain situations and temperature ranges in which differences due to estimation method become more pronounced. If the temperatures experienced transcend growth boundary values, e.g. maximum or minimum temperatures for growth, estimates of the predicted growth by the two methods can differ significantly and lead to different frequency distributions of predicted growth. They are unlikely to be important for prediction of the growth of *Listeria monocytogenes* in RTE foods, however, because the lower (= 0°C) or upper (= 45°C) temperature thresholds for *L. monocytogenes* are unlikely to be experienced in normal refrigerated storage.

Thus, the results of that study (Ross, 1999) suggest that the use of the average temperature approach can provide a reasonable prediction of the extent of the growth of *L. monocytogenes* under real conditions of storage and distribution.

A3.9.2 Upper and lower limits

When distributions of temperatures are defined, they should reflect reality, i.e. the distributions should be truncated at realistic values. Similarly, when the range of temperature defined in the exposure model exceeds the minimum and optimum or maximum temperatures for growth of the organism, the growth model used must model the response of *L. monocytogenes*, i.e. the decline in growth rate as temperatures increase above that optimal for growth rate; and the cessation of growth at temperatures above or below the limits for growth.

Further pitfalls may occur in the use of unbounded temperature distributions. If the temperature distribution exceeds the range of the predictive model, nonsense predictions can occur, and may not be revealed by the simulation software used. While the effects might be subtle, they are likely to increase the range of uncertainty in the final model prediction.

A3.9.3 Lag time response

Microbial lag time is dependent both on the environment and its effect on growth rate, and the amount of "work" the cell has to do before it can initiate growth. This has presented problems for modellers, because models are developed under sets of constant conditions and it has been difficult to relate one set of conditions to another. The use of the relative lag time (RLT) concept, and RLT distributions, provides a way to overcome these problems in developing exposure assessments.

Ross (1999), Mellefont, McMeekin and Ross (2003) and Mellefont and Ross (2003) combined lag time data from experiments deliberately intended to induce long lag times with the published observations of other workers to investigate the distribution of lag times that are observed. When the lag time is expressed as an equivalent number of generation times of the organism in the same environment, i.e. lag time divided by generation time, or RLT, the distribution of RLTs observed has a sharp peak in the range 3–6. Augustin and Carlier (2000) also collated relative lag time distributions. The results are shown in Figures A3.3 and suggest that in many situations there is a practical upper limit to the lag time duration.

The number of generations of growth is predicted from the time and environmental conditions. The relative lag time is sampled from the RLT distribution and deducted from the predicted growth. If the predicted generations of growth do not exceed the lag time, no growth is predicted. If it does, the growth predicted to have occurred is given by the difference between the predicted generations of growth less the RLT.

A3.9.4 Jameson Effect

There has been very little work done to include in predictive models factors that contribute to the "Jameson effect" (Stephens et al., 1997), i.e. the suppression of growth of all micro-organisms in the food by high *total* microbial loads. In some products, this effect may greatly reduce the health risk from *L. monocytogenes* predicted on the basis of models currently available. Example 4 in this report (cold-smoked fish) introduces a method for inclusion of the Jameson effect in exposure assessment modelling. It models the increase in spoilage or other microorganisms, or both, on the product simultaneously with the growth of *L. monocytogenes*. If the predicted growth of other microbiota is predicted to exceed 10^9 CFU/g, the predicted growth of *L. monocytogenes* is modified accordingly. Full details are given in the example.

A3.9.5 Physiological state of cells

Environmental and physiological factors during food processing or present in foods are reported to affect the infectivity or virulence, or both, of *L. monocytogenes* (Buchanan et al., 1994; Zemser and Martin, 1998). These have been reviewed (Rees et al., 1995; Archer, 1996; Rowan, 1999), and also specifically in relation to *L. monocytogenes* in foods (Lou and Yousef, 1999). Conversely other workers (Conte et al., 1994; Gahan and Hill, 1999) have found little effect of environmental conditions on virulence.

Harsh environments in foods that stress the microbial cell produce a response that makes the cell more resistant to subsequent stressful or potentially lethal conditions, extending the survival of the cell under those conditions. Cells that are in stationary phase will also have increased tolerance to potentially lethal environments. This phenomenon has been suggested as increasing the chance that pathogenic bacteria, including *L. monocytogenes*, will survive passage through the acid environment of the stomach, thereby effectively increasing their virulence.

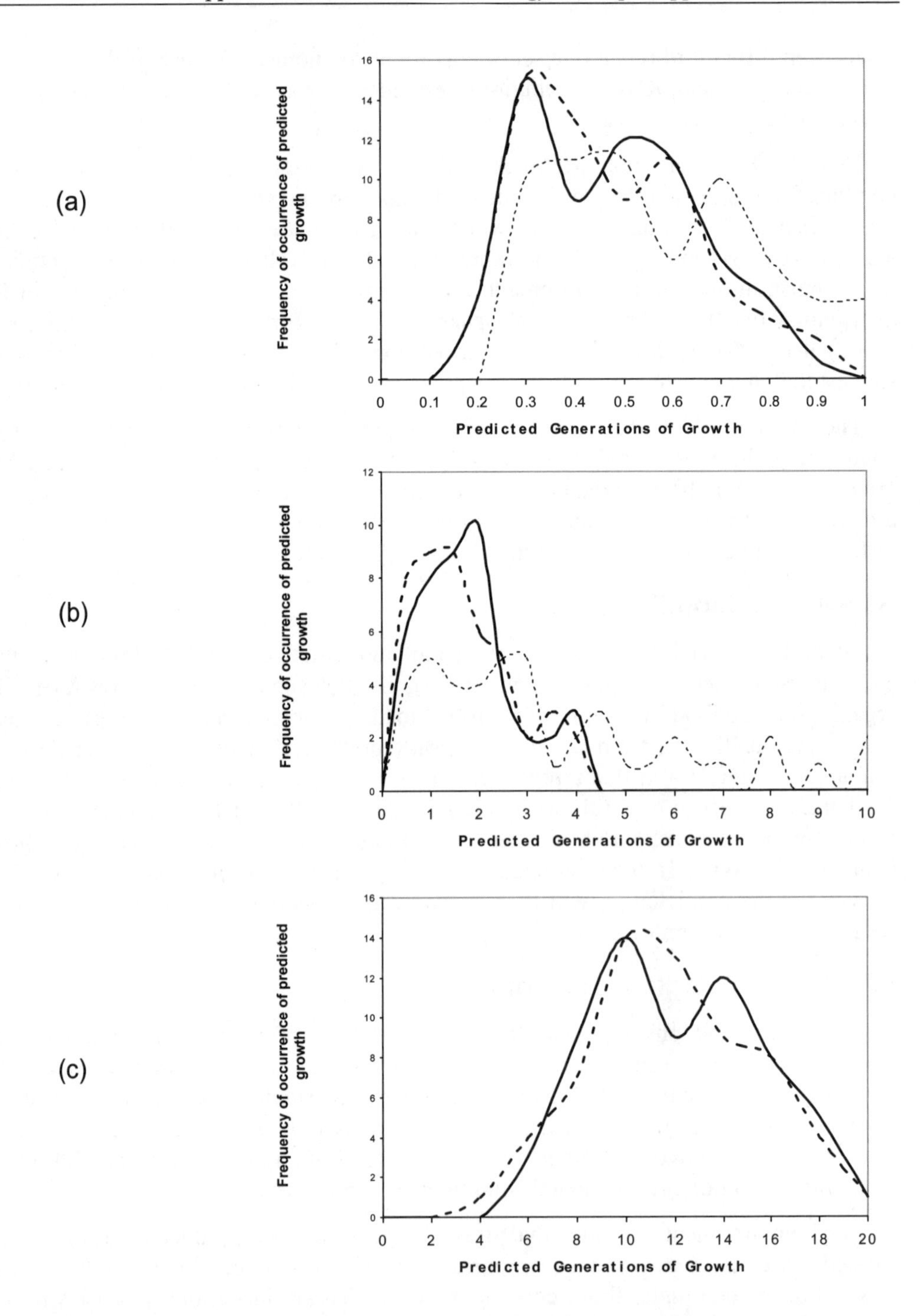

Figures A3.2 Graphs showing the distribution of the predicted number of generations of growth of pseudomonads during (a) transport from retail to home, (b) foodservice, and (c) domestic storage (home refrigerators) The heavy dashed line represents the predictions based on the average temperature; the solid line represents predictions based on time-temperature function integration; and the light dotted line represents predictions based on the maximum temperature recorded.

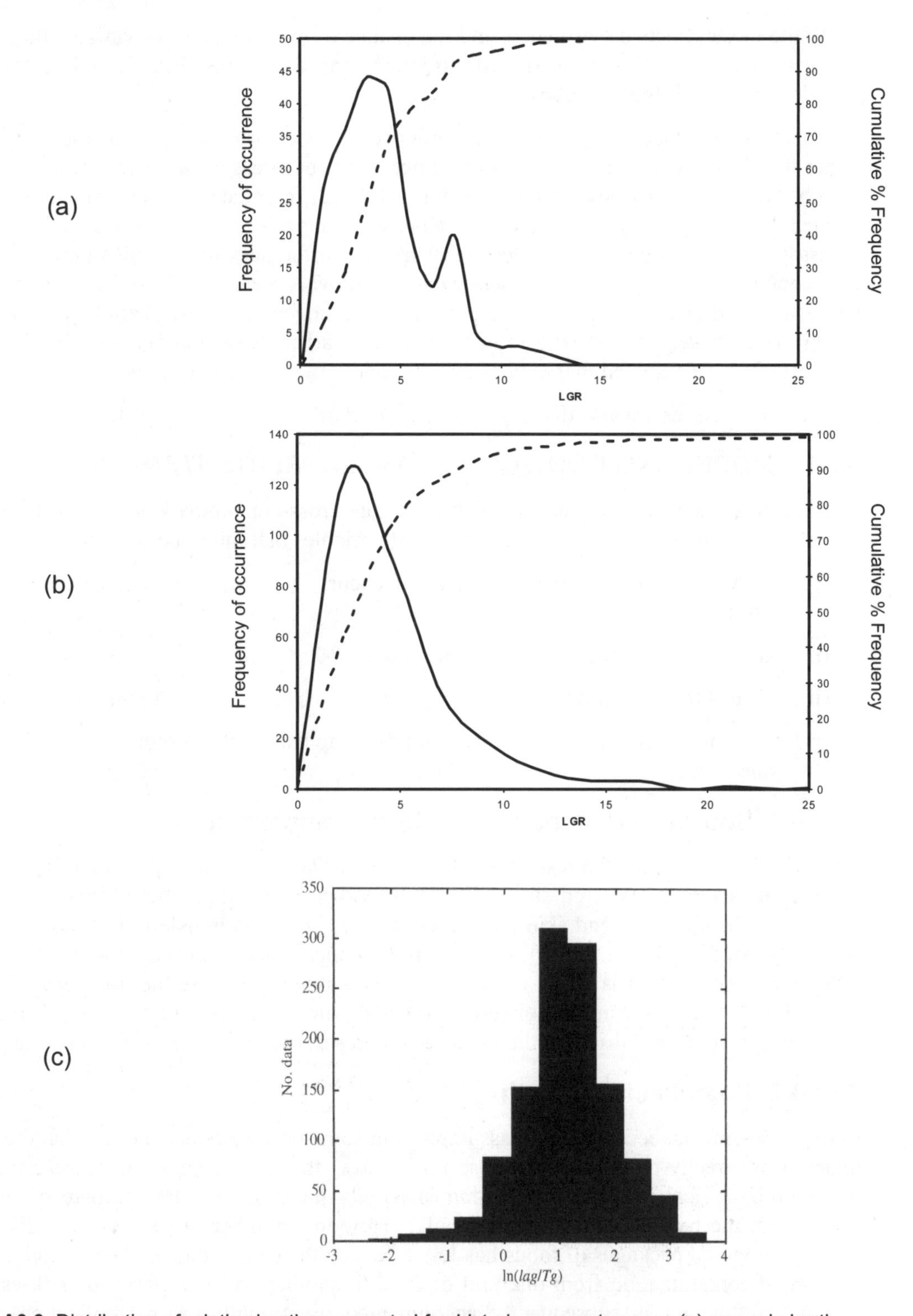

Figure A3.3 Distribution of relative lag times reported for *Listeria monocytogenes*: (a) grown in broth under laboratory conditions (collated in Ross, 1999); (b) in foods (collated in Ross, 1999); and (c) collated by Augustin and Carlier (2000) for all sources and plotted as ln(RLT). In figures (a) and (b), the dotted lines represent the cumulative frequency.

While these effects on virulence, and the chance of infection, are recognized, they are not well characterized and may be specific to strains and conditions (Buncic and Avery, 1996; Buncic, Avery and Rogers, 1996).

There is also uncertainty about the virulence of foodborne strains of *L. monocytogenes*. Other than that most cases of foodborne illness have been associated with serotypes 4a and 1/2a/b, specifically virulent strains cannot yet be differentiated. Notermans et al. (1998) studied the infectivity of more than 20 foodborne strains of *L. monocytogenes,* using a mouse bioassay and chick embryo test. Despite observing differences in virulence between strains, they concluded that almost all *L. monocytogenes* serovars present in foods have clear virulent properties, and should be considered potentially pathogenic, a view shared by McLauchlin (1996). Conversely, some exposure assessments (Farber, Ross and Harwig, 1996; Bemrah et al., 1998) have assumed that only 1–10% of foodborne *L. monocytogenes* are pathogenic.

The above issues are also discussed in detail in Part 2 of the main report.

A3.10 MODELLING CONTAMINATION AND RE-CONTAMINATION

There is little data available upon which to enable cross-contamination, or its effects, to be modelled quantitatively. There are a number of variables that might be considered:

(i) If contact with contaminated material occurs, how often does cross-contamination result?

(ii) At what point in the food chain does it occur?

(iii) What is the potential for growth on fomites, such as cutting equipment?

(iv) How much material is transferred and does the nature of the source affect the amount transferred?

A3.10.1 Source and amount of material transferred

FAO/WHO (2002) cites the results of Zhao et al. (1998), who developed a model system to enumerate bacteria transferred during common food preparation practices. Zhao et al. (1998) found that chicken meat and skin inoculated with 10^6 bacteria transferred 10^5 to a chopping board and hands and then 10^3–10^4 to vegetables chopped on the unclean board. It should be noted that chicken skin is likely to be wet and this might facilitate the transfer of bacteria, suspended in a surface film of moisture, compared with what might be transferred from RTE foods, which are often "drier" to the touch, e.g. cheeses, processed meats, smoked fish, etc.

A3.10.2 Potential for growth

In risk assessments, because of the assumption in some dose-response models that the risk of infection is directly proportional to dose for the low- to medium-dose range, the estimate of the microbiological risk to a population is largely governed by the estimate of the total numbers of the pathogen in the food supply. How that number of pathogens is distributed among individual packages of foods has less effect on the risk estimate. Accordingly, simple transfer of contamination from one unit of food to another will not affect the risk estimate, unless that transfer is subsequently accompanied by growth (i.e. multiplication) of the pathogen on a contact surface that contaminates uncontaminated material or growth in the (now) contaminated product itself.

In many RTE products, *L. monocytogenes* cells that contaminate the product will remain effectively immobilized at the site of contamination unless there is free liquid in the package, or other modes of transfer, to transport them to other parts of the package. Consequently, *L. monocytogenes* in some foods may be highly localized and exist as discrete pockets of contamination. This may limit the potential for growth of the organism, as nutrients are utilized and wastes accumulate at that site of contamination. Transfer to a new environment provides new potential for growth and potential increase in risk.

Grau (1993) traced the flow of *L. monocytogenes* through meat processing plants and found it to be transferred to many sites, such as trolleys, door handles and the surface packaging of finished products by contact contamination and cross-contamination. In these cases, paucity of nutrients and moisture may inhibit the growth of the organism, and limit the numbers of *L. monocytogenes* transferred to any product.

L. monocytogenes is known to colonize processing plants and, in particular, wet areas in plants. In these areas, if organic matter is present, growth can be expected to occur given sufficient time. Investigations in the United States of America suggest that listeriosis outbreaks often arise when virulent strains "colonize" a production line (Tompkin, 2002). Sites of colonization include hard-to-clean processing equipment. Hollow rollers on production lines are also known to deteriorate and crack, allowing water, nutrients and bacteria to colonize the interior. These niches are very difficult to clean, and provide a reservoir of pathogenic contaminants.

On equipment that is in direct contact with food and becomes fouled with food, growth would be expected to be occur. The amount of growth that could occur would be determined by the product composition, the temperature of that part of the plant, and the time before the contamination was removed by cleaning. As an example, if the processing line were operating at 15°C and slicing a processed meat product (e.g. pH 6.2, a_w 0.975, 100 ppm nitrite), *L. monocytogenes* growing in a residual material on contact surfaces could double in numbers approximately every 5 hours.

A3.10.3 Point in food chain at which contamination occurs

As stated above, there is no increase in risk as a consequence of cross-contamination unless there is increased potential for microbial growth as a result. The amount of increase will depend on the product, its storage conditions, and the time between the contamination event and consumption. If the integrity of the chill chain between the point of production and consumption were uniform, the potential consequences of contamination would be expected to be greater for contamination at the point of production than at the point of retail sale or in the consumer's home. This is because of the increased time available for growth to high numbers before consumption.

A3.10.4 Likelihood of transfer

Even if uncontaminated material comes into contact with contaminated material, the probability of cross-contamination is not absolute, but would be expected to depend on the concentration of pathogens, and their distribution on, or in, the contaminated material.

A3.11 RELATIVE RATE FUNCTIONS

Growth rate modelling can be simplified enormously by using relative rate functions, particularly when combined with square root-type models, or cardinal parameter models. The simple square root model (Ratkowsky et al., 1982) describes the effect of temperature on the growth rate of almost all bacteria. The square root model is:

$$\sqrt{\mu} = b(T - T_{min})$$

where μ is the rate of growth
T is temperature (°C)
b is a constant to be fitted related to the maximum growth rate of the organism
T_{min} is the temperature at which the growth rate is predicted to be zero.

It should be noted that T_{min} is a notional temperature, and is usually several degrees below the minimum temperature at which growth is observed to occur. It should also be noted that the simple square root model above applies only to the sub-optimal temperature region for growth, up to ~35–37°C for *L. monocytogenes.*

If:

- temperature is the only factor affecting the growth rate of a bacterium in a food that varies during the storage and distribution of the product (i.e. if pH, water activity, etc., are constant),
- T_{min} is known for the organism, and
- the growth rate in a product of interest is known at one temperature,

then the growth rate of the organism in that product at any other temperature can be derived using the following relationship, based on the simple square root model (McMeekin et al., 1993):

$$\mu_T = \mu_{REF} * \frac{(T - T_{\min})}{(T_{REF} - T_{\min})}$$

where μ_{REF} is the known growth rate at some temperature T_{REF},
μ_T is the unknown growth rate at some temperature T,
and the other parameters are as previously defined.

For example, FDA/FSIS (2001) collated data for the growth rate at 5°C of *L. monocytogenes* in many RTE food products. Growth at a temperature other than 5°C was calculated using the relationship $\sqrt{\mu_T} = \sqrt{\mu_5}\left(T - T_{\min}\right)\left(5 - T_{\min}\right)^{-1}$. This approach has been adopted in several of the exposure assessments in Part 4 of the main report.

The use of the relative rate function is a simplification. As conditions become less favourable for microbial growth, e.g. due to decreased water activity or increased acidity, the difference between T_{min} and the minimum temperature at which growth is possible increases. This was discussed above in relation to models for growth limits under multiple hurdles to growth (e.g. Tienungoon et al., 2000). Bajard et al. (1996) suggested that the simple square root model does not describe the growth rate response to temperature of *L. monocytogenes* as well as it does for other organisms. Nonetheless, in the context of the other sources of uncertainty that arise in microbial risk assessments, these are considered to be relatively minor deficiencies.

A3.12 PREDICTIVE MICROBIOLOGY MODELS

A summary of some currently available predictive models for *L. monocytogenes* is presented in Table A3.1.

Table A3.1 Summary of predictive models available for the growth, survival and inactivation of *Listeria monocytogenes* in foods.

1.	2.	3. (°C)	4. (a_w or %)	5. pH	6.	7. (µg/ml)	8.	9.	10.
GROWTH									
Y	broth	5–37	0.5–4.5%	4.5–7.5	–	0–1000	aerobic/ anaerobic	N	[1]
Y	food	3–35	2–8%	4.5–7.5	–	–	aerobic	Y	[2]
Y	broth	5–35	0.5–8%	4.6–7.4	–	–	aerobic	Y	[3]
N	broth	4–20 1–20	–	4.5–7.0 4.3–7.2	acetic 0–10 000 lactic 0–20 000	–	aerobic	Y	[4]
Y	broth	9	1.0–4.0%	5.5–6.5	lactic 0–0.6%; acetic 0–0.6%	70	aerobic	Y	[5]
Y	broth	2–46	–	–	–	–	aerobic	N	[6]
Y	broth	4–20	0.5–8%	4.5–7.0	–	–	CO_2: 0–100%, balance N_2	Y	[7]
Y	food	3, 7, 11	0.5	–	–	–	Air: 0.03% CO_2, 78.03% N_2, 20.99% O_2; Modified atmosphere #1: 76% CO_2, 13.3% N_2, 10.7% O_2; Modified atmosphere #2: 80% CO_2, 20% N_2	N	[8]
Y	meat broth	4–30	0.992–0.960 (a_w)	5.4–7	–	–	aerobic	Y	[9]
N	broth	20–35	2–10%	4–8.5	–	–	–	N	[10]
N	broth	1.0–35	0.5–11.5%	4.0–7.2	–	0–200	–	N	[11]
Y	lean beef and fatty beef tissue	0–30.6	–	5.46–6.98	–	–	aerobic	Y	[12]
Y	lean beef fat beef	0–43 0–31	~0.99	5.6–6.7	–	–	aerobic	Y	[13]
Y	broth	3–37	0.5–13%	4.2–7.3	lactic 0–450 mM	–	aerobic	Y	[14]
Y	broth	4–37	0.5–13%	5.6–7	lactic 0, 200 mM		aerobic	Y	[15]
N	roast beef	-1.5 & 3	–	6.1	–	–	vac. pack and saturated CO_2	N	[16]
N	broth	5–30	0.5–8%	4.6–7.4	–	0–400	–	N	[17]
Y	broth	5–35	0.95–0.997	#1 4.6–6.7 #2 4.6–7.4	–	–		Y	[18]
Y	broth	5–37	0.5 & 4.5	6.0 & 7.5	–	0–1000	aerobic and anaerobic	N	[19]

1.	2.	3. (°C)	4. (a_w or %)	5. pH	6.	7. (µg/ml)	8.	9.	10.
Dynamic growth									
Y	fluid whole milk		4, 6, 8, 10, 15, 20, 25, 30, 35	–	–	–	–	Y	[20]
Repair of heat injury									
Y	broth	4–43	0.5–10.0	4.2–9.6	–	–	–	N	[21]
SURVIVAL/GROWTH LIMITS/GROWTH INITIATION									
Probability of growth initiation in defined period of time									
N	broth	4–30	0.5–12.5	>5.9	–	See Note (1)	–	N	[22]
Survival and ongrowth									
		5, 10, 30	0–18	4.19–4.83	–	–	aerobic	N	[23]
Growth limits									
	broth	3.1–35.8 3.1–36.4	0.5–13% 0.5–13%	3.9–7.3 3.9–7.7	lactic 0–500 lactic 0–450	–	aerobic	Y	[24]
Effect of heat stress									
Y	broth	53–60	–	–	–	–	Stationary phase cells	N	[25]
INACTIVATION									
Thermal									
Y	milk (bovine)	60.5–69.5 for 3–60 secs (HTST pasteurization process)	–	–	–	–	–	Y	[26]
Y	food	55, 60, 65	–	5, 6, 7	–	–	milkfat 0, 2.5, 5%	Y	[27]
Y	food	55–65	0–6%	4–8	–	–	sodiumpyro-phosphate 0–0.3%	Y	[28]
Y	food (infant formula)	55, 60, 65	0, 2, 4%	5, 6, 7	–	–	physiological states (lag, exponential, stationary) of test cultures	Y	[29] using data from [41]
Heating rate and thermal inactivation									
Y	broth	50–64	–	–	–	–	Sodiumpyro-phosphate 0–0.3%	N	[30]
Heat resistance									
Y	broth	50, 60, 65	–	–	–	–	Physiological state of cells (end of log phase cells; heat shocked cells; cells resistant to prolonged heat)	Y	[31]
Y	buffer	50, 55, 60	–	–	–	–	–	Y	[32]
Y	broth	30, 10, 5	0–18%	4.19–4.83	–	–	–	Y	[33]
Non-thermal									
N	broth	4 to 42	0.5–19%	3.3–7.3	lactic 0–2%	0–200		N	[34]

1.	2.	3. (°C)	4. (a_w or %)	5. pH	6.	7. (µg/ml)	8.	9.	10.
N	broth	5 to 42	0.5–19%	3.3–7.4	lactic 0–2%	0–200	O_2 levels reduced, N_2 flushed vessels	N	[35]
Y	broth	28	–	4–7	lactic 0–18%; acetic 0–12%	–	–	Y	[36]
Y	broth	4, 19, 28	–	3–4.5	acetic: 0–2.0%; ascorbic: 0–2.0%	–	aerobic	N	[37]
Y	broth	4–42	0.5–19%	–	lactic 0–1%	0–200	–	Y	[38]
COMBINED									
Growth survival death									
Y	broth	4–12	2–4%	6.2	–	–	phenol: 5, 12.5, 20 ppm	Y	[39]
Biotic interactions									
Y	broth	10	2%	5–5.8	0–5 mM protonated lactic acid	–	*Lactococcus lactis* (non-nisin producing)	Y	[40]

KEY TO COLUMNS: (1) Model given? Y = Yes; N = No. (2) Medium. (3) Temperature (°C). (4) Water activity (a_w) or salt (NaCl) percentage. (5) pH. (6) Organic acids. (7) Nitrite, expressed as µg/ml (= ppm). (8) Other. (9) Validation data? Y = Yes; N = No. (10) Source (see below).

NOTES: (1) methyl paraben 0–2%; sodium propionate 0.3%; sodium benzoate 0.1%; potassium sorbate 0.3%; inoculum size 0.01–100 000 CFU/ml; *Listeria* spp. (*L. monocytogenes, L. innocua, L. seeligeri* and *L. ivanovii*).

SOURCES: [1] Buchanan and Phillips, 1990. [2] Murphy, Rae and Harrington, 1996. [3] Wijtzes et al., 1993. [4] George, Richardson and Peck, 1996. [5] Nerbrink et al., 1999. [6] Duh and Schaffner,1993. [7] Fernandez, George and Peck, 1997. [8] Zhao, Wells and Marshall,1992. [9] Lebert, Bégot and Lebert, 1998. [10] McClure, Roberts and Otto Oguru, 1989. [11] McClure et al., 1997. [12] Grau and Vanderlinde, 1993. [13] Grau and Vanderlinde, 1992. [14] Tienungoon, 1998. [15] Ross, 1993. [16] Hudson, Mott and Penny, 1994. [17] McClure, Kelly and Roberts, 1991. [18] McClure, Zwietering and Roberts, 1993. [19] Buchanan, Stahl and Whiting, 1989. [20] Alavi et al., 1999. [21] Chawla, Chen and Donnelly, 1996. [22] Razavilar and Genigeorgis, 1998. [23] Cole, Jones and Holyoak, 1990. [24] Tienungoon et al., 2000. [25] Breand et al., 1998. [26] Piyasena, Liou and McKellar, 1998. [27] Chabra et al., 1999. [28] Juneja and Eblen, 1999. [29] Xiong et al., 1999. [30] Stephens, Cole and Jones, 1994. [31] Augustin, Carlier and Rozier, 1998. [32] Linton et al., 1995. [33] Cole, Jones and Holyoak, 1990. [34] Buchanan and Golden, 1995. [35] Buchanan, Golden and Phillips, 1997. [36] Buchanan and Golden, 1995. [37] Golden, Buchanan and Whiting, 1995. [38] Buchanan et al., 1994. [39] Farber, Cai and Ross, 1996. [39] Membre, Thurette and Catteau, 1997. [40] Breidt and Fleming, 1998. [41] Linton et al., 1996.

A3.13 REFERENCES CITED IN APPENDIX 3

Alavi, S.H., Puri, V.M., Knabel, S.J., Mohtar, R.H. & Whiting, R.C. 1999. Development and validation of a dynamic growth model for *Listeria monocytogenes* in fluid whole milk. *Journal of Food Protection*, **62**: 170–176

Alber, S.A. & Schaffner, D.W. 1992. Evaluation of data transformations used with the Square Root and Schoolfield models for predicting bacterial growth rate. *Applied and Environmental Microbiology*, **58**: 3337–3342.

Archer, D.L. 1996. *Listeria monocytogenes* – the science and policy. *Food Control*, **7**: 181–182.

Armas, A.D., Wynne, A. & Sutherland, J.P. 1996. Validation of predictive models using independently published data. Poster/Abstract. 2nd International Conference of Predictive Microbiology. Hobart, Tasmania,18–22 February 1996.

Augustin, J.C. & Carlier, V. 2000. Mathematical modelling of the growth rate and lag time for *Listeria monocytogenes. International Journal of Food Microbiology*, **56**: 29–51.

Augustin, J.C., Carlier, V. & Rozier, J. 1998. Mathematical modelling of the heat resistance of *Listeria monocytogenes. Journal of Applied Microbiology*, **84**: 185–191.

Bajard, S., Rosso, L., Fardel, G. & Flandrois, J.P. 1996. The particular behaviour of *Listeria monocytogenes* under sub-optimal conditions. *International Journal of Food Microbiology,* **29**:201 - 211.

Baranyi, J., Ross, T., McMeekin, T.A. & Roberts, T.A. 1996. Effects of parameterization on the performance of empirical models used in 'Predictive Microbiology'. *Food Microbiology*, **13**: 83–91.

Baranyi, J., Pin, C. & Ross, T. 1999. Validating and comparing predictive models. *International Journal of Food Microbiology,* **48**: 159–166.

Begot, C., Lebert, I. & Lebert, A. 1997. Variability of the response of 66 *Listeria monocytogenes* and *Listeria innocua* strains to different growth conditions. *Food Microbiology*, **14**: 403–412.

Bemrah, N., Sana, M., Cassin, M.H., Griffiths, M.W. & Cerf, O. 1998. Quantitative risk assessment of human listeriosis from consumption of soft cheese made from raw milk. *Preventative Veterinary Medicine*, **37**: 129–145.

Breand, S., Farde, G., Flandrois, J.P., Rosso, L. & Tomassone, R. 1998. Model of the influence of time and mild temperature on *Listeria monocytogenes* non-linear survival curves. *International Journal of Food Microbiology,* **40**: 185–195.

Breidt, F. & Fleming, H.P. 1998. Modeling of the competitive growth of *Listeria monocytogenes* and *Lactococcus lactis* in vegetable broth. *Applied and Environmental Microbiology*, **64**: 3159–3165.

Buchanan, R.L. & Golden, M.H. 1995. Model for the non-thermal inactivation of *Listeria monocytogenes* in a reduced oxygen environment. *Food Microbiology*, **12**: 203–212.

Buchanan, R.L. & Phillips, J.G. 1990. Response surface model for predicting the effects of temperature, pH, sodium chloride content, sodium nitrite concentration and atmosphere on the growth of *Listeria monocytogenes*. *Journal of Food Protection*, **53**: 370–376.

Buchanan, R.L. & Whiting, R.C. 1997. Risk assessment - A means for linking HACCP plans and public health. *Journal of Food Protection*, **61**: 1531–1534.

Buchanan, R.L., Golden, M.H. & Phillips, J.G. 1997. Expanded models for the non-thermal inactivation of *Listeria monocytogenes. Journal of Applied Microbiology*, **82**: 567–577.

Buchanan, R.L., Golden, M.H. Whiting, R.C., Phillips, J.G. & Smith, J.L. 1994. Non-thermal inactivation models for *Listeria monocytogenes*. *Journal of Food Science*, **59**: 179–188.

Buchanan, R.L., Stahl, H.G. & Whiting, R.C. 1989. Effects and interactions of temperature, pH, atmosphere, sodium chloride and sodium nitrite on the growth of *Listeria monocytogenes*. *Journal of Food Protection*, **52**: 884–851.

Buncic, S. & Avery, S.M. 1996. Relationship between variations in pathogenicity and lag phase at 37°C of *Listeria monocytogenes* previously stored at 4°C. *Letters in Applied Microbiology*, **23**: 18–22.

Buncic, S., Avery, S.M. & Rogers, A.R. 1996. Listeriolysin O production and pathogenicity of non-growing *Listeria monocytogenes* stored at refrigeration temperature. *International Journal of Food Microbiology*, **31**: 133–147.

Cassin, M.H., Lammerding, A.M., Todd, E.C.D., Ross, W. & McColl, R.S. 1998. Quantitative risk assessment for *Escherichia coli* O157:H7 in ground beef hamburgers. *International Journal of Food Microbiology*, **41**: 21–44.

Chabra, A.T., Carter, W.H., Linton, R.H. & Cousin, M.A. 1999. A predictive model to determine the effects of pH, milkfat, and temperature on thermal inactivation of *Listeria monocytogenes. Journal of Food Protection,* **62**: 1143–1149.

Chawla, C.S., Chen, H. & Donnelly, C.W. 1996. Mathematically modeling the repair of heat-injured *Listeria monocytogenes* as affected by temperature, pH, and salt concentration. *International Journal of Food Microbiology,* **30**: 231–242.

Cole, M.B., Davies, K.W., Munro, G., Holyoak, C.D. & Kilsby, D.C. 1993. A vitalistic model to describe the thermal inactivation of *Listeria monocytogenes*. *Journal of Industrial Microbiology*, **12**: 232–239.

Cole, M.B., Jones, M.V. & Holyoak, C. 1990. The effect of pH, salt concentration and temperature on the survival and growth of *Listeria monocytogenes. Journal of Applied Bacteriology*, **69**: 63–72.

Conte, M.P., Longhi, C., Petrone, G., Polidoro, M., Valenti, P. & Seganti, L. 1994. *Listeria monocytogenes* infection of caco-2 cells - role of growth temperature. *Research Microbiology*, **145**: 677–682.

Dalgaard, P. 1997. Predictive microbiological modelling and seafood quality. pp. 431–443, *in:* J. Luten, T. Børresen and J. Oehlenschläger (eds). *Seafood from Producer to Consumer, Integrated Approach to Quality.* Amsterdam, The Netherlands: Elsevier.

Dalgaard, P. 2000. Fresh and lightly preserved seafood. pp. 110–139, *in:* C.M.D. Man and A.A. Jones (eds). *Shelf life evaluation of foods*. 2nd edition. Maryland, USA: Aspen Publishing.

Dalgaard, P., Ross, T., Kamperman, L., Neumeyer, K. & McMeekin, T.A. 1994. Estimation of bacterial growth rates from turbidimetric and viable count data. *International Journal of Food Microbiology*, **23**: 391–404.

Dalgaard, P., Mejlholm, O. & Huss, H.H. 1997. Application of an iterative approach for development of a microbial model predicting the shelf-life of packed fish. *International Journal of Food Microbiology*, **38**: 169–179.

Duh, Y.-H. & Schaffner, D.W. 1993. Modeling the effect of temperature on the growth rate and lag time of *Listeria innocua* and *Listeria monocytogenes*. *Journal of Food Protection,* **56**: 205–210.

FAO/WHO. 2002. Risk assessments of *Salmonella* in eggs and broiler chickens. *FAO/WHO Microbiological Risk Assessment Series*, No. 2. 300p.

Farber, J.M. 1986. Predictive modeling of food deterioration and safety. pp. 57–90, *in:* M.D. Pierson and N.J. Stern (eds). *Foodborne Microorganisms and their Toxins.* New York NY: Marcel Dekker.

Farber, J.M., Cai, Y. & Ross, W.H. 1996. Predictive modelling of the growth of *Listeria monocytogenes* in CO_2 environments. *International Journal of Food Microbiology*, **32**: 133–144.

Farber, J.M., Ross, W.H. & Harwig, J. 1996. Health risk assessment of *Listeria monocytogenes* in Canada. *International Journal of Food Microbiology*, **30**: 145–156.

FDA/FSIS [U.S. Food and Drug Administration/Food Safety and Inspection Agency (USDA)]. 2001. Draft Assessment of the relative risk to public health from foodborne *Listeria monocytogenes* among selected categories of ready-to-eat foods. Center for Food Safety and Applied Nutrition (FDA) and Food Safety Inspection Service (USDA) (Available at: www.foodsafety.gov/~dms/lmrisk.html). [Report published September 2003 as: Quantitative assessment of the relative risk to public health from foodborne *Listeria monocytogenes* among selected categories of ready-to-eat foods. Available at: www.foodsafety.gov/~dms/lmr2-toc.html].

Fehlhaber, K. & Krüger, G. 1998. The study of *Salmonella enteriditis* growth kinetics using rapid automated bacterial impedance technique. *Journal Applied Microbiology*, **84**: 945–949.

Fernandez, P.S., George, S.M. & Peck, M.W. 1997. Predictive model of the effect of CO_2, pH, temperature and NaCl on the growth of *Listeria monocytogenes*. *International Journal of Food Microbiology*, **37**: 37–45.

Gahan, C.G.M. & Hill, C. 1999. The relationship between acid stress responses and virulence in *Salmonella typhimurium* and *Listeria monocytogenes*. *International Journal of Food Microbiology*, **50**: 93–100.

George, S.M., Lund, B.M. & Brocklehurst, T.F. 1988. The effect of pH and temperature on initiation of growth of *Listeria monocytogenes*. *Letters in Applied Microbiology*, **6**: 153–156.

George, S.M., Richardson, L.C.C. & Peck, M.W. 1996. Predictive models of the effect of temperature, pH and acetic and lactic acids on the growth of *Listeria monocytogenes*. *International Journal of Food Microbiology,* **32**: 73–90.

Golden, M.H., Buchanan, R.L. & Whiting, R.C. 1995. Effect of sodium acetate or sodium propionate with EDTA and ascorbic acids on the inactivation of *Listeria monocytogenes*. *Journal of Food Safety,* **15**: 53–65.

Grau, F.H. 1993. Processed meats and *Listeria monocytogenes*. pp. 13–24, *in*: Prevention of Listeria in Processed Meats. Proceedings of a series of workshops. CSIRO Division of Food Science and Technology, Meat Research Laboratory, Queensland, Australia.

Grau, F.H. & Vanderlinde, P.B. 1992. Occurrence, numbers, and growth of *Listeria monocytogenes* on some vacuum-packaged processed meats. *Journal of Food Protection*, **55**: 4–7.

Grau, F.H. & Vanderlinde, P.B. 1993. Aerobic growth of *Listeria monocytogenes* on beef lean and fatty tissue: equations describing the effects of temperature and pH. *Journal of Food Protection*, **56**: 96–101.

Hudson J.A, Mott, S.J. & Penny N. 1994. Growth of *Listeria monocytogenes, Aeromonas hydrophil*a and *Yersinia enterocolitica* on vacuum and saturated carbon dioxide controlled atmosphere-packaged sliced roast beef. *Journal of Food Protection,* **57**: 204–208.

ICMSF. 1996. *Microorganisms in Foods, Microbiological Specifications of Food Pathogens*. Vol. 5. London: Blackie Academic and Professional. 513p.

Jarvis, B. 1989. *Statistical Aspects of the Microbiological Analysis of Foods*. Amsterdam, The Netherlands: Elsevier.

Juneja, V.K. & Eblen, B.S. 1999. Predictive thermal inactivation model for *Listeria monocytogenes* with temperature, pH, NaCl, and sodium pyrophosphate as controlling factors. *Journal of Food Protection,* **62**: 986–993.

Lebert, I., Bégot, C. & Lebert, A. 1998. Development of two *Listeria monocytogenes* growth models in a meat broth and their application to beef meat. *Food Microbiology*, **15**: 499–509.

Linton, R.H., Carter, W.H., Pierson, M.D. & Hackney, C.R. 1995. Use of a modified Gompertz equation to model non-linear survival curves for *Listeria monocytogenes* Scott A. *Journal of Food Protection,* **58**: 946–954.

Linton, R.H., Carter, W.H., Pierson, M.D., Hackney, C.R. & Eifert, J.D. 1996. Use of a modified Gompertz equation to predict the effects of temperature, pH and NaCl on the inactivation of *Listeria monocytogenes* Scott A heated in infant formula. *Journal of Food Protection*, **59**: 16–23.

Lou, Y. & Yousef, A.E. 1999. Characteristics of *Listeria monocytogenes* important to food processors. pp. 131–225, *in:* Ryser & Marth, 1999, q.v.

McDonald, K. & Sun, D.-W. 1999. Predictive food microbiology for the meat industry: a review. *International Journal of Food Microbiology,* **52**: 1–27.

McClure, P.J, Kelly, T.M. & Roberts, T.A. 1991. The effects of temperature, pH, sodium chloride and sodium nitrite on the growth of *Listeria monocytogenes. International Journal of Food Microbiology*, **14**: 77–92.

McClure, P.J., Roberts, T.A. & Otto Oguru, P. 1989. Comparison of the effects of sodium chloride, pH and temperature on the growth of *Listeria monocytogenes* on gradient plates and in liquid medium. *Letters in Applied Microbiology*, **9**: 95–99.

McClure, P.J., Zwietering, M.H. & Roberts, T.A. 1993. Modelling bacterial growth of *Listeria monocytogenes* as a function of water activity, pH and Temperature. *International Journal of Food Microbiology*, **18**: 139–149.

McClure, P.J., Beaumont, A.L., Sutherland, J.P. & Roberts, T.A. 1997. Predictive modelling of growth of *Listeria monocytogenes*. The effects on growth of NaCl, pH, storage temperature and $NaNO_2$. *International Journal of Food Microbiology*, **3**: 221– 232.

McDonald, K. & Sun, D.-W. 1999. Predictive food microbiology for the meat industry: a review. *International Journal of Food Microbiology,* **52**: 1–27.

McLauchlin, J. 1996. The relationship between *Listeria* and listeriosis. *Food Control*, **7**: 187–193.

McMeekin, T.A., Olley, J., Ross, T. & Ratkowsky, D.A. 1993. *Predictive Microbiology. Theory and Application*. Taunton, UK: Research Studies Press. 340p.

Mellefont, L.A., McMeekin, T.A. & Ross, T. 2003. The effect of abrupt osmotic shifts on the lag phase duration of foodborne bacteria. *International Journal of Food Microbiology,* **83**: 281–293.

Mellefont, L.A. & Ross, T. 2003. The effect of abrupt shifts in temperature on the lag phase duration of *Escherichia coli* and *Klebsiella oxytoca. International Journal of Food Microbiology,* **83**: 295–305.

Membre, J.M., Thurette, J. & Catteau, M. 1997. Modelling the growth, survival and death of *Listeria monocytogenes*. *Journal of Applied Microbiology*, **82**: 345–350.

Murphy, P.M., Rae, M.C. & Harrington, D. 1996. Development of a predictive model for growth of *Listeria monocytogenes* in a skim milk medium and validation studies in a range of dairy products. *Journal of Applied Microbiology*, **80**: 557–564.

Nerbrink, E., Borch, E., Blom, H. & Nesbakken, T. 1999. A model based on absorbance data on the growth rate of *Listeria monocytogenes* and including the effect of pH, NaCl, Na-lactate and Na-acetate. *International Journal of Food Microbiology,* **47**: 99–109.

Notermans, S., Dufrenne, J., Teunis, P. & Chackraborty, T. 1998. Studies on the risk assessment of *Listeria monocytogenes*. *Journal of Food Protection*, **61**: 244–248.

Peleg, M. & Cole, M.B. 1998. Reinterpretation of microbial survival curves. *Critical Reviews in Food Science and Nutrition,* **38**: 353–380.

Piyasena, P., Liou, S. & McKellar, R.C. 1998. Predictive modelling of inactivation of *Listeria* spp. in bovine milk during high-temperature short-time pasteurization. *International Journal of Food Microbiology*, **39**: 167–173.

Ratkowsky, D.A. 1992. Predicting response times in predictive food microbiology. Department of Primary Industry, Fisheries & Energy, Tasmania, Research and Development Unit, Biometrics Section, Occasional Paper No. 1992/1.

Ratkowsky, D.A., Olley, J., McMeekin, T.A. & Ball, A. 1982. Relationship between temperature and growth rate of bacterial cultures. *Journal of Bacteriology*, **149**: 1–5.

Ratkowsky, D.A., Ross, T., McMeekin, T.A. & Olley, J. 1991. Comparison of Arrhenius-type and Belehradek-type models for prediction of bacterial growth in foods. *Journal of Applied Bacteriology,* **71**: 452–459.

Ratkowsky, D.A., Ross, T., Macario, T.W. & Kamperman, L. 1996. Choosing probability distributions for modelling generation time variability. *Journal of Applied Bacteriology*, **80**: 131–137.

Razavilar, V. & Genigeorgis, C. 1998. Prediction of *Listeria* spp. growth as affected by various levels of chemicals, pH, temperatures and storage time in a model broth. *International Journal of Food Microbiology*, **40:** 149–157.

Rees, C.E.D., Dodd, C.E.R., Gibson, P.T., Booth, I.R. & Stewart, G.S.A.B. 1995. The significance of bacteria in stationary phase to food microbiology. *International Journal of Food Microbiology*, **28**: 263–275.

Ross, T. 1993. A philosophy for the development of kinetic models in predictive microbiology. Ph.D. Thesis, University of Tasmania, Hobart, Australia.

Ross, T. 1996. Indices for performance evaluation of predictive models in food microbiology. *Journal of Applied Bacteriology*, **81**: 501-508.

Ross, T. 1999. Predictive food microbiology models in the meat industry. Meat and Livestock Australia, Sydney, Australia. 196p.

Ross, T., Baranyi, J. & McMeekin, T.A. 1999. Predictive Microbiology and Food Safety. pp. 1699–1710, *in:* R. Robinson, C.A. Batt and P. Patel (eds). *Encyclopaedia of Food Microbiology*. London: Academic Press.

Ross, T. & McMeekin, T.A. 1994. Predictive microbiology - a review. *International Journal of Food Microbiology,* **23**: 241–264.

Rowan, N.J. 1999. Evidence that inimical food-preservation barriers alter microbial resistance, cell morphology and virulence. *Trends in Food Science Technology*, **10**: 261–270.

Ryser, E.T., & Marth, E.H. (eds). 1999. *Listeria, Listeriosis, and Food Safety.* 2nd edition, revised and expanded. New York NY: Marcel Dekker. 738p.

Stephens, P.J., Cole, M.B. & Jones, M.V. 1994. Effect of heating rate on the thermal inactivation of *Listeria monocytogenes. Journal of Applied Bacteriology*, **77**: 702–708.

Stephens, P.J., Joynson, J.A., Davies, K.W., Holbrook, R., Lappin-Scott, H.M. & Humphrey, T.J. 1997. The use of an automated growth analyser to measure recovery times of single heat injured *Salmonella* cells. *Journal of Applied Microbiology,* **83**: 445–455.

Sutherland, J.P., Bayliss, A.P. & Roberts, T.A. 1994. Predictive modelling of the growth of *Staphylococcus aureus*: The effects of temperature, pH and sodium chloride. *International Journal of Food Microbiology,* **21**: 217–236.

te Giffel, M.C. & Zwietering, M.H. 1999. Validation of predictive models describing the growth of *Listeria monocytogenes. International Journal of Food Microbiology*, **46**: 135–149.

Tienungoon, S. 1998. Some aspects of the ecology of *Listeria monocytogenes* in salmonid aquaculture. Ph.D. Thesis, University of Tasmania, Hobart, Australia.

Tienungoon, S., Ratkowsky, D. A., McMeekin, T. A. & Ross, T. 2000. Growth limits of *Listeria monocytogenes* as a function of temperature, pH, NaCl and lactic acid. *Applied and Environmental Microbiology*, **66**: 4979–4987.

Tompkin, R.B. 2002. Control of *Listeria monocytogenes* in the food-processing environment. *Journal of Food Protection,* **65**: 709–725.

Walls, I. & Scott, V.N. 1997. Validation of predictive mathematical models describing the growth of *Listeria monocytogenes*. *Journal of Food Protection*, **60**: 1142–1145.

Wijtzes, T., McClure, P.J., Zwietering, M.H. & Roberts, T.A. 1993. Modelling bacterial growth of *Listeria monocytogenes* as a function of water activity, pH and temperature. *International Journal of Food Microbiology*, **18**: 139–149.

Xiong, R., Xie, G., Edmondson, A.S., Linton, R.H. & Sheard, M.A. 1999. Comparison of the Baranyi model with the modified Gompertz equation for modelling thermal inactivation of *Listeria monocytogenes* Scott A. *Food Microbiology*, **16**: 269–279.

Zhao, Y., Wells, J.H. & Marshall, D.L. 1992. Description of log phase growth for selected microorganisms during modified atmosphere storage. *Journal of Food Process Engineering*, **15**: 299–317.

Zhao, P., Zhao, T., Doyle, M., Rubino, J., & Meng, J. 1998. Development of a model for evaluation of microbial cross-contamination in the kitchen. *Journal of Food Protection*, **61**: 960–963.

Zwietering, M.H., Cuppers, H.G.A.M., deWit, J.C. & van't Riet, K. 1994. Evaluation of data transformation and validation of a model for the effect of temperature on bacterial growth. *Applied and Environmental Microbiology*, **60**: 195–203.

Zemser, R.B. & Martin, S.E. 1998. Heat stability of virulence-associated enzymes from *Listeria monocytogenes* SLCC 5764. *Journal of Food Protection*, **61**: 899–902.

Appendix 4. Prevalence and incidence of *Listeria monocytogenes* in Fermented Meat Products

A4.1 REPORTED PREVALENCE AND INCIDENCE

The prevalence and incidence *of Listeria monocytogenes* in fermented meat products (FMPs) as reported in the literature is summarized in Table A4.1. It is to be noted, as mentioned in Section A4.2, below, that some products in the list in Table A4.1 might have a single name but represent very different products and processes in different countries. The authors have not attempted to distinguish these in the risk assessment modelling, but have instead treated all prevalence and concentration data as representative of all FMPs.

Table A4.1 Reported prevalence and incidence *of Listeria monocytogenes* in fermented meat products

Product Description	Positive (samples or proportion)	Samples	% positive	Conc.	Location of Survey	Ref.
Fermented sausages	up to 0.20	5	20.00%		Various countries	[1] [2]
Fermented sausages					Austria	[3]
Fermented sausages	4	21	19.05%		Yugoslavia	[4]
Raw sausage	16	20	80.00%		Brazil	[5]
Fermented sausage	0.22 to 0.83				Spain	[6]
Dry sausages	0.22 to 0.83	18	44.00%		Various countries	[7] [8] [9] [10] [11]
Fermented sausages	6	30	20.00%		Canada	[12]
Raw sausage	13	25	52.00%		UK	[13]
Mettwurst with onion, fresh	1	11	9.09%		Germany	[14]
Sausages	2	8	25.00%		Hungary	[15]
Mettwurst with onion	27	245	11.00%		Germany	[16]
Spreadable, fermented	43	381	11.30%		Germany	[16]
Sliceable, fermented	11	228	4.80%		Germany	[16]
Raw sausage	30	120	25.00%	<100 CFU/g	Germany	[17]
Mettwurst, coarse	6	30	20.00%	<1000 CFU/g	Germany	[18]
Mettwurst, fresh	18	30	60.00%	<1000 CFU/g	Germany	[18]
Raw sausage, salami type	5	30	16.67%	<100 CFU/g	Germany	[18]
Beef sausage	0	1	0.00%		UK	[19]
Sausage	0	3	0.00%		UK	[19]
Raw fresh sausages	4	98	4.08%		France	[20]
Raw sausage	12	68	17.65%		Germany	[21]
Mettwurst, fresh	22	132	16.67%		Germany	[22]

Product Description	Positive (samples or proportion)	Samples	% positive	Conc.	Location of Survey	Ref.
Raw sausage, sliced	2	126	1.59%		Germany	[22]
Salsiccia	6	52	11.54%		Italy	[23]
Fermented sausages, salami type	0	70	0.00%		Norway	[24]
Ground/minced muscle (dry fermented sausages)	36	308	11.69%		Belgium	[25]
Fermented sausages	up to 0.20	5	20.00%	less than in non-fermented RTE cooked meats		[26]
Salami		128	10.00%		UK	[27]
Salami		67	16.00%		UK	[28]
Salami		59	5.00%	20 CFU/g	Switzerland	[29]
Mettwurst		14	0.00%		Switzerland	[30]
Dry cured		136	10.00%		Hungary	[31]
Fermented		21	10.00%		Hungary	[31]
Smoked		23	13.00%		Hungary	[31]
Cervelat		44	0.00%		South Africa	[32]
Vacuum-packed salami		19	0.00%		Australia	[33]
Salami		132	40.00%		Australia	[34]
Uncooked, preserved meat products						
1994/5 data	77	328	23.50%	1.8% > 10 (& <100) CFU/g; 0.6% >100 CFU/g		
1997 data from retail-level processors	19	132	14.40%	13.6%>10 (& <100) CFU/g; 0.8% > 100 CFU/g	Denmark	[35]
1998 data from retail-level processors	37	225	16.50%	14.7%>10 (& <100) CFU/g; 1.8% > 100 CFU/g		

SOURCES: [1] Breer and Schopfer, 1989. [2] Farber, Sanders and Johnston, 1989. [3] Breuer and Prandl, 1988. [4] Buncic, 1991. [5] Destro, Serrano and Kabuki, 1991. [6] Encinas et al., 1999. [7] Farber, Sanders and Johnston, 1989. [8] Nicolas and Vidaud, 1985. [9] McClain and Lee, 1988. [10] Breuer and Prandl, 1988. [11] Schmidt et al., 1988. [12] Farber, Sanders and Johnston, 1989. [13] Gilbert, Hall and Taylor, 1989. [14] Karches and Teufel, 1988. [15] Kiss et al., 1996. [16] 1991–2 data supplied to FAO/WHO by BgVV, Germany, in response to a call for data, 2000. [17] Leistner and Schmidt, 1992. [18] Leistner, Schmidt and Kaya, 1989. [19] MacGowan et al., 1994. [20] Nicolas and Vidaud, 1985. [21] Noack and Jockel, 1993. [22] Ozari and Stolle, 1990. [23] Pacini et al., 1995. [24] Rørvik and Yndestad, 1991. [25] Uyttendaele, Troy and Debevere, 1999. [26] WHO, 1988. [27] Velani and Gilbert, 1990. [28] Gilbert, 1991. [29] Trüssel, 1989. [30] Trüssel, 1989. [31] Kovacs-Domjan, 1991. [32] Vorster, Greebe and Nortje, 1993. [33] Grau and Vanderlinde, 1992. [34] Varabioff, 1992. [35] Nørrung, Andersen and Schlundt, 1999.

A4.2 PRODUCTION METHODS AND STYLES OF FERMENTED MEATS

A4.2.1 Introduction to fermented meat products

Fermented meats, including salami, have been manufactured for centuries (Lücke, 1985; Leistner, 1995; Ricke and Keeton, 1997). European sausages have been produced since the Middle Ages and per capita production and consumption of fermented meat products (FMPs) is still greatest in Europe. European migrants to North America, and elsewhere, took their FMPs methods and styles with them to their new homelands, where new variations evolved, i.e. these traditional products in some cases were changed to suit conditions in the New World. It is important, then, to recognize that FMPs products from the Old and New Worlds that have the same name may often differ in composition and processing. For example, all "Mettwurst' and "Teewurst" in the United States of America is cooked, and NaCl levels in United States of America products are normally higher than their European counterparts, due to regulations for the control of the parasite, *Trichinella.* United States of America producers typically use nitrite only, with no nitrate added (B. Tompkin, pers. comm., 2001).

Dry sausages include chorizo (Spanish, smoked, highly spiced), frizzes (similar to pepperoni, but not smoked), pepperoni (not cooked, air dried), Lola or Lolita and Lyons sausage (mildly seasoned pork with garlic), and Genoa salami (Italian, usually made from pork but may have a small amount of beef; in the preparation process it is moistened with wine or grape juice and seasoned with garlic).

Chinese-style fermented sausages, with pork as the main ingredient, are also common in Asia and date back thousands of years (Leistner, 1995; Yu and Chou, 1997). The Thai fermented sausage Nham is also receiving attention in the scientific literature (ASCA, 1986; Petchsing and Woodburn, 1990).

Most FMPs products have long shelf lives due to the combination of acidification (through fermentation), removal of oxygen, addition of compounds that favour the growth of some microbes while retarding the growth of others, and, ultimately, the removal of water.

Semi-dry sausages are usually heated in a smokehouse to fully cook the product and partially dry it. Semi-dry sausages are semi-soft sausages with good keeping qualities due to their lactic acid fermentation. "Summer Sausage" (another word for cervelat) is the general classification for mildly seasoned, smoked, semi-dry sausages like mortadella and Lebanon bologna.

Unless otherwise noted the information in these sections is drawn from Lücke (1985), Leistner (1995), Lücke (1995) and Ricke and Keeton (1997).

A4.2.1 Processing

The fundamental steps involved in the production of FMPs are:

- chopping and mixing of ingredients, and filling into casing;
- fermentation; and
- drying (or maturation).

A4.2.3 Ingredients

Meat and Fat

From a product quality perspective, the type of meat used in FMPs is important. It is less important for the microbiological safety of the product, unless some types of meat are more highly contaminated with pathogens than others. The proportion of meat to fat, and the type of fat, is not important microbiologically except in the sense that the proportion of fat affects the amount of free water in the product. Of the lean muscle in the mix, about 70–75% by weight is water. It is the *concentration* of the additives in the aqueous (water) phase of the food that is important for understanding the microbiology of the product. More fat in the mixture means that there is less lean meat, which in turn means less water. As a guide, for a product containing 30% (by weight) fat, water makes up only about 53% of the weight of the batch. Thus, the effective concentration of any water-soluble additives is about twice that predicted simply on the basis of its weight compared to the overall weight of the batch. During maturing of FMPs, weight losses of 20–30% occur in "semi-dry" FMPs, and even more for "dry"-style products. This is due to loss of water only, and further increases the effective concentration of the water-soluble components, so that the final concentration can be up to four times the apparent level added to the mixture expressed on a weight-for-weight basis.

Table A4.2 Typical physico-chemical properties of styles of finished FMPs products.

Category	Final pH	Lactic acid (%)	Moisture : protein ratio	Moisture loss	Moisture(1)	Comments
Dry sausages	5.0–5.3 (<5.3)	0.5–1.0	<2.3:1	25–50	<35	See Note (2)
Cervelat			1.9:1		32–38	Shelf-stable
Cappicola			1.3:1		23–29	Shelf-stable
German Dauerwurst	4.7–4.8		1.1:1		25–27	Shelf-stable
German salami	4.7–4.8		1.6:1	1	34–35	Shelf-stable
Peperoni	4.5–4.8	0.8–1.2	1.6:1	35	25–32	Shelf-stable
Italian salami, hard or dry			1.9:1	30	32–38	Shelf-stable
Genoa salami	4.9	0.79	2.3:1	28	33–39	Shelf-stable
Thüringer, dry	4.9	1.0	2.3:1	28	46–50	Shelf-stable
Semi-dry sausages	4.7–5.1 (<5.3)	0.5–1.3	>2.3<3.7:1	8–15	45–50	See Note (3)
Lebanon bologna	4.7	1.0–1.3	2.5:1	10–15	56–62	Refrigerate
Cervelat, soft			2.6:1	10–15		Refrigerate
Salami, soft			2.3–3.7:1	10–15	41–51	Refrigerate
Summer sausage	<5.0	1.0	3.1:1	10–15	41–52	Refrigerate
Thüringer, soft			3.7:1		46–50	Refrigerate
For comparison						
Dried beef			2.04:1	29		
Beef jerky			0.75:1	>50	28–30	
Air-dried sausage			2.1:1			

NOTES: (1) Water activity ranges for dry and semi-dry sausages are <0.85 to 0.91, and 0.90 to 0.94, respectively. European Council Directive 77/99/EEC (Health problems affecting the production and marketing of meat products and certain other products of animal origin) requests a_w of <0.91 or pH <4.5 for dry sausages to be shelf-stable, or a combined a_w and pH of 0.95 and <5.2, respectively. (2) Heat processed (optional, but see note (4)); dried or aged after fermentation for moisture loss; smoked.(3) Heat processed (but see Note (4)); typically smoked; packaged after processing and chilling.(4) USDA/FSIS Title 9 CFR may be amended to require specified time and temperature heating combinations after fermentation, or verification that processing conditions destroy all pathogenic micro-organisms.SOURCES: Various authors, cited in Ricke and Keeton, 1997.

Salt

Typically, 2.5–3.3% NaCl (w/w) is added to FMPs mixes. The water activity (a_w) of the product decreases during processing as the product loses water. This leads to effective concentrations in the typical finished semi-dry product of about 7.5–12% salt, corresponding to water activities in the range 0.95–0.92. Lower water activities (~0.85) are achieved in southern European style dry sausages (Calicioglu et al., 1997; Ricke and Keeton, 1997). Water activity values can be translated to aqueous phase NaCl concentration by reference to calibration tables or curves, e.g. Chirife and Resnik (1984).

Sugar, pH and organic acids

Sugars (0.4–0.8%) are added to the mixture as a carbon source for the fermentative bacteria. These bacteria, usually lactic acid bacteria, metabolize the sugars, producing lactic acid in the process, which is released into the FMPs. In a review of lactic acid bacterial fermentation and the principal antimicrobial factors produced by lactic acid bacteria, Adams and Nicolaides (1997) concluded that the principal antimicrobial factor is the ability of all lactic acid bacteria to produce organic acids and decrease the pH of foods in which they grow.

The biochemistry of conversion of simple sugars (e.g. glucose) results in almost twice as much lactic acid being produced as the concentration of simple sugars added. For more complex sugars, a smaller ratio of lactic acid to sugar results due to incomplete utilization of the carbohydrate. Other organic acids are also produced, but at much lower levels. The presence of lactic acid reduces the pH of the product during fermentation, typically to the range 4.6–5.0. The range of lactic acid concentrations in the final product is shown in Table A4.2. The range corresponds to total effective lactic acid concentrations (i.e. in the water phase) of from about 100 (e.g. semi-dry) to 500 mM (pepperoni).

Other additives

Other ingredients of FMPs may include a variety of spices, and nitrite or nitrate. Spices, including pepper, paprika, garlic, mace, pimento and cardamom, may be added, but their primary role is sensory. The redox potential (E_h) of FMPs is low. After mixing, the unfermented product is stuffed into casings. This effectively removes some oxygen. The predominant spoilage organisms of raw, aerobically stored meat, will be included in the mix and quickly consume residual oxygen. The presence of ascorbate and sugars also contributes to the creation of a low redox potential in the sausage.

A4.2.4 Production of "safe" FMPs

Production of safe FMPs requires prevention of the growth of pathogens during the fermentation step and maximizing death of surviving pathogens during maturation and storage. Some processors (especially in North America) include a heating step after fermentation that is intended to inactivate pathogens, including *Salmonella*, pathogenic *E. coli* and *Trichinella spiralis*. The initial stages of the fermentation process can permit growth of enteric pathogens such as *Salmonella*, *E. coli* and *Staphylococcus aureus*. The rapid acidification of the medium by the starter culture is considered a critical control point for minimization or prevention of pathogen growth (Bacus, 1997).

A4.3 PHYSICO-CHEMICAL PARAMETERS OF FMPs

See Table A4.3 and Figure A 4.1

Table A4.3 Typical composition and processing parameters for various FMPs styles

	Semi-dry sausage	**Dry sausage**	
		Northern European type	Southern European type
Examples	Summer sausages German cervelat Bologna sausages	German salamis Danish salamis	Italian salamis (Milanese; Calabrese) Saucissons secs Spanish chorizos
Raw mixture			
Meat : fat	Pork or beef, lean and fat	Lean pork : lean beef : fat pork (1 : 1 : 1)	Lean pork : fat pork (2 : 1)
Sugars (e.g. glucose, lactose, sucrose)	0.3–1.5%	0.3–0.8%	0–0.4%
Nitrate	–	–	<300 ppm
Nitrite	0–150 ppm	20–200 ppm	20–200 ppm
NaCl	2–2.5%	2–2.5%	2–2.5%
Seasoning (e.g. pepper, garlic, cardamom)	++	++	++
Starter cultures (10^6 CFU/g)	Yes	Yes	Yes
Lactobacillus sakei, L. curvatus, L. plantarum	+	++	+++
Pediococcus acidilactici, P. pentosaceus	+++	+++	+
Staphylococcus carnosus, S. xylosus	–	++ or –	++
Kocuria varians	–	++ or –	+
Penicillium chrysogenum, P. nalviogense	+	–	++
Debaryomyces hansenii, Candida lipolytica	+	–	++
Fermentation period (time/temperature/ relative humidity)	15–20 h/27–41°C/90% (USA) 18–48 h/20–32°C/85–95% (Germany)	18–48 h/20–30°C/58–95%	Day 1 – 22–24°C/94–96% Day 2 – 20–22°C/90–92% Day 3 – 18–20°C/85–88% or 2–3 d at 22–25°C/90–95%
Drying period (time/temperature/relative humidity)	2–3 d/10°C/68–72% (USA) 10–25 days (Germany)	1–3 weeks/12–15°C/75–80%	4–6 weeks/12–15°C/75–78%, or 8–14 weeks (traditional)
Method of production			
Smoking	Yes	Yes	No
Cooking	Yes	No	No
Product caracteristics			
Final pH	4.4–5	4.6–5.1	5.1–5.5
Final a_w	0.93–0.98	0.92–0.94	0.85–0.86
Water content	40–50%	30–40%	20–30%
Moisture : protein ratio (w/w)	2.3–3.7	2–2.3	1.6–1.9

KEY: + = occasionally used; ++ = frequently used; +++ = regullarly used; – = not used.

SOURCE: Reprinted from Montel, M.C. Fermented meat products. pp. 745–753, *in:* R.K. Robinson, C.A. Batt and P.D. Patel (eds). *Encyclopaedia of Food Microbiology.* Copyright (2000), with permission from Elsevier.

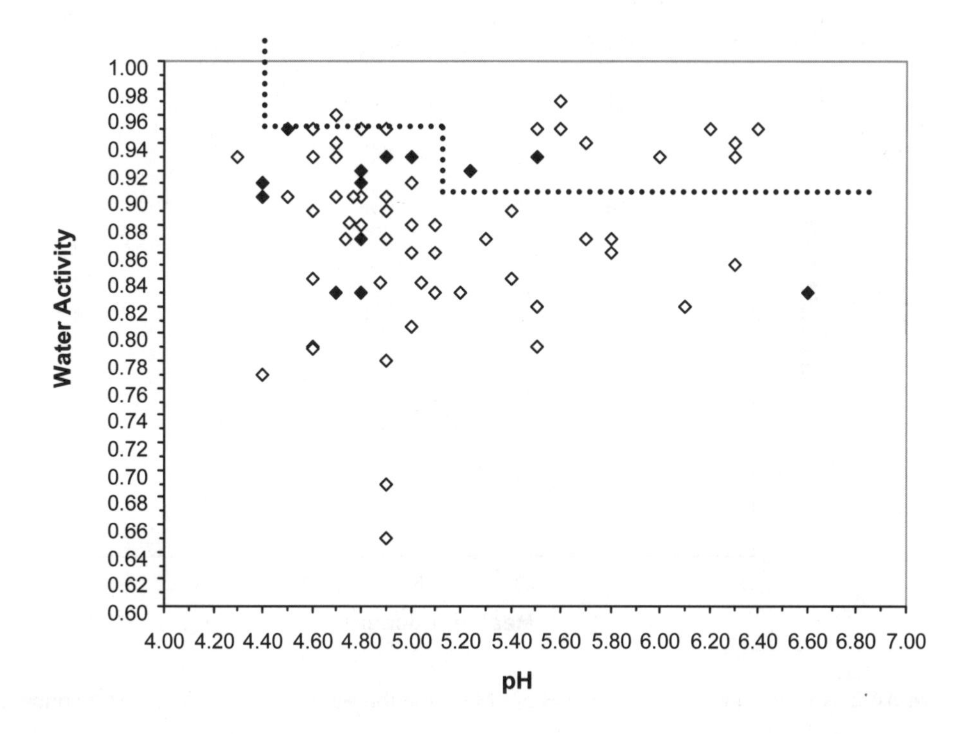

Figure A4.1 Collation of final pH and water activity of fermented meat products available in North America (open diamonds) and Australia (closed diamonds). pH and water activity limits below which the product is considered safe, in the absence of other inhibitors of microbial growth, are also shown (dotted lines)

SOURCE: After Ross et al., in press.

A4.4 ESTIMATION OF LACTIC ACID CONCENTRATION IN FMPs

The primary determinant of pH in meat and fermented meat is lactic acid. Natural levels in post-mortem meat are up to 125 mM (Gill, 1982) for meat in the pH range 5.5–6.5.

pH values, lactic acid and moisture content levels presented in Table A4.1 were tabulated. The lactic acid level was converted to lactic acid (using the Henderson-Haselbalch equation) in the aqueous phase and pH plotted against lactic acid concentration. Data for pH and lactic acid concentration in meat were also included in the tabulation, and the data plotted (*see* Figure A4.2).

The simple regression through the data is described by the line:

$$\text{lactic acid (ppm)} = 50 + (((6.6\text{-pH})/2.3) \times 300)$$

The model was generated based on the assumption that the lactic acid concentration in meat at pH 6.6 (highest pH reported in salami in Figure A4.1) is 50 mM, and that the lactic acid concentration in the salami with the lowest pH (4.3) is ~350 mM (*see* Section A4.2). It was further assumed that pH was directly related to lactic acid concentration in the range pH 6.6–4.3.

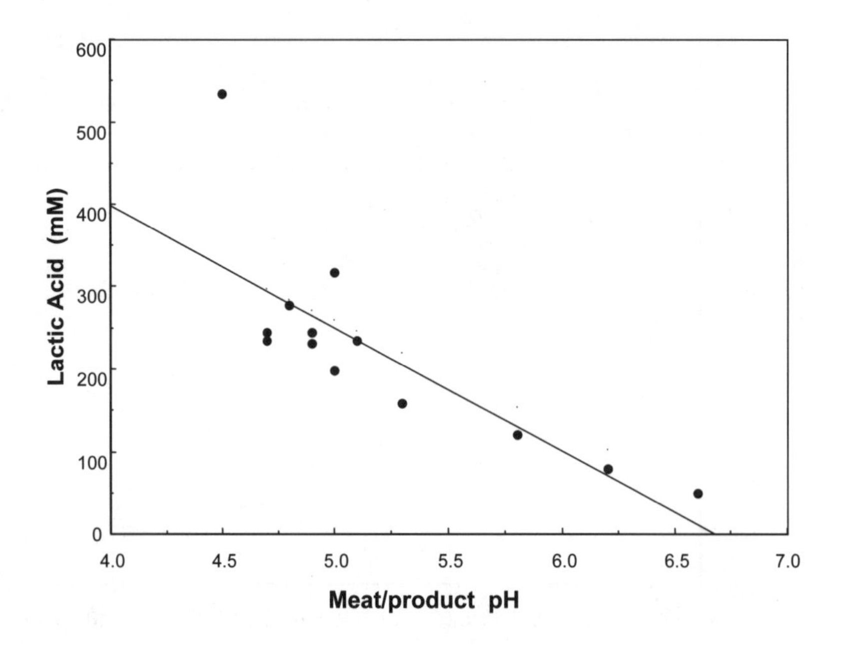

Figure A4.2 Relationship between the pH of FMPs and the aqueous phase lactic acid concentraton.

A4.5 DATA ILLUSTRATING THE REMAINING SHELF-LIFE OF AUSTRALIAN FMPs AT THE TIME OF PURCHASE BY DOMESTIC CONSUMERS

An ad hoc survey was conducted of retail outlets in Hobart, Australia, including supermarkets and small stores and delicatessens, of nominal remaining shelf-life of fermented meats on retail display. The survey involved 28 samples of 13 different products from 3 Australian producers. The survey was part of a larger survey of all processed meats, involving >700 samples.

By comparing the "use-by" (or "expiration") date with the survey data, it is possible to infer the nominal shelf-life remaining. From the "use-by" date and manufacture date it is possible to infer the total shelf-life. The survey revealed that the mean stated shelf-life of Australian FMPs is 140 (±70) days, but samples included only three producers.

It is assumed that the survey represents a snapshot of the remaining shelf-life that a purchaser would have available to them. Full details of the survey are presented in Ross et al. (in press).

A summary is shown in Figure A4.3.

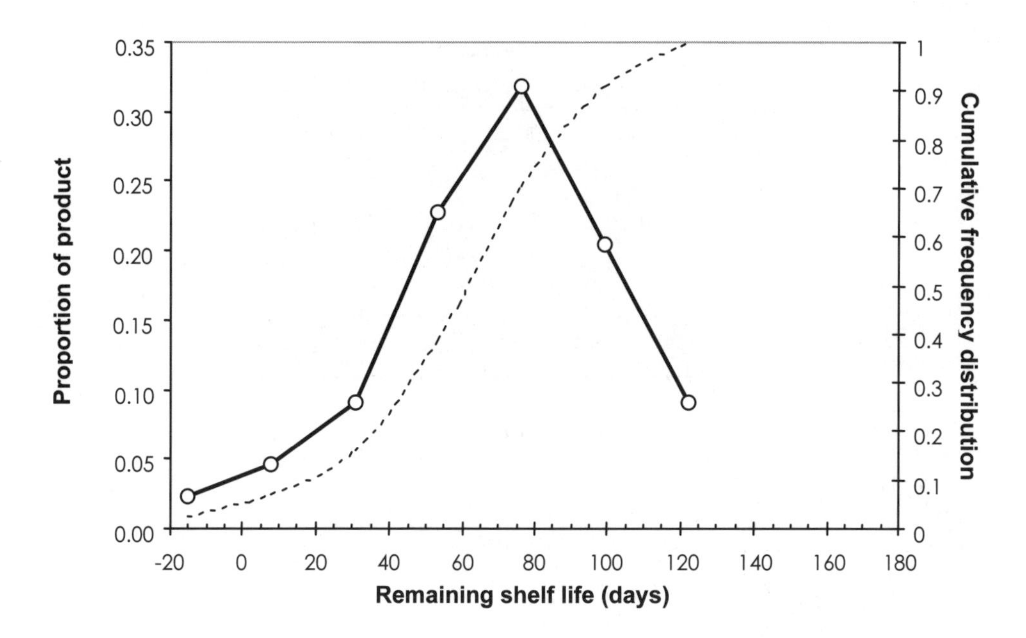

Figure A4.3 A histogram of nominal remaining shelf-life of Australian fermented meats on retail display (heavy lines) and the cumulative frequency curve derived from that data. The sample size was 24, comprising 13 different products from three producers.

A4.6 SERVING SIZE DISTRIBUTION

See Figure A4.4.

A4.7 NATIONAL CONSUMPTION OF FMPs

Differences in consumption between nations is expected and reported indirectly (Holdsworth et al., 2000), but practically no data quantifying national FMPs production were found. Typically, FMPs statistics are included in total processed meat statistics. However, some data were extracted from supermarket sales records and nutritional surveys, as described below.

United States of America

FDA/FSIS (2001) estimated an average of 6.41 servings of FMPs per annum for a population of 271 000 000 (NGS, 1999). This consumption is relatively low compared to other developed nations, and expert opinion (B. Tompkin, pers. comm., 2001) also suggests that this estimate of consumption is unrealistically low. The 50th percentile serving size is 46 g. This equates to 295 g/person-year, or a total national consumption of 82 265 tonne/year. It is noted that total consumption of "deli meats" in the United States of America is 12 times higher than FMPs consumption.

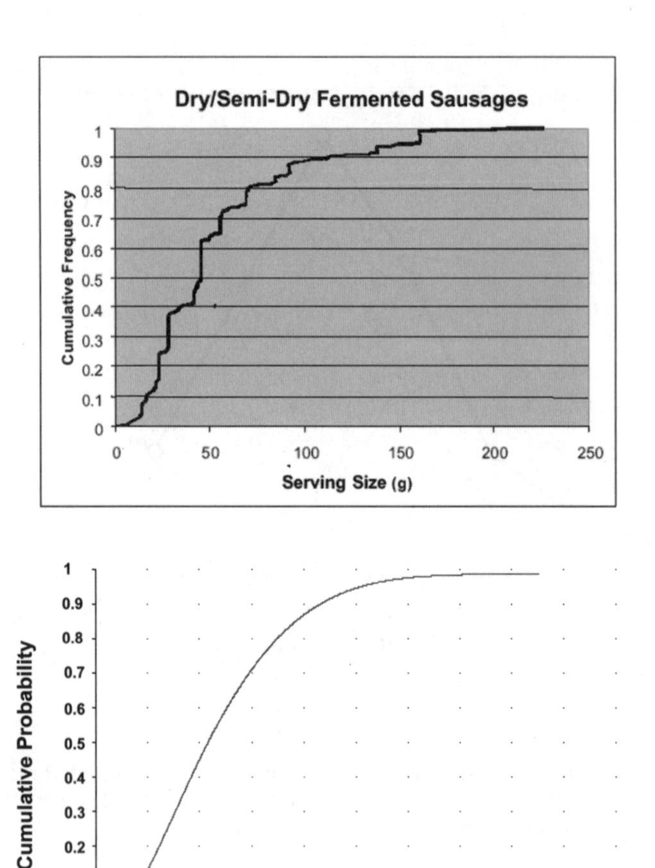

Figure A4.4 Comparison of serving size distribution from United States of America data (FDA/FSIS, 2001) (top) and that generated by the model used in this study (bottom).

Australia

National survey data (ABS, 1995) suggests that annual per capita consumption of processed meat is between 5 and 19 kg, but this could include sausages and other meat products intended to be cooked prior to consumption. From diverse Australian production and sales data reviewed by Ross et al. (in press), it was estimated that total production of FMPs in Australia was 7795–32 379 tonne/year, equivalent to an annual per capita consumption of 400–1680 g. It is noted that this also suggests that in Australia the total FMPs consumption is one tenth to one twelfth of total processed meats, as reported for the United States of America. The Australian population is about 19 800 000 (ABS, 2002).

Canada

National consumption of FMPs in Canada is estimated as 912 g/person/year, based on 5% of consumers daily eating approximately 50 g. Canada has a population of about 31 000 000 (NGS, 1999).

Germany

van Schothorst (1997) suggests that per capita annual average national consumption of processed meats in Germany is 28.5 kg. Assuming that 10% of this consumption is FMPs, (analogous to Australian and United States of America estimates), an estimate of 2.85 kg FMPs/person/year is made. However, German survey data from 1986 (G. Klein, pers. comm., 2000) indicate that per capita annual consumption of semi-dry fermented salami averaged only 723 g in West Germany, lower than that estimated by comparison with other nations. The reason for this large difference in estimates is currently unresolved. The population of Germany is about 81 000 000 (NGS, 1999).

Finland

FFDIF (2000) reported that in 1998 and 1999 national annual per capita consumption of processed meat averaged 32 kg. National consumption included about 7000 tonnes dry sausage, 120 000 tonnes of other sausage, and 38 000 tonnes of hams and other processed meats. The population of Finland is approximately 5 170 000 (NGS, 1999). If it is assumed that dry sausage refers only to FMPs, then consumption is estimated at 1.35 kg/person/year. If the assumption from other nation's data is used, i.e. that 10% of processed meat consumption is FMPs, estimated annual per capita consumption is about 3.2 kg, and for consistency this was the estimated value used in the present risk assessment.

A4.8 REFERENCES CITED IN APPENDIX 4

ABS [Australian Bureau of Statistics]. 1995. Australian National Nutrition Survey 1995. Australian Government Publishing Service, Canberra, Australia.

ABS. 2002. *Australia Now. Population. Population clock.* Downloaded 14/10/2002 from www.abs.gov.au/Ausstats/ABS

Adams, M.R. & Nicolaides, L. 1997. Review of the sensitivity of different foodborne pathogens to fermentation. *Food Control,* **8**: 227–239.

ASCA [Association for Science Cooperation in Asia]. 1986. A Concise Handbook of indigenous fermented foods in the ASCA Countries. Edited by S. Saono, R.R. Hull and B. Dhamcharee. Association for Science Cooperation in Asia, Australian Government Publishing Service, Canberra.

Bacus, J. 1997. Processing procedures to control *Salmonella* and *E. coli* in fermented sausage products. *Food Australia*, **49**: 543–547.

Breer, C. & Schopfer, K. 1989. Listerien in Nahrungsmitteln [*Listeria* in foodstuffs]. *Schweizerische Medizinische Wochenschrift*, **119**: 306–311.

Breuer, V. J. & Prandl, O. 1988. Nachweis von Listerien und deren Vorkommen in Hackfleisch und Mettwursten in Osterreich [The detection and incidence of *Listeria*s in ground beef and mettwurst in Austria]. *Archiv fur Lebensmittelhygiene,* **39**: 28–30.

Buncic, S. 1991. The incidence of *Listeria monocytogenes* in slaughtered animals, in meat and in meat products in Yugoslavia. *International Journal of Food Microbiology*, **12**: 173–180.

Calicioglu, M., Faith, N.G., Buege, D.R. & Luchansky, J.B. 1997. Viability of *Escherichia coli* O157:H7 in fermented semidry low-temperature-cooked beef summer sausage. *Journal of Food Protection*, **60**: 1158–1162.

Chirife, J. & Resnik, S.L. 1984. Unsaturated solutions of sodium chloride as reference sources of water activity at various temperatures. *Journal of Food Science,* **49**: 1486–1488.

Destro, M.T., Serrano, A.D. & Kabuki, D.Y. 1991. Isolation of *Listeria* species from some Brazilian meat and dairy products. *Food Control*, **2**: 110–112.

Encinas, J.P., Sanz, J.J., Garcia-Lopez, M.-L. & Otero, A. 1999. Behaviour of *Listeria* spp. in naturally contaminated *chorizo* (Spanish fermented sausage). *International Journal of Food Microbiology*, **46**: 167–171.

Farber, J.M., Sanders, G.W. & Johnston, M.A. 1989. A survey of various foods for the presence of *Listeria* species. *Journal of Food Protection*, **52**: 456–458.

FDA/FSIS [U.S. Food and Drug Administration/Food Safety and Inspection Agency (USDA)]. 2001. Draft Assessment of the relative risk to public health from foodborne *Listeria monocytogenes* among selected categories of ready-to-eat foods. Center for Food Safety and Applied Nutrition (FDA) and Food Safety Inspection Service (USDA) (Available at: www.foodsafety.gov/~dms/lmrisk.html). [Report published September 2003 as: Quantitative assessment of the relative risk to public health from foodborne *Listeria monocytogenes* among selected categories of ready-to-eat foods. Available at: www.foodsafety.gov/~dms/lmr2-toc.html].

FFDIF [Finnish Food and Drink Industries Federation]. 2000. Facts about Finnish Food Industry. Downloaded from: http://www.etl.fi/english/stat/pdf/facts.pdf (22 October 2002).

Gilbert, R.J. 1991. Occurrence of *Listeria monocytogenes* in foods in the United Kingdom. pp. 82–88, *in:* Proceedings of the International Conference on Listeriosis and Food Safety. Laval, France, cited in J.M. Farber and P.I. Peterkin, 1999. Incidence and behaviour of *Listeria monocytogenes* in meat products. pp. 505–564, *in:* Ryser & Marth, 1999, q.v.

Gilbert, R.J., Hall, S.M. & Taylor, A.G. 1989. Listeriosis update. *U.K. PHLS Microbiology Digest*, **6**: 33–37.

Gill, C.O. 1982. Microbial interaction with meats. pp. 225–264, *in:* M.H. Brown (ed). *Meat Microbiology*. UK: Applied Science Publishers.

Grau, F.H. & Vanderlinde, P.B. 1992. Occurrence, numbers, and growth of *Listeria monocytogenes* on some vacuum-packaged processed meats. *Journal of Food Protection*, **55**: 4–7.

Holdsworth, M., Gerber, M., Haslam, C., Scali, J., Bearsdworth, A., Avallone, M.H. & Sherratt, E. 2000. A comparison of dietary behaviour in Central England and a French Mediterranean region. *European Journal of Clinical Nutrition*, **54**: 530–539.

Karches, H. & Teufel, P. 1988. *Listeria monocytogenes*. Vorkommen in Hackfleisch bei Kuhl- und Gefrierlagerung. *Fleischwirtschaft*, **68**: 1388–1392.

Kiss, R., Papp, N. E., Vamos, G. Y. & Rodler, M. 1996. *Listeria monocytogenes* isolation from food in Hungary. *Acta Alimentaria*, **25**: 83–91.

Kovacs-Domjan, H. 1991. Occurrence of Listeria infection in meat industrial raw-materials and end products. *Magyar Allatorvosok Lapja*, **46**: 229–233.

Leistner, L. 1995. Stable and safe fermented sausages world-wife. pp. 160-175, *in:* G. Campbell-Platt and P.E. Cook, (eds). *Fermented Meats.* UK: BlackieAcademic and Professional.

Leistner, L. & Schmidt, U. 1992. Aktuelles uber Listerien in Fleisch. *Mitteilungsblatt der Bundesanst. fur Fleischforshung, Kulmbach,* **31**: 197–206.

Leistner, L., Schmidt, U. & Kaya, M. 1989. Listerien bei Fleisch und Fleischerzeugnissen. [Listeria in meat and meat products]. *Mitteilungsblatt der Bundesanst. fur Fleischforshung, Kulmbach,* **104**: 192–199.

Lücke, F-K. 1985. Fermented sausages. pp. 41-83, *in:* B.J.B. Wood (ed). *Microbiology of Fermented Foods*. London: Elsevier Applied Science Publ.

Lücke, F-K. 1995. Fermented meats. pp. F-1–F-23, *in: LFRA Microbiology Handbook: Meat Products*. Leatherhead, Surrey, UK: Food Research Association.

MacGowan, A.P., Bowker, K., McLauchlin, J., Bennett, P.M. & Reeves, D.S. 1994. The occurrence and seasonal changes in the isolation of *Listeria* spp. in shop bought foodstuffs, human faeces, sewage and soil from urban sources. *International Journal of Food Microbiology*, **21**: 325–334.

McClain, D. & Lee, W.H. 1988. Development of USDA FSIS method for isolation of Listeria monocytogenes from raw meat and pooultry. *Journal of the Association of Official Analytical Chemists,* **71**: 660–664.

Montel, M.C. 2000. Fermented meat products. pp. 745–753, *in:* R.K. Robinson, C.A. Batt and P.D. Patel (eds). *Encyclopaedia of Food Microbiology*. San Diego CA: Academic Press.

NGS [National Geographic Society]. 1999. The National Geographic desk reference. Edited by R.M. Downs, F.A Day, P.L. Knoxw, P.H. Meserve and B. Warf. Washington DC: National Geographic Society. 700p.

Nicolas, J.A. & Vidaud, N. 1985. Contamination des viandes et des produits de charcuterie par *Listeria monocytogenes* en Haute-Vienne. *Sciences des aliments*, **5**: 175–180.

Noack, D.J. & Joeckel, J. 1993. *Listeria monocytogenes*, occurrence and significance in meat and meat products and experience with recommendations for its detection and assessment. *Fleischwirtschaft*, **73**: 581–584.

Nørrung, B., Andersen, J.K. & Schlundt, J. 1999. Incidence and control of *Listeria monocytogenes* in foods in Denmark. *International Journal of Food Microbiology,* **53**: 195–203.

Ozari, R. von & Stolle, F. A. 1990. Zum Vorkommen von *Listeria monocytogenes* in Fleisch und Fleischerzeugnissen einschliesslich Geflugelfleisch des Handels. *Archiv. fur Lebensmittelhygiene*, **41**: 47–50.

Pacini, M., Quagli, E., Galassi, R., Marinari, M. & Marzotto, G. 1995. Profilo microbiologico di alcuni prodotti alimentari di largo consumo. *Annali di Microbiologia ed Enzimologia*, **45**: 37–49.

Petchsing, U. & Woodburn, M.J. 1990. *Staphylococcus aureus* and *Escherichia coli* in nham (Thai-style fermented pork sausage). *International Journal of Food Microbiology,* **10**: 183–192.

Ricke, S.C. & Keeton, J.T. 1997. Fermented meat, poultry, and fish products. pp. 610–628, *in:* M.P. Doyle, L.R. Beuchat and T.J. Montville (eds). *Food Microbiology Fundamentals and Frontiers*. Washington DC: ASM Press.

Rørvik, L.M. & Yndestad, M. 1991. *Listeria monocytogenes* in foods in Norway. *International Journal of Food Microbiology*, **13**: 97–104.

Ross, T., Fazil, A., Paoli, G. & Sumner, J. In press. *Listeria monocytogenes* in Australian processed meat products: risks and their management. Meat and Livestock Australia, Sydney, NSW, Australia.

Ryser, E.T., & Marth, E.H. (eds). 1999. *Listeria, Listeriosis, and Food Safety.* 2nd edition, revised and expanded. New York NY: Marcel Dekker. 738p.

Schmidt, U., Seeliger, H.P.R., Glenn, E., Langer, B. & Leistner, L. 1988. Listerienfunde in rohen fleischerzeugnissen [Listeria findings in raw meat products]. *Fleischwirtschaft*, **68**: 1313–1316.

Trüssel, M. 1989. The incidence of *Listeria* in the production of cured and air-dried beef, salami and mettwurst. *Schweizer Archiv fur Tierheilkunde*, **131**: 409–421.

Uyttendaele, M., de Troy, P. & Debevere, J. 1999. Incidence of *Salmonella, Campylobacter jejuni, Campylobacter coli* and *Listeria monocytogenes* in poultry carcasses and different types of poultry products for sale on the Belgian retail market. *Journal of Food Protection*, **62**: 735–740.

van Schothorst, M. 1997. Practical approaches to risk assessment. *Journal of Food Protection,* **60**: 1439–1443.

Varabioff, Y. 1992. Incidence of *Listeria* in smallgoods. *Letters in Applied Microbiology*, **14**: 167–169.

Velani, S. & Gilbert, R.J. 1990. *Listeria monocytogenes* in prepacked ready-to-eat sliced meals. *UK PHLS Microbiology Digest*, **7**: 56.

Vorster, S.M., Greebe, R.P. & Nortje, G.L. 1993. The incidence of *Listeria* in processed meats in South Africa. *Journal of Food Protection*, **56**: 169–172.

WHO. 1988. Foodborne listeriosis. Report of the WHO Working Group. *Bulletin of the World Health Organization*, **66**: 421–428.

Yu, C.F., & Chou, C.C. 1997. Fate of *Escherichia coli* O157:H7 in Chinese-style sausage during the drying step of the manufacturing process as affected by the drying condition and curing agent. *Journal of the Science of Food and Agriculture*, **74**: 551–556.

Appendix 5. Background for the cold-smoked fish assessment

A5.1 ESTIMATE OF GLOBAL PRODUCTION AND NATIONAL AND INDIVIDUAL CONSUMPTION OF COLD-SMOKED FISH

A5.1.1 Scope

The most abundant type of cold-smoked fish is cold-smoked salmon. Due to a paucity of data, cold-smoked fish consumption estimates were based on data describing global production of cold-smoked salmon.

A5.1.2 National and global consumption characteristics

National consumption was initially estimated from data in Globefish (1996)[6] detailing national production and imports/exports of cold-smoked salmon, as shown in Table A5.1. Various other sources of consumption estimates are also shown, and it is noted that the estimates from different sources are not completely consistent, which leads to uncertainty in the estimates.

National population data (NGS, 1999) were combined with the national consumption (calculated as "disappearance" data), to determine per capita consumption. Those data are also shown in Table A5.1.

In Germany and Denmark, hot-smoked product constitutes only a negligible or very small proportion of smoked salmon consumption (P.K. Ben Embarek, pers. comm., 2000; G. Klein, pers. comm., 2000). Similarly, in Australia, hot-smoked salmon products constitute ~10% of production and consumption (Walsh, 1999). Conversely, the contribution of other types of cold-smoked fish is not included in the estimates. Recognizing this limitation, the data are nonetheless used as proxy values for total cold-smoked fish consumption.

From that data, there are various approaches available to calculate the annual per-person consumption of cold-smoked fish and its variability and uncertainty. If the total population of the nations is considered against the total production, the average consumption is 90.2 g/person/year. Per capita consumption in individual nations appears to vary between 8 and 1000 g/person/year, with a median value of 138 g/consumer/year. The average of the estimates of national per-person annual consumption is, however, 231 g. This estimate

6 Globefish have published an updated report on Salmon - A Study of Global Supply and Demand (Globefish, 2003). This provides more recent data on national production and imports/exports of cold-smoked salmon. However, due to limited time and resources it was not possible to incorporate the more recent data into this risk assessment.

differs from the global average obtained if each national data set has equal weight in the average global consumption estimate. If each national consumption estimate is weighted according to the population size, the global average is calculated to be 146 g. The difference between this and the original global estimates arises because data for Canada, Chile, Germany and West Germany, and Norway could not be used because one element of the needed data was missing; see Table A5.1. The median of the remaining national estimates is 144 g/person/year.

Table A5.1 Production, import and export of cold-smoked salmon, and estimated per capita consumption.

	Production (*P*) (tonne)	Import (*I*) (tonne)	Export (*E*) (tonne)	Consumption (inferred from *P+I - E*) (tonne)	Population (million)	Consumption per person-year (gram)	Note
Australia	980	700	–	1 167	18.53	90.7	[1]
Austria	0	848	0	848	8.09	104.9	
Belgium	2 775	2 324	297	4 802	10.23	469.6	
Canada	501	–	501	0	30.59	–	
Chile	1 074	–	1 074	0	15.02	–	
Denmark	15 786	1 406	13 102	4 090	5.24	781.9	[2]
Denmark	–	–	–	–	–	202	[3]
Faeroe islands	407	–	407	0	–	–	
France	11 059	2 941	1 059	12 941	59.07	219.1	
Germany	5 063	7 279	693	11 649	81.95	142.1	
W. Germany	–	–	–	–	–	47.8	[4]
Germany	–	–	–	–	–	1 000	[5]
Ireland	357	–	357	0	3.73	–	
Italy	2 500	5 100	0	7 600	57.72	131.7	
Italy (1998)	–	–	–	9 000	57.72	145.5	[6]
Japan	7 853	765	0	8 618	126.75	68.0	
Netherlands	3 668	0	0	3 668	15.80	232.2	
Norway	2 446	0	2 446	0	4.46	–	
Others	3 954	5615	2 916	6 653	–	–	
Spain	0	313	0	313	39.42	7.9	
Sweden	–	–	–	2 400	8.86	271.0	[7]
Switzerland	0	656	0	656	7.12	92.1	
UK	11 000	146	3 247	7 899	59.36	133.1	
USA	5 116	1133	164	6 085	270.93	22.5	
Total or average	**73 562**	**28 526**	**26 263**	**79 389**	**880.56**	**90.2**	

SOURCES: Data from Globefish (1996) unless otherwise noted.

NOTES: [1] Estimated by Ross and Sanderson, 2000. [2] P.K. Ben Embarek, pers. comm., 2000. [3] Danish food Authority *via* P.K. Ben Embarek, pers. comm., 2000, [4] 1986 data, G. Klein, pers. comm., 2000, based on 1998 population, including East Germany. [5] Buchanan et al., 1997. [6] AC Nielsen data supplied to FAO, 2000. [7] Lindqvist and Westöö, 2000.

A5.1.3 Serving size estimates

Individual consumption frequency by nation

There is limited data available on the number of consumers who eat cold-smoked salmon products. West German data from 1986 (G. Klein, pers. comm., 2000,) reports that 311 of 23 131 interviewees (1.34%) consumed cold-smoked salmon on the survey day and that the mean serving size among eaters was 9.77 g/day, but with an upper 95th percentile of 28.60 g. From the same data source, differences among population sub-groups were revealed but are not used explicitly in this assessment. Using data for consumption of all smoked seafoods, there was no significant difference in serving size by geographical region (north or south Germany) or age group (more or less than 60 years).

In Australia, cold-smoked salmon is considered a luxury food. National consumption was estimated by Ross and Sanderson (2000) at approximately 0.15–0.20 kg/person/year, roughly equivalent to 1% of consumers per day eating a 60 g serving of cold-smoked salmon, or all members of the population eating 3 to 4 servings per year.

Canadian data (CFPNS, 1992–1995) shows that cold-smoked salmon is consumed infrequently in that country. Approximately 5% of consumers on the survey day ate cold-smoked fish products, which included kippered Atlantic herring; cold-smoked Chinook (spring) salmon; smoked haddock; Chinook (lox) salmon; and smoked cod. About half of these were smoked salmonid products. It should be noted that smoked cod is normally cooked before consumption, but represents 8% of eating occasions in the data.

Smoked fish meal size data were estimated by FDA/FSIS (2001) using data from CSFII and NHANES. The data were modified for this case study by removal of data (mostly for smoked oysters) that did not relate to smoked fish products. The average serving size based on age and gender is shown in Figure A5.1. It should be noted that there were few data available – the number of data represented by each bar in the figure varies from 1 to 8.

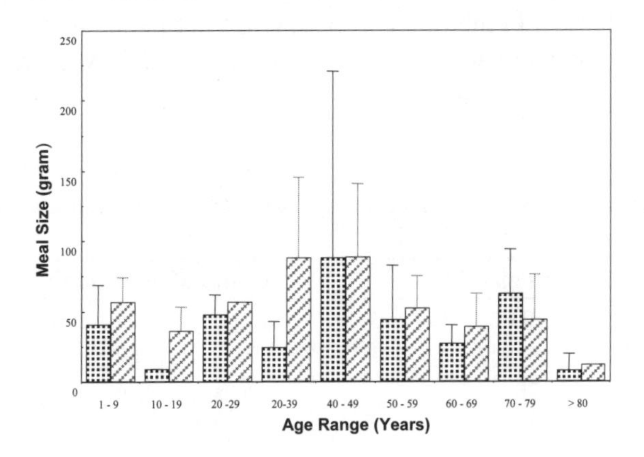

Figure A5.1 Cold-smoked fish serving size as a function of age and gender based on United States of America data (FDA/FSIS, 2001).

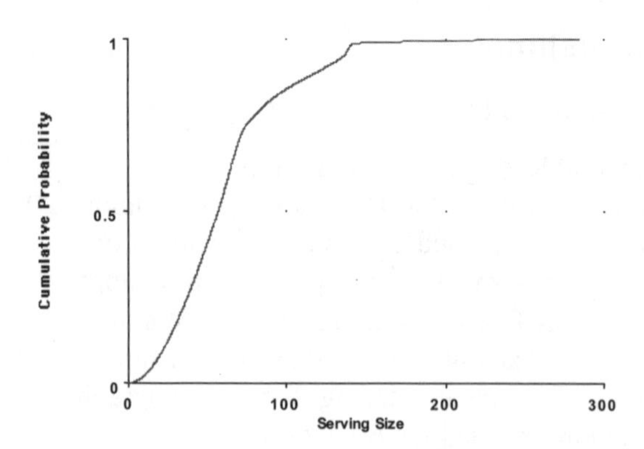

Figure A5.2 Modelled cumulative probability distribution of serving sizes.

Those data indicate that serving size for cold-smoked fish products varies between 1 and 357 g with a median value of ~50 g per serving, and an average value of 58 g/serving. Similarly, using Canadian data (CFPNS, 1992–1995), consumption amounts were aggregated over the smoked fish foods considered, up to the amount consumed on all occasions in a day. The observations are skewed, with median value at 61 g, and long upper tail extending to approximately 225 g, representing approximately the 97.5th percentile.

The above serving size data was used to estimate per capita frequency of consumption from the annual per person consumption estimates in Table A5.1. Those data were used as the basis for the meal size distribution that was used in the model, which is shown in Figure A5.2 and was described in the model using a CumDist function based on the values shown in Table A5.2. The median value of the distribution is 57 g, and the mean value is 63 g.

For the nations considered, the consumption frequency is in the range 0.15–18 servings per person per year, with most in the range 2–5 servings per person per year. The distribution of the number of meals per consumer per year in the model is described empirically by Beta(0.5, 2.5, 0, 18), as shown in Figure A5.3.

Table A5.2 Values used to generate the distribution of serving sizes of cold-smoked fish used in the simulation model.

Serving size (g)	**Cumulative probability**
0	0.00
57	0.50
75	0.75
136	0.95
142	0.99
284	1.00

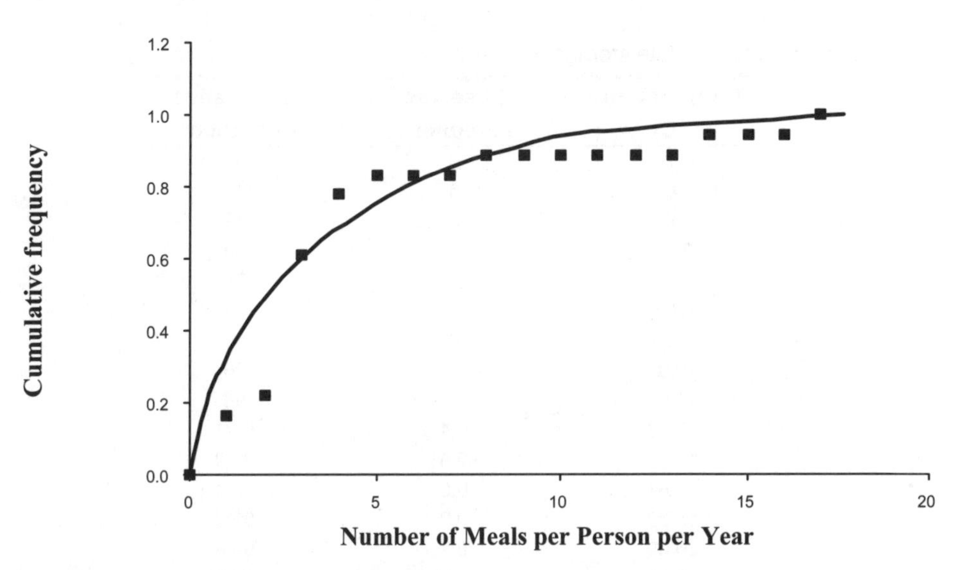

Figure A5.3 Modelled cumulative probability distribution of number of cold-smoked fish servings compared to the observed data.

A5.1.4 Reality check

From the values described above, global total annual consumption in the nations considered can be estimated from the simulation model from the distribution of the product of serving frequency estimate and serving size estimate. This is a useful check on the performance of the simulation model. The median modelled global consumption is 62 300 tonnes and the mean modelled global consumption is 118 000 tonnes. The latter value is ~50% higher than the consumption of cold-smoked salmon estimated from the data in Table A5.1. The basis of this difference is not known with certainty, but may derive from the fact that serving size estimates in the model are derived from all types of smoked fish whereas consumption is based on smoked salmon data only.

A5.2 DESCRIPTION OF STORAGE TEMPERATURES

The storage temperature distribution (Section 4.5.3.4) is derived from Audits International (2000) data for refrigerated cabinets used for storage of cold-smoked fish at retail.

The data used is tabulated below (Table A5.3) and was fitted empirically to a Beta distribution. Comparison of the original data and the fitted distribution is shown in Figure A5.4.

Table A5.3 Data used to simulate storage temperature.

Temperature (°C)	Observed % Frequency	Cumulative % Frequency
-1.67	5	5
0.00	5	10
1.67	8	18
3.33	21	39
5.00	26	65
6.67	16	81
8.33	8	89
10.00	7	96
11.67	2	98
13.33	0.4	98.4
15.00	0.5	98.9
16.67	0.5	99.4
18.33	0.2	99.6
20.00	0.2	99.8
21.67	0.2	100

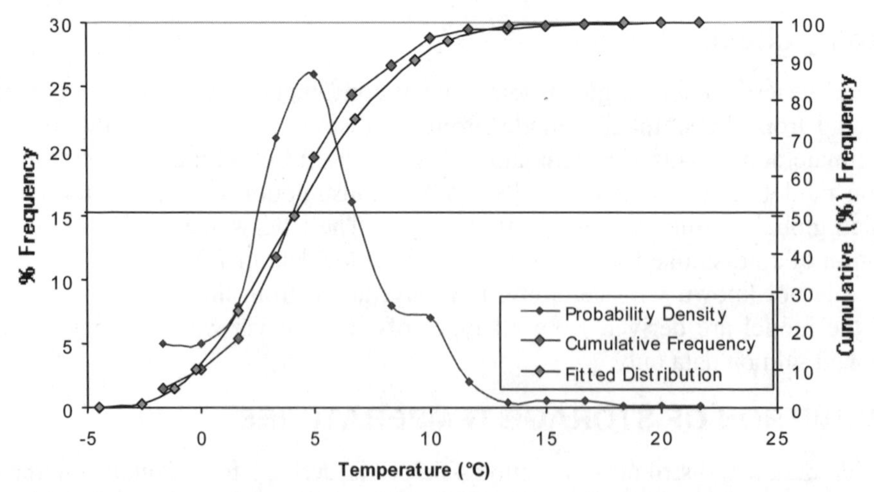

Figure A5.4 Comparison of observed and fitted temperature distribution data.

A5.3 DETAILS OF GROWTH MODELLING

A5.3.1 Physico-chemical parameters

Ranges of physico-chemical parameters of cold-smoked fish that could affect growth of *L. monocytogenes* were presented in Section 4.5.3.5. In the simulation model, these features were described by the following distributions:

- water activity Normal(0.98, 0.0027)
- pH Normal(6.2, 0.05)
- lactic acid concentration (mM) was estimated from pH by the empirical relationship: $1105.0 - 162.50 \times \text{pH}$.

Other growth inhibiting factors (e.g. phenol, spices, etc.) can be included in the model predictions by a simple multiplicative constant.

A5.3.2 *L. monocytogenes* growth rate model

The *L. monocytogenes* growth rate model is derived from Tienungoon (1998). It is a square root type model. The model was further developed, described and evaluated against independent literature data by Ross (1999) and found to have $B_f = 0.88$ and $A_f = 1.94$ (measures of predictive model performance – see Ross, 1996, and Baranyi, Pin and Ross, 1999), which was as good as or better than other published models for *L. monocytogenes* growth rate. In the modelling, the growth rate prediction of the model was multiplied by 0.9 to compensate for the bias of the model. This correction was implemented using the "Other growth inhibiting factors" input in the simulation model.

To calculate growth, physico-chemical parameters sampled from the distributions described above are first "filtered" through the growth/no-growth model of Tienungoon et al. (2000) to determine whether the scenario sampled represents a product that will allow growth of *L. monocytogenes.* If growth is predicted to be possible, the extent of growth is modelled using the sampled storage time and the growth rate model, including a correction for lag time.

The prediction of *L. monocytogenes* growth is further filtered by applying an upper limit to the population density (CFU/g product) predicted to be achievable. Including the effects of lactic acid bacteria in the model is expected to preclude this being necessary in most model iterations, but in those (rare) scenarios where *L. monocytogenes* growth is modelled not to be constrained by any other factor, it will eventually limit its own growth, i.e. achieve its maximum population density (MPD). FDA/FSIS (2001) reviewed the available literature and noted that *L. monocytogenes* rarely achieves levels in cold-smoked salmon as high as it does in pure culture in laboratory broth. It is probable, however, that many of those observations are due to the effects of other bacteria in the foods, which are modelled in this assessment. MPD was therefore set at 3×10^9 CFU/g, a level representative of otherwise ideal conditions for those scenarios in which no other factors constrain *L. monocytogenes* growth.

A5.3.3 Lag time

Lag time data specific for *L. monocytogenes* in cold-smoked fish were not found in the literature. Ross (1999) collated data for lag times from the published literature and expressed these as relative lag times (see Section 3.5.3.3). *L. monocytogenes* relative lag times in foods were in the range 0–40, with a peak value near 2.5. Lag times in laboratory broths had a similar range, but the peak value was nearer to 4.5. Figure A5.5 presents this data.

Dalgaard and Jørgensen (1998) state that *L. monocytogenes* cells that contaminate cold-smoked fish are likely to be damaged due to the effects of processing. Other studies (Rørvik and Yndestad, 1991; Rørvik et al., 1997, 2000), however, suggest that most contamination of cold-smoked fish arises after smoking, from contamination sources in processing plants. Because of this ambiguity, two distributions were assessed to gauge the importance of assumptions about lag time distributions. These were termed "short" (Beta(3, 30, 0, 35)) and "long" (Beta(6, 35, 0, 35)) relative lag times, and produce the distributions shown in Figure A5.6. The overall growth model and the interaction of factors governing the extent of growth are depicted in Figure A5.7.

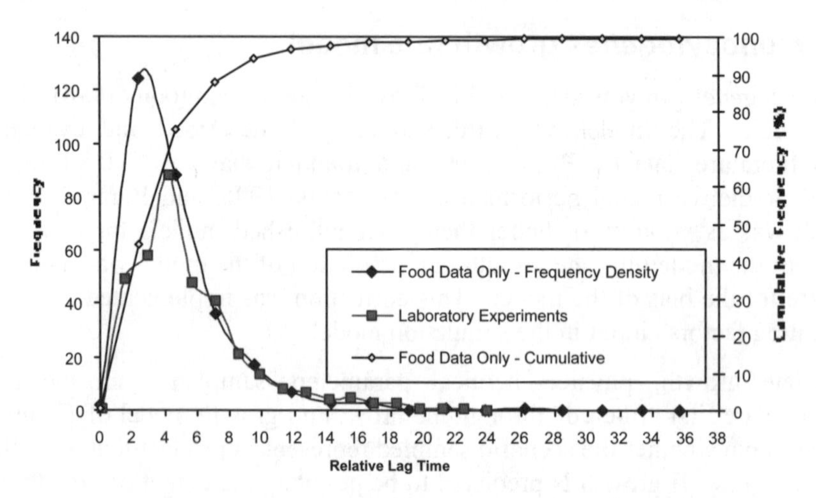

Figure A5.5 Relative lag time data for *Listeria monocytogenes* reported in published literature.
SOURCE: After Ross, 1999.

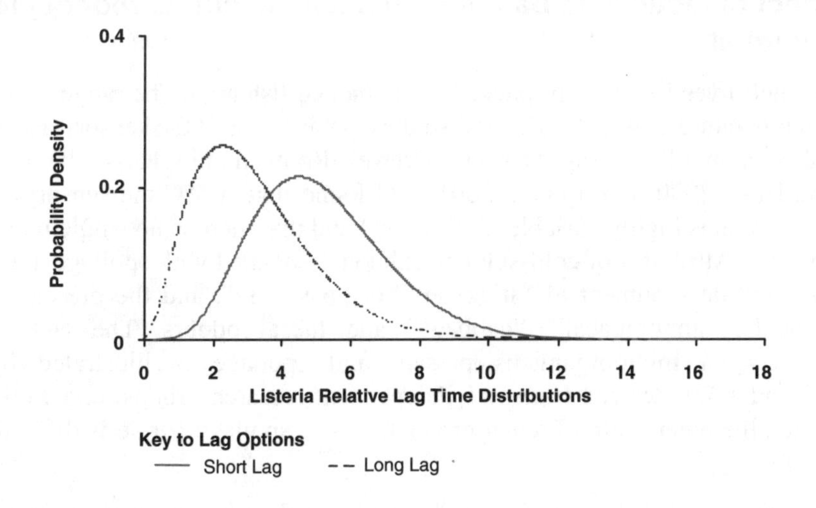

Figure A5.6. Outputs from the simulation model showing the two relative lag time distributions used to model the effects of lag time on risk of listeriosis from cold-smoked fish.

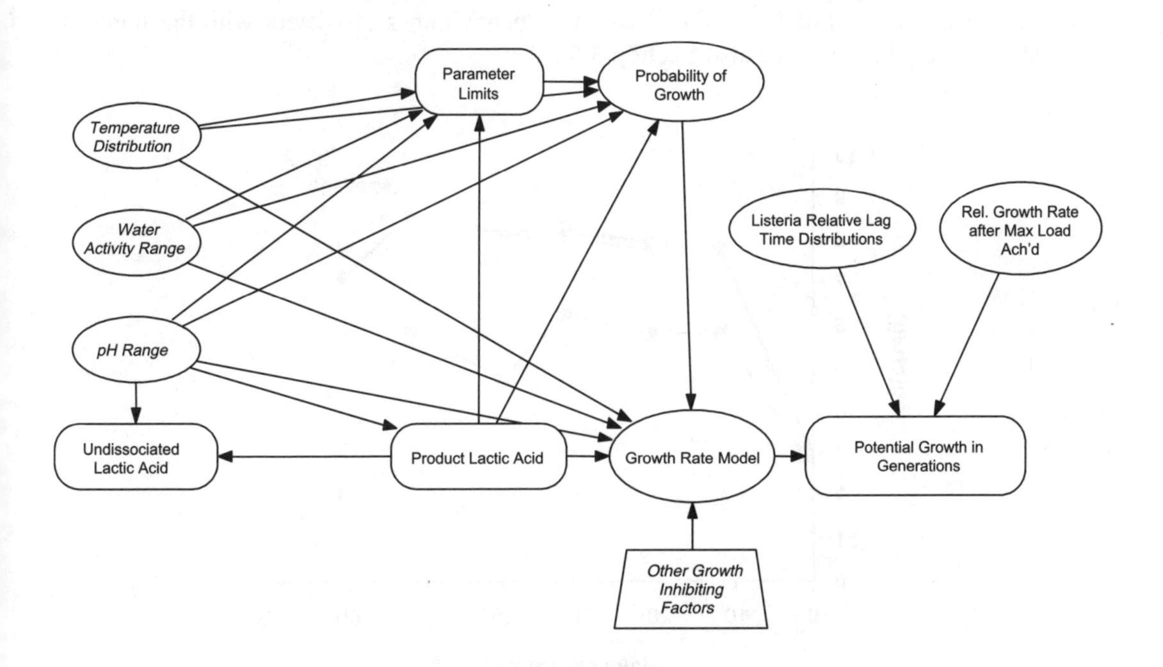

Figure A5.7 Influence diagram taken from the simulation model used in this risk assessment and showing the interaction of factors used to estimate growth of *Listeria monocytogenes* during storage of cold-smoked fish.

A5.3.4 Effect of lactic acid bacteria on shelf life and *L. monocytogenes* growth potential

The nominal shelf lives for vacuum-packed cold-smoked fish are in the range of 3 to 6 weeks at a storage temperature of 4–5°C. Several studies have assessed the sensory acceptability of cold-smoked salmon (Truelstrup Hansen, Drewes Røntved and Huss, 1998; Jørgensen, Dalgaard and Huss, 2000; Leroi et al., 2001) and found that at 5°C the sensory shelf-life of cold-smoked salmon is highly variable (3–9 weeks) and that there is no single indicator of the onset of spoilage. Attributes of cold-smoked salmon associated with spoilage are a softening of the texture and development of "stickiness" or "pastiness", and the presence of "sour", "bitter", "rancid", "ammoniacal", "cabbage" and faecal odours. The lack of a clear relationship between microorganisms present and spoilage is illustrated by data in Figures A5.8 and A5.9, derived from two independent research groups, one working with a Danish product, the other with a French product. Thus, spoilage per se is difficult to model mechanistically.

In raw and processed meats and fish chilled and stored under vacuum, lactic acid bacteria become the dominant population and preserve the product with a "hidden" fermentation (Stiles, 1996). Thus, of particular note in Figures A5.8 and A5.9 is the cessation of growth of any component of the population when the total psychrotrophic count appears to achieve a stationary phase at a level of 10^7–10^8 CFU/g. This behaviour is consistent with the Jameson effect (Stephens et al., 1997; see also Section 3.5.3.1).

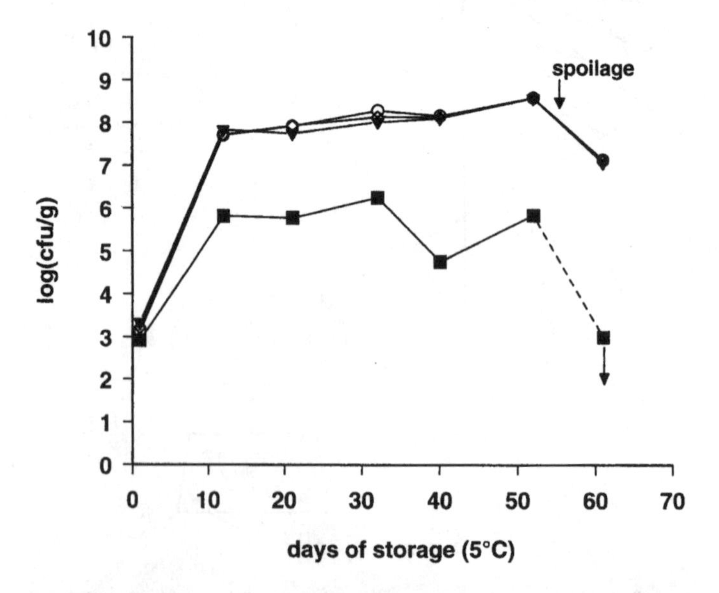

Figure A5.8 Microbiological changes in Danish-produced vacuum-packed cold-smoked salmon (4.6% water phase salt (WPS)) during storage at 5°C. Total count (○); total psychrotrophic count (▼); lactic acid bacteria (◊) and Enterobacteriaceae (■). Arrow indicates the time of sensory rejection. (Data of Truelstrup Hansen, Gill and Huss, 1995, reproduced from Gram and Huss, 1996).

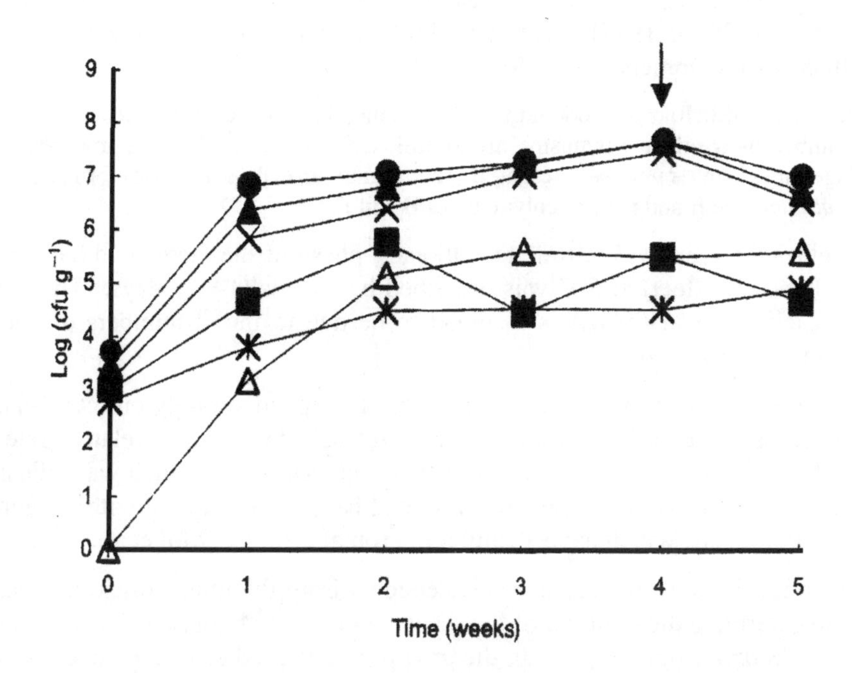

Figure A5.9 Microbiological changes in French-produced vacuum-packed cold-smoked salmon during storage at 5°C. Total psychrotrophic count (●); total lactic acid bacteria (▲); lactobacilli (△); Enterobacteriaceae (■); *Brocothrix thermosphacta* (X) and yeast (⋊). The arrow indicates the time of sensory rejection (Reproduced from Leroi et al., 2001).

The Jameson effect can be likened to a race to reach stationary phase. The winner is that sub-group within the total microbial population that first achieves stationary phase.

When that happens, the race is over and all other contestants finish the race (i.e. they also enter stationary phase) at that point in time, although unpublished data (L.A. Mellefont, B. Davidson and T. Ross, Univ. of Tasmania, pers. comm., 2002) indicate that in some cases growth is not completely inhibited, but is nevertheless greatly reduced.

The Jameson effect has relevance for estimation of the risk from microbiological hazards in cold-smoked fish products. As Figures A5.8 and A5.9 show, and as has been reported also for vacuum-packed meats (Mol et al., 1971; Egan, Ford and Shay, 1980; Korkeala et al., 1989), spoilage of vacuum-packed meat and fish does not usually occur until well after the total count has reached stationary phase. In cold-smoked salmon, that occurs after one to two weeks under recommended storage conditions. Thus, growth of pathogens in the product may only be possible for 25–50% of the full shelf-life (use-by period) of the product.

A5.3.5 Modelling the effect of lactic acid bacteria

The mechanism of the Jameson effect is not yet fully understood. It may be due to competition for nutrients, production of toxic end products, or production of specific antibiotics against other bacteria. Under some circumstances, a pathogen may be numerically dominant at the time of production and, under improper storage, may grow fast enough to reach stationary phase before any other element of the population on the food (e.g. see Nilsson, Huss and Gram, 1997). This possibility is explicitly recognized in the Seafood HACCP Alliance's recommendations (SHA, 1997):

> "In cold-smoked fish, it is important that the product does not receive so much heat that the number of spoilage organisms are significantly reduced. This is true because spoilage organisms must be present to inhibit the growth and toxin formation of *C. botulinum* type E and nonproteolytic types B and F."

Based on Figures A5.8 and A5.9, time to stationary phase of the lactic acid bacteria (i.e. the onset of the Jameson effect) at 5°C was described by Normal(14, 2) days. The variation in the time to reach stationary phase is assumed to be due to initial numbers of bacteria and specific product composition.

Because the time to stationary phase is expected to depend strongly on the temperature of storage, it is adjusted according to the temperatures selected using a relative rate function with a T_{min} of 0°C. This value was used as a first approximation, which was adjusted based on temperature–growth rate responses of lactic acid bacteria associated with vacuum-packed processed meat products, which have a similar microbial ecology (Mol et al., 1971).

The time to reach the stationary phase is deducted from the total storage time sampled in an iteration to determine the duration of the second, constrained, phase of growth. During the first phase of *L. monocytogenes* growth, the growth rate is predicted to be unconstrained and predicted by the growth rate model for the temperature and product characteristics sampled during that iteration. After that time, however, growth is predicted to be reduced by some factor. It could be complete inhibition (as described in Figures A5.8 and A5.9), but other data (L.A. Mellefont, B. Davidson and T. Ross, Univ. of Tasmania, pers. comm., 2002, unpublished data) that suggest that *L. monocytogenes* might continue to grow slowly. Thus, the specific growth behaviour is uncertain and has been left as an assumption whose influence can be tested (see Section 4.5.5).

Nominal storage life also has to be adjusted for storage conditions because higher temperatures will cause premature spoilage. Conversely, there are reports (Ben Embarek, pers. comm., 2001) that in some countries 2–3-month shelf lives are realized, apparently without consumer rejection. Accordingly, several scenarios have been modelled. However, in each case a filter is applied so that if the storage time at 5°C exceeds 10 weeks (or its equivalent calculated at other temperatures) the product is considered completely spoiled and no further growth occurs. This is achieved by truncating the predictions of growth based on shelf-life at the growth levels that could have occurred at the equivalent of 70 days at 5°C. These predictions are not, however, removed from the simulation.

These interactions are shown as an influence diagram in Figure A5.10.

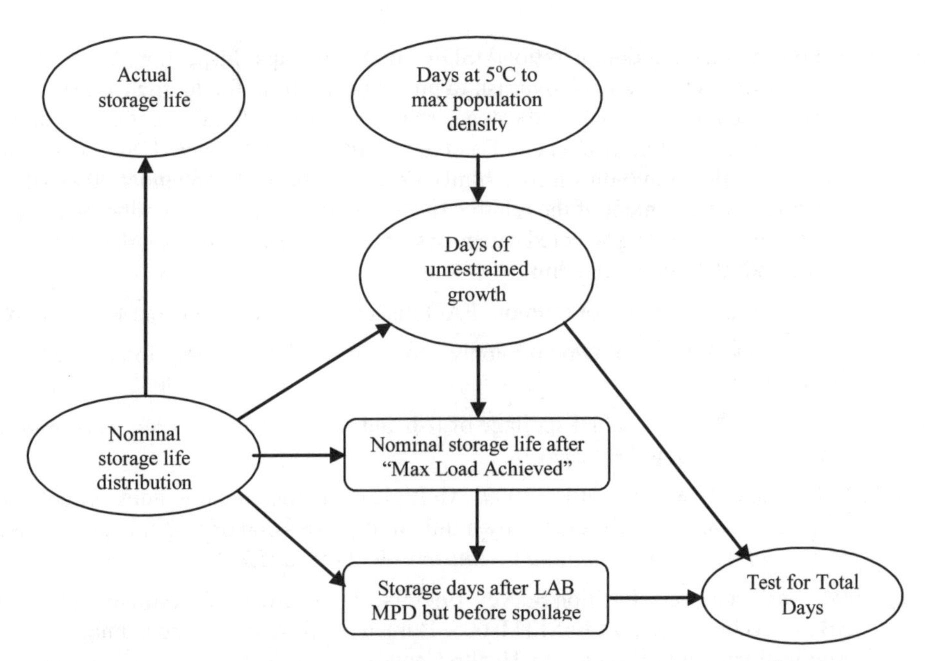

Figure A5.10 Influence diagram derived from the simulation model, showing the interaction of nominal storage life and spoilage due to bacterial growth, as well as the estimation of the time required for lactic acid bacteria to reach the stationary phase and for the Jameson effect to constrain *Listeria monocytogenes* growth.

A5.4 REFERENCES CITED IN APPENDIX 5

Audits International. 2000. 1999 U.S. Food Temperature Evaluation. Design and Summary Pages. Audits International and U.S. Food and Drug Administration. 13p.

Baranyi, J., Pin, C. & Ross, T. 1999. Validating and comparing predictive models. *International Journal of Food Microbiology,* **48**: 159–166.

Buchanan, R.L., Damert, W.G., Whiting, R.C. & van Schothorst, M. 1997. Use of epidemiologic and food survey data to estimate a purposefully conservative dose-response relationship for *Listeria monocytogenes* levels and incidence of listeriosis. *Journal of Food Protection.* **60**: 918–922.

CFPNS [Canadian Federal-Provincial Nutrition Surveys]. 1992–1995. Bureau of Biostatistics and Computer Applications, Food Directorate, Health Canada. See also: Karpinski & Nargundkar, 1992; Junkins & Karpinski, 1994; Junkins, 1994; Junkins &Laffey, 2000; Junkins, Laffey & Weston, 2001.

Dalgaard, P. & Jørgensen, L.V. 1998. Predicted and observed growth of *Listeria monocytogenes* in seafood challenge tests and in naturally contaminated cold-smoked salmon. *International Journal of Food Microbiology,* **40**: 105–115.

Egan, A.F., Ford, A.L. & Shay, B.J. 1980. A comparison of *Mocrobacterium thermosphactum* and lactobacilli as spoilage organisms of vacuum-packaged sliced luncheon meats. *Journal of Food Science,* **45**: 1745–1748.

FDA/FSIS [U.S. Food and Drug Administration/USDA Food Safety and Inspection Agency]. 2001. Draft Assessment of the relative risk to public health from foodborne *Listeria monocytogenes* among selected categories of ready-to-eat foods. Center for Food Safety and Applied Nutrition (FDA) and Food Safety Inspection Service (USDA) (Available at: www.foodsafety.gov/~dms/lmrisk.html). [Report published September 2003 as: Quantitative assessment of the relative risk to public health from foodborne *Listeria monocytogenes* among selected categories of ready-to-eat foods. Available at: www.foodsafety.gov/~dms/lmr2-toc.html].

Globefish. 1996. The World Market for Salmon. FAO Fishery Industries Division, Rome, Italy.

Globefish. 2003. Salmon- A Study of Global Supply and Demand. FAO Fishery Industries Division, Rome, Italy.

Gram, L. & Huss, H.H. 1996. Microbial spoilage of fish and fish products. *International Journal of Food Microbiology*, **33**: 121–138.

Jørgensen, L.V., Dalgaard, P. & Huss, H.H. 2000. Multiple compound quality index for cold-smoked salmon (*Salmo salar*) developed by multivariate regression of biogenic amines and pH. *Journal of Agricultur and Food Chemistry*, **48**: 2448–2453.

Junkins, E. 1994. Saskatchewan Nutrition Survey 1993/94. Methodology for estimating usual intake. BBCA Technical Report E451311-005. Bureau of Biostatistics and Computer Applications, Food Directorate, Health Canada.

Junkins, E. & Karpinski, K. 1994. Enquéte québécoise sur la nutrition. Méthodologie pour estimer l'apport habituel, les statistiques sommaires et les erreurs-types. Bureau of Biostatistics and Computer Applications, Food Directorate, Health Canada.

Junkins, E. & Laffey, P. 2000. Alberta Nutrition Survey 1994. Methodology for estimating usual intake. BBCA Technical Report E451311-006. Bureau of Biostatistics and Computer Applications, Food Directorate, Health Canada.

Junkins, E., Laffey, P. & Weston, T. 2001. Prince Edward Island Nutrition Survey 1995. Methodology for estimating usual intake. BBCA Technical Report E451311-007. Bureau of Biostatistics and Computer Applications, Food Directorate, Health Canada.

Karpinski, K. & Nargundkar, M. 1992. Nova Scotia Nutrition Survey. Methodology Report. BBCA Technical Report E451311-001. Bureau of Biostatistics and Computer Applications, Food Directorate, Health Canada.

Korkeala, H., Alanko, T., Makela, P. & Lindroth, S. 1989. Shelf-life of vacuum-packed cooked ring sausages at different chill temperatures. *International Journal of Food Microbiology*, **9**: 237–347.

Leroi, F., Joffraud, J.J., Chevalier F. & Cardinal, M. 2001. Research of quality indices for cold-smoked salmon using a stepwise multiple regression of microbiological counts and physico-chemical parameters. *Journal of Applied Microbiology*, 90: 578–587.

Lindqvist, R. & Westöö, A. 2000. Quantitative risk assessment for *Listeria monocytogenes* in smoked or gravad salmon/rainbow trout in Sweden. *International Journal of Food Microbiology,* **58**: 181–196.

Mol, J.H.H., Hietbrink, J.E.A., Mollen, H.W.M. & van Tinteren, J. 1971. Observations on the microflora of vacuum-packed sliced cooked meat products. *Journal of Applied Bacteriology*, **34**: 377–397.

NGS [National Geographic Society]. 1999. The National Geographic desk reference. Edited by R.M. Downs, F.A Day, P.L. Knoxw, P.H. Meserve and B. Warf. Washington DC: National Geographic Society. 700p.

Nilsson, L., Huss, H.H. & Gram, L. 1997. Inhibition of *Listeria monocytogenes* on cold-smoked salmon by nisin and carbon dioxide atmosphere. *International Journal of Food Microbiology,* **38**: 217–228.

Rørvik, L.M. & Yndestad, M. 1991. *Listeria monocytogenes* in foods in Norway. *International Journal of Food Microbiology*, **13**: 97–104.

Rørvik, L.M., Aase, B., Alvestad, T. & Caugant, D.A. 2000. Molecular epidemiological survey of *Listeria monocytogenes* in seafoods and seafood-processing plants. *Applied and Environmental Microbiology*, **66**: 4779–4784.

Rørvik, L.M., Skjerve, E., Knudsen, B.R. & Yndestad, M. 1997. Risk factors for contamination of smoked salmon with *Listeria monocytogenes* during processing. *International Journal of Food Microbiology,* **37**: 215–219.

Ross, T. 1996. Indices for performance evaluation of predictive models in food microbiology. *Journal of Applied Bacteriology*, **81**: 501-508.

Ross, T. 1999. Predictive food microbiology models in the meat industry. Meat and Livestock Australia, Sydney, Australia. 196p.

Ross, T. & Sanderson, K. 2000. A risk assessment of selected seafoods in New South Wales. SafeFood NSW, Sydney, Australia. 275p.

SHA [Seafood HACCP Alliance]. 1997. Compendium of fish and fishery product processes, hazards and controls. National Seafood HACCP Alliance for Training and Education. Available at httt://seafood.ucdavis.edu/haccp/compendium/compend.htm

Stephens, P.J., Joynson, J.A., Davies, K.W., Holbrook, R., Lappin-Scott, H.M. & Humphrey, T.J. 1997. The use of an automated growth analyser to measure recovery times of single heat injured *Salmonella* cells. *Journal of Applied Microbiology,* **83**: 445–455.

Stiles, M.E. 1996. Biopreservation by lactic acid bacteria. *Antonie van Leeuwenhoek*, **70**: 331–345.

Tienungoon, S. 1998. Some aspects of the ecology of *Listeria monocytogenes* in salmonid aquaculture. Ph.D. Thesis, University of Tasmania, Hobart, Tasmania, Australia.

Tienungoon, S., Ratkowsky, D.A., McMeekin, T.A. & Ross, T. 2000. Growth limits of *Listeria monocytogenes* as a function of temperature, pH, NaCl and lactic acid. *Applied and Environmental Microbiology*, **66**: 4979–4987.

Truelstrup Hansen, L., Drewes Røntved, S. & Huss, H.H. 1998. Microbiological quality and shelf life of cold-smoked salmon from three different processing plants. *Food Microbiology*, **15**: 137–150.

Truelstrup Hansen, L., Gill, T. & Huss, H.H. 1995. Effects of salt and storage temperature on chemical, microbiological and sensory changes in cold-smoked salmon. *Food Research International*, **28**: 123–130.

Walsh, P. 1999. Consultancy for Researching the Business Profile of the NSW Seafood Industry and Food Safety Hazards of Seafood in NSW. Safe Food Production NSW, Sydney, NSW. 151p.

FAO/WHO MICROBIOLOGICAL RISK ASSESSMENT SERIES

1 Risk assessments of *Salmonella* in eggs and broiler chickens: Interpretative Summary. 2002
2 Risk assessments of *Salmonella* in eggs and broiler chickens. 2002
3 Hazard characterization for pathogens in food and water: Guidelines. 2003
4 Risk assessment of *Listeria monocytogenes* in ready-to-eat foods: Interpretative Summary. 2004
5 Risk assessment of *Listeria monocytogenes* in ready-to-eat foods: Technical Report. 2004
6 *Enterobacter sakazakii* and micro-organisms in powdered infant formula: Meeting Report. 2004